MATERNAL IMMUNIZATION

MATERNAL IMMUNIZATION

Edited by

Elke E. Leuridan
Marta C. Nunes
Christine E. Jones

Academic Press is an imprint of Elsevier
125 London Wall, London EC2Y 5AS, United Kingdom
525 B Street, Suite 1650, San Diego, CA 92101, United States
50 Hampshire Street, 5th Floor, Cambridge, MA 02139, United States
The Boulevard, Langford Lane, Kidlington, Oxford OX5 1GB, United Kingdom

Notices

Knowledge and best practice in this field are constantly changing. As new research and experience broaden our understanding, changes in research methods, professional practices, or medical treatment may become necessary.

Practitioners and researchers must always rely on their own experience and knowledge in evaluating and using any information, methods, compounds, or experiments described herein. In using such information or methods they should be mindful of their own safety and the safety of others, including parties for whom they have a professional responsibility.

To the fullest extent of the law, neither the Publisher nor the authors, contributors, or editors, assume any liability for any injury and/or damage to persons or property as a matter of products liability, negligence or otherwise, or from any use or operation of any methods, products, instructions, or ideas contained in the material herein.

Library of Congress Cataloging-in-Publication Data
A catalog record for this book is available from the Library of Congress

British Library Cataloguing-in-Publication Data
A catalogue record for this book is available from the British Library

ISBN 978-0-12-814582-1

For information on all Academic Press publications
visit our website at https://www.elsevier.com/books-and-journals

Publisher: Andre G. Wolff
Acquisition Editor: Linda Versteeg-buschman
Editorial Project Manager: Megan Ashdown
Production Project Manager: Vignesh Tamil
Cover Designer: Miles Hitchen

Typeset by SPi Global, India

Contents

3 Immunobiological aspects of vaccines in pregnancy: Maternal perspective

Helen Y. Chu, Arnaud Marchant

4 Immunobiological aspects of vaccines in pregnancy: Infant perspective

Christopher R. Wilcox, Christine E. Jones

5 Global considerations on maternal vaccine introduction and implementation

Michelle L. Giles, Pauline Paterson, Flor M. Munoz, Heidi Larson, Philipp Lambach

II

Vaccines with current recommendations for use in pregnancy

6 Tetanus

Inci Yildirim, Saad B. Omer

7 Influenza

Deshayne B. Fell, Milagritos D. Tapia, Marta C. Nunes

8 Pertussis

Kirsten Maertens, Kathryn Edwards, Elke E. Leuridan

9 Vaccination in pregnancy in specific circumstances

Anna Calvert, Miranda Quenby, Paul T. Heath

III

Future vaccines for use in pregnancy

14 Malaria

Patrick E. Duffy, Sara Healy, J. Patrick Gorres, Michal Fried

IV

Conclusion

15 Conclusion

Elke E. Leuridan, Marta C. Nunes, Christine E. Jones

Contributors

Kristina Adachi Department of Pediatrics, Division of Infectious Diseases, University of California, Los Angeles David Geffen School of Medicine, Los Angeles, CA, United States

Anna Calvert Vaccine Institute & Paediatric Infectious Diseases Research Group, St Georges, University of London, London, United Kingdom

Helen Y. Chu Departments of Medicine and Epidemiology, University of Washington, Seattle, WA, United States

Patrick E. Duffy Laboratory of Malaria Immunology and Vaccinology, National Institute of Allergy and Infectious Diseases, National Institutes of Health, Bethesda, MD, United States

Linda O. Eckert Department of Obstetrics and Gynecology; Department of Global Health, University of Washington, Seattle, WA, United States

Kathryn Edwards Division of Infectious Diseases, Vanderbilt Vaccine Research Program, Department of Pediatrics, Vanderbilt University School of Medicine, Nashville, TN, United States

Janet A. Englund Department of Pediatrics, Pediatric Infectious Diseases, Seattle Children's Hospital, University of Washington, Seattle, WA, United States

Deshayne B. Fell School of Epidemiology and Public Health, University of Ottawa; Children's Hospital of Eastern Ontario Research Institute; ICES, Ottawa, Canada

Michal Fried Laboratory of Malaria Immunology and Vaccinology, National Institute of Allergy and Infectious Diseases, National Institutes of Health, Bethesda, MD, United States

Michelle L. Giles Department of Obstetrics and Gynaecology, Monash University, Melbourne, VIC, Australia

J. Patrick Gorres Laboratory of Malaria Immunology and Vaccinology, National Institute of Allergy and Infectious Diseases, National Institutes of Health, Bethesda, MD, United States

Sara Healy Laboratory of Malaria Immunology and Vaccinology, National Institute of Allergy and Infectious Diseases, National Institutes of Health, Bethesda, MD, United States

Paul T. Heath Vaccine Institute & Paediatric Infectious Diseases Research Group, St Georges, University of London, London, United Kingdom

Christine E. Jones Faculty of Medicine and Institute for Life Sciences, University of Southampton and University Hospital Southampton NHS Foundation Trust, Southampton, United Kingdom

Alisa Kachikis Department of Obstetrics and Gynecology, University of Washington, Seattle, WA, United States

Gaurav Kwatra Medical Research Council: Respiratory and Meningeal Pathogens Research Unit; Department of Science and Technology/National Research Foundation: Vaccine Preventable Diseases Unit, Faculty of Health Sciences, University of the Witwatersrand, Johannesburg, South Africa; Department of Clinical Microbiology, Christian Medical College, Vellore, India

Philipp Lambach World Health Organization, Geneva, Switzerland

Heidi Larson Department of Infectious Disease Epidemiology, The London School of Hygiene & Tropical Medicine, London, United Kingdom

Elke E. Leuridan Centre for the Evaluation of Vaccination, Vaccine & Infectious Diseases Institute, Faculty of Medicine and Health Sciences, University of Antwerp, Antwerp, Belgium

Shabir A. Madhi Medical Research Council: Respiratory and Meningeal Pathogens Research Unit; Department of Science and Technology/National Research Foundation: Vaccine Preventable Diseases Unit, Faculty of Health Sciences, University of the Witwatersrand, Johannesburg, South Africa

Kirsten Maertens Centre for the Evaluation of Vaccination, Vaccine & Infectious Diseases Institute, Faculty of Medicine and Health Sciences, University of Antwerp, Antwerp, Belgium

Arnaud Marchant Institute for Medical Immunology, Université libre de Bruxelles, Brussels, Belgium

Flor M. Munoz Department of Pediatrics, Section of Infectious Diseases, Baylor College of Medicine, Houston, TX, United States

Karin Nielsen-Saines Department of Pediatrics, Division of Infectious Diseases, University of California, Los Angeles David Geffen School of Medicine, Los Angeles, CA, United States

Marta C. Nunes Medical Research Council: Respiratory and Meningeal Pathogens Research Unit, Faculty of Health Sciences; Department of Science and Technology/National Research Foundation: Vaccine Preventable Diseases Unit, University of the Witwatersrand, Johannesburg, South Africa

Saad B. Omer Division of Infectious Diseases, Department of Pediatrics, Emory University; Department of Epidemiology; Hubert Department of Global Health, Rollins School of Public Health, Atlanta, GA, United States

Pauline Paterson Department of Infectious Disease Epidemiology, The London School of Hygiene & Tropical Medicine, London, United Kingdom

Natalie Quanquin Department of Pediatrics, Division of Infectious Diseases, Children's Hospital Los Angeles, Los Angeles, CA, United States

Miranda Quenby MONASH University, Melbourne, VIC, Australia

Mark R. Schleiss Division of Pediatric Infectious Diseases and Immunology, Center for Infectious Diseases and Microbiology Translational Research, University of Minnesota Medical School, Minneapolis, MN, United States

Milagritos D. Tapia Center for Vaccine Development, Bamako, Mali; University of Maryland, School of Medicine, Center for Vaccine Development and Global Health, Baltimore, MD, United States

Christopher R. Wilcox NIHR Clinical Research Facility, University Hospital Southampton NHS Foundation Trust, Southampton, United Kingdom

Nicholas Wood National Centre for Immunisation Research and Surveillance, The Children's Hospital at Westmead; The University of Sydney Children's Hospital Westmead Clinical School, Sydney, NSW, Australia

Inci Yildirim Division of Infectious Diseases, Department of Pediatrics, Emory University; Department of Epidemiology, Rollins School of Public Health, Atlanta, GA, United States

Editors' biography

Dr. Elke E. Leuridan is an Associate Professor at the Vaccine & Infectious Diseases Institute, Centre for the Evaluation of Vaccination, University of Antwerp, Belgium. She completed her M.D. degree from the Catholic University of Leuven, Belgium and obtained a PhD at the University of Antwerp, working on vaccine preventable diseases and maternal antibodies. As postdoctoral researcher, she supervises several projects on all aspects of pertussis vaccination during pregnancy in Belgium and abroad (Vietnam, Thailand). She is the promoter of several PhD students on the topic, is involved in the training of medical students and supervises yearly several master and advanced master theses. She is member of the Belgian National Health Council, working group vaccination, and participates in the decision-making process on current and future immunization programs in Belgium. Since 2015, she combines working in research with general practice.

Dr. Marta C. Nunes is a Reader, Associate Professor, at the Vaccine Preventable Diseases Unit, Faculty of Health Sciences, University of the Witwatersrand, South Africa. Marta developed her PhD work at the Department of Neurology & Neuroscience, Weill Medical College of Cornell University, New York, USA and obtained her PhD from the University of Lisbon, Medical College, Portugal. After her post-doctoral training at the Institut Pasteur in Paris, France, where she developed different projects aimed at understanding the cell biology of *Plasmodium falciparum* and to identify molecular candidates for malaria vaccines for pregnant women, she moved to South Africa in 2009.

To approach the problem of infant morbidity and mortality related to infections, she is exploring the potential of intervening through vaccination of pregnant women to protect the women and their babies against infections. This also includes the evaluation of this intervention in reducing adverse birth outcomes.

Dr. Chrissie E. Jones is an Associate Professor in Pediatric Infectious Diseases at the University of Southampton and University Hospital Southampton NHS Trust in the UK. During her PhD she carried out a mother-infant study looking at the effect of maternal HIV and tuberculosis infection on infant responses to vaccination. This has led to her current work on vaccines in pregnancy in order to optimize protection from birth against vaccine preventative infections. She leads clinical trials of vaccines in pregnancy. She leads the Wessex Congenital Infection service and has a clinical and academic interest in Cytomegalovirus, in particular primary prevention in pregnancy to reduce the risk of congenital cytomegalovirus infection. She is the Chair of the Committee for Education for the European Society of Paediatric Infectious Diseases.

PART I

Concepts of maternal immunization

CHAPTER

1

The history of maternal immunization

Alisa Kachikis[a], *Linda O. Eckert*[a,b], *Janet A. Englund*[c]

[a]Department of Obstetrics and Gynecology, University of Washington, Seattle, WA, United States, [b]Department of Global Health, University of Washington, Seattle, WA, United States, [c]Department of Pediatrics, Pediatric Infectious Diseases, Seattle Children's Hospital, University of Washington, Seattle, WA, United States

Introduction

Maternal immunization, or the vaccination of pregnant women, for prevention of maternal, fetal and neonatal morbidity and mortality, has emerged as an exciting and rapidly expanding area of vaccine-preventable diseases research and clinical practice in the last decades. The intricacies of the maternal and neonatal immune system, the transplacental transfer of antibodies to the fetus and potential inhibition of subsequent active infant immunization have added an additional layer of complexity to vaccine research. Nevertheless, vaccines are currently in development specifically for use in maternal immunization such as those against respiratory syncytial virus (RSV) and Group B *Streptococcus* (GBS) [1]. While maternal immunization strategies are not a new phenomenon and have a history similar to vaccine strategies overall, their current expanding application more closely mirrors trends in public policy regarding research in pregnancy. Consideration of past lessons learned, current work in vaccinology, as well as policy regarding inclusion of reproductive age and pregnant women in research may better predict future directions and successes for immunization in pregnancy.

Maternal Immunization
https://doi.org/10.1016/B978-0-12-814582-1.00001-2

The need from a historical perspective: Pregnancy

Improvements in obstetrical care, medical knowledge, access to better nutrition, increases in standard of living, and access to health care have all contributed to improved survival of mothers and their infants in the 21st century. The increased risk for morbidity and mortality during pregnancy, delivery, and post-partum is recorded throughout history. Worldwide, maternal morbidity rates (MMR) per 100,000 deliveries within 42 days of childbirth were between 800 and 1000 per 100,000 live births in the early 1800s and have now fallen to less than 10 deaths per 100,000 live births in high-income countries. Substantial decreases have been also documented in many developing countries, with MMR in 2015 at about 216 deaths per 100,000 live births worldwide [2,3]. Strides are being made in some middle-income countries such as India (MMR 174 per 100,000 in 2015) [3], but less so in certain low-income countries and in those with armed conflict such as the Democratic Republic of Congo (MMR 693 per 100,000 in 2015) [4].

Pregnant women have been documented to have increased susceptibility to vaccine-preventable diseases compared to non-pregnant women resulting in increased maternal, fetal, and neonatal morbidity and mortality. Historical records on smallpox infections in the 19th century report increased case fatality rates and adverse outcomes among pregnant women [5,6]. The increased susceptibility of pregnant women and their fetuses to viral illness was again demonstrated in the measles outbreaks in the Faroe Islands in 1846 and in Greenland in 1951 [7–13]. During the influenza pandemics of 1918 and 1957, high rates of mortality were reported among pregnant women. One report of influenza among 1350 pregnant women in 1918 in the United States (US) showed mortality rates of 27%, while another study during the same time period of 86 pregnant women with influenza infection in Chicago reported a 45% mortality rate [11,12]. During the 1957 pandemic, influenza was listed as the leading cause of death for pregnancy-associated deaths in Minnesota; half of the women of reproductive age who died due to influenza were pregnant at the time [13]. Adverse pregnancy outcomes including high rates of miscarriage and preterm birth were reported among pregnant women during the influenza pandemic of 1918 [11,12]. In addition, concern for increase in congenital defects of fetuses in pregnant women affected by the Asian influenza pandemic of 1957 and seasonal influenza have been reported [13–18]. Over several influenza seasons in the 1970s–1990s, pregnant women were significantly more likely to be hospitalized and to present for medical visits than non-pregnant women [19,20]. This finding was reaffirmed during the H1N1 influenza pandemic (2009), when pregnant women and their fetuses had an increased risk for morbidity and mortality compared to women who were not pregnant [21–23].

The need from a historical perspective: Early childhood

Infant mortality rates have also been decreasing worldwide over the past century. In urban settings in the US and Europe in the late 1800s, up to 30% of children died before their first birthday compared with current infant rates in developing countries of 30.5 deaths per 1000 live births [24,25]. Global childhood mortality rates in children less than 5 years of age have fallen from 18.4% in 1960 to 4.3% in 2015. In 2016, 75% of all deaths in children under 5 years still occurred in the first year of life, demonstrating the need for continued improvement. The highest rates of childhood mortality today occur in the neonatal period, or first month of life, due in large part to complications of birth, prematurity, infections, and congenital anomalies [26]. Although infant survival has dramatically improved through improved perinatal care, increased emphasis on the prenatal visits for women, medically-attended deliveries, and infant follow-up is ongoing internationally with support from the World Health Organization (WHO) and other partners. Immunization is playing a significant role in this effort to improve early childhood survival.

History of immunization in pregnant women

Reports of vaccination date back as far as the 17th and 18th century with the use of smallpox inoculation (also known as variolation) to prevent smallpox infection in China, Turkey and the African continent prior to its spread to Europe and America [27,28]. Historic records published in the 19th century demonstrated that compared to women who had not received smallpox vaccine prior to pregnancy, women who had been vaccinated had at least partial protection against smallpox during pregnancy [5]. In 1879, Burckhardt reported his uncontrolled case series on "intrauterine vaccination" from Basel, Switzerland, in which he showed that infants of women who were re-immunized with smallpox vaccine during pregnancy were refractory to the smallpox vaccine which was administered 4–6 days after birth [29].

In 1904, Polano published his findings on transplacental transfer of tetanus and diphtheria toxoid antibody [30].

In the 20th century, as vaccine research focus shifted to larger scale population-based projects, maternal immunization research also became more population and laboratory-based. Clinical trials on the administration of whole cell pertussis vaccine in pregnancy were conducted in the 1930s and 1940s with variable results. These studies did demonstrate safety in the infant and the potential to transmit protective antibodies to the neonate, but remarkably, did not report on reactogenicity or adverse effects in the pregnant study subjects [31,32]. In the 1940s and 1950s research on the diphtheria vaccine demonstrated diphtheria-specific antibodies in cord blood and the impact of passive immunity in neonates on infant vaccine immune response [33–35]. In the 1950s and 1960s, administration of influenza and polio vaccines during pregnancy became routine given high prevalence of poliomyelitis, known adverse effects of influenza infection in pregnancy, and the benefit of protecting the pregnant woman against both polio and influenza. During that timeframe, safety of live attenuated polio vaccine and inactivated influenza vaccine during pregnancy was demonstrated by long term prospective studies involving 3000 or more women and their infants [36]. Work on vaccines in pregnancy continued globally, with the subsequent publication of an important landmark study in 1961 on tetanus disease in New Guinea. Schofield et al. demonstrated that the risk of neonatal tetanus, an important cause of neonatal mortality globally, was significantly decreased after maternal inoculation with multiple doses of tetanus vaccine during pregnancy [37,38]. Outbreaks of polio in Finland and Israel and meningococcal disease in Brazil were so widespread in the 1970s–1990s that mass immunizations for the entire population of these countries was recommended, which included pregnant women. In the 1970s, 90 million people including pregnant women received the meningococcal A vaccine supplied by Sanofi Pasteur [39]. Follow up studies on the effects of these vaccinations in pregnancy did not demonstrate adverse effects in mother, fetus or infant [40–44]. Nevertheless, by the 1980s, except for certain high-risk groups, maternal immunization in many countries stopped given concerns centered around vaccine safety, efficacy and preparation components for pregnant women. However, this shift in perception of maternal immunization was based on hypothetical potential risk, rather than distinct adverse vaccination outcome data (see "Historical perspectives on women's inclusion in research" section).

In more recent years, early epidemiology reporting increased risks to pregnant women due to pandemic 2009 influenza A/H1N1 infection stimulated research conducted during that influenza pandemic which ultimately demonstrated the beneficial effect of maternal influenza immunization for mothers, fetuses, and infants [21,22]. Maternal pertussis vaccination has also been shown to decrease the morbidity and mortality of pertussis in infants in recent pertussis epidemics by boosting transplacental antibody transfer to the fetus. Another mechanism of infant protection is through the reduction of maternal susceptibility to infection and thereby decreasing the risk of neonatal postnatal exposure [45]. An example of the progression of maternal vaccination is shown in Fig. 1.

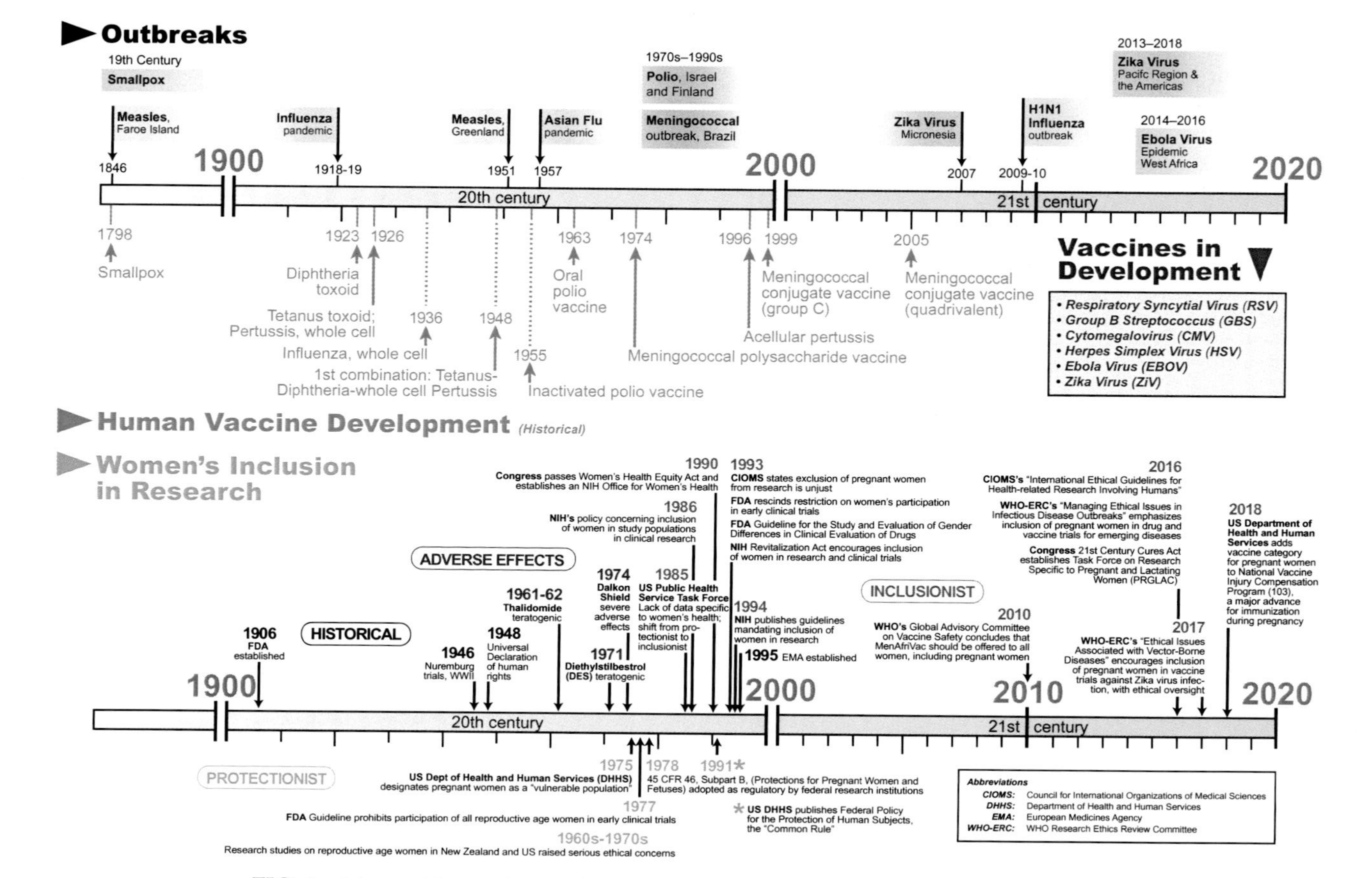

FIG. 1 Maternal immunization development and research through history [28,46–77].

(Continued)

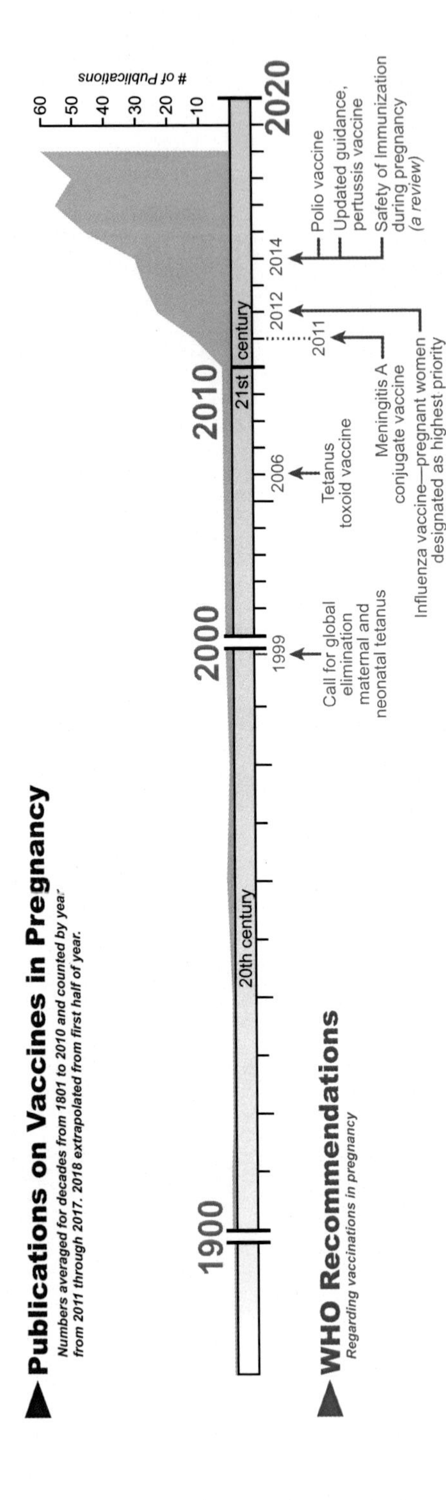
Publications on Vaccines in Pregnancy
Numbers averaged for decades from 1801 to 2010 and counted by year from 2011 through 2017. 2018 extrapolated from first half of year.
# of Publications
60
50
40
30
20
10
1900
2000
2010
2020
20th century
21st century
WHO Recommendations
Regarding vaccinations in pregnancy
1999
Call for global elimination maternal and neonatal tetanus
2006
Tetanus toxoid vaccine
2011
Meningitis A conjugate vaccine
2012
Influenza vaccine—pregnant women designated as highest priority
2014
Polio vaccine
Updated guidance, pertussis vaccine
Safety of Immunization during pregnancy (a review)

FIG. 1, CONT'D

Examples of recent work in select vaccines

Tetanus toxoid vaccine

Following the demonstration of decreased maternal and neonatal tetanus with tetanus toxoid vaccine administration during pregnancy in the 1960s, several global initiatives have been put in place including the resolution to eliminate neonatal tetanus via maternal immunization set forth by the WHO in 1989. This was endorsed in several subsequent global fora and was broadened in 1999 to include the goal to also eliminate maternal tetanus [50]. Other partners in this effort include the United Nations Children's Fund (UNICEF), the United Nations Population Fund (UNFPA), the United States Agency for International Development (USAID), the Global Alliance for Vaccines and Immunizations (Gavi), Save the Children, the Bill & Melinda Gates Foundation, Kiwanis International, and other international and national stakeholders [78]. The overall numbers of cases and countries with maternal and neonatal tetanus (MNT) continues to decrease, with MNT successfully eliminated in North and South America and Europe. However, as of January 2018, 15 countries have yet to eliminate MNT (see Fig. 2) [78]. In addition to maternal immunization, strategies that are currently being used or may be increasingly utilized in the future include tetanus toxoid vaccine boosters to females and males both in childhood and adolescence in order to ensure that women are protected before pregnancy and to decrease the tetanus incidence in the general population [79].

Diphtheria vaccine

Research conducted on the diphtheria toxoid vaccine—particularly by Barr et al. in the mid 20th century—explored not only diphtheria vaccine efficacy in neonates but also demonstrated the effect of maternal antibodies on subsequent neonatal immune response to active immunization (see Fig. 3) [33,34]. Maternal immunization with diphtheria vaccine is not currently used for neonatal protection against diphtheria, but since the formulation of tetanus toxoid and acellular pertussis vaccine contains the vaccine against diphtheria as well (i.e., the tetanus, diphtheria and acellular pertussis vaccine [TdaP]), it is routinely administered in pregnancy [47,81]. Completing the primary series of diphtheria vaccine (3 injections) with one booster in neonates, children or adults has been found to incur long-term immune protection against the disease [81].

Influenza vaccine

Following the 2009 A/H1N1 influenza pandemic and research demonstrating decreased morbidity and mortality of pregnant women and neonates following immunization, the WHO designated pregnant

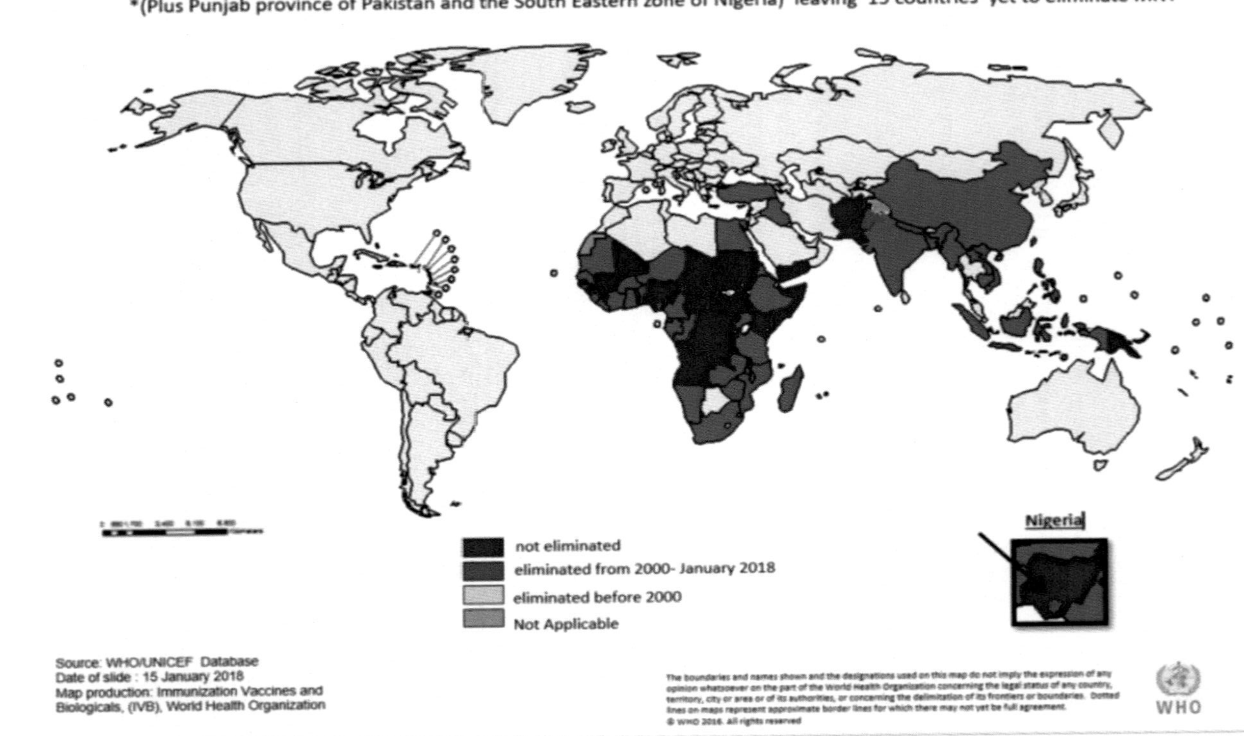

FIG. 2 Elimination of maternal and neonatal tetanus (MNT) by 2018. *Reprinted with permission World Health Organization. 44 Countries eliminated MNT between 2000 & January 2018: unicef.org; 2018. [Available from: https://www.unicef.org/health/index_43509.html].*

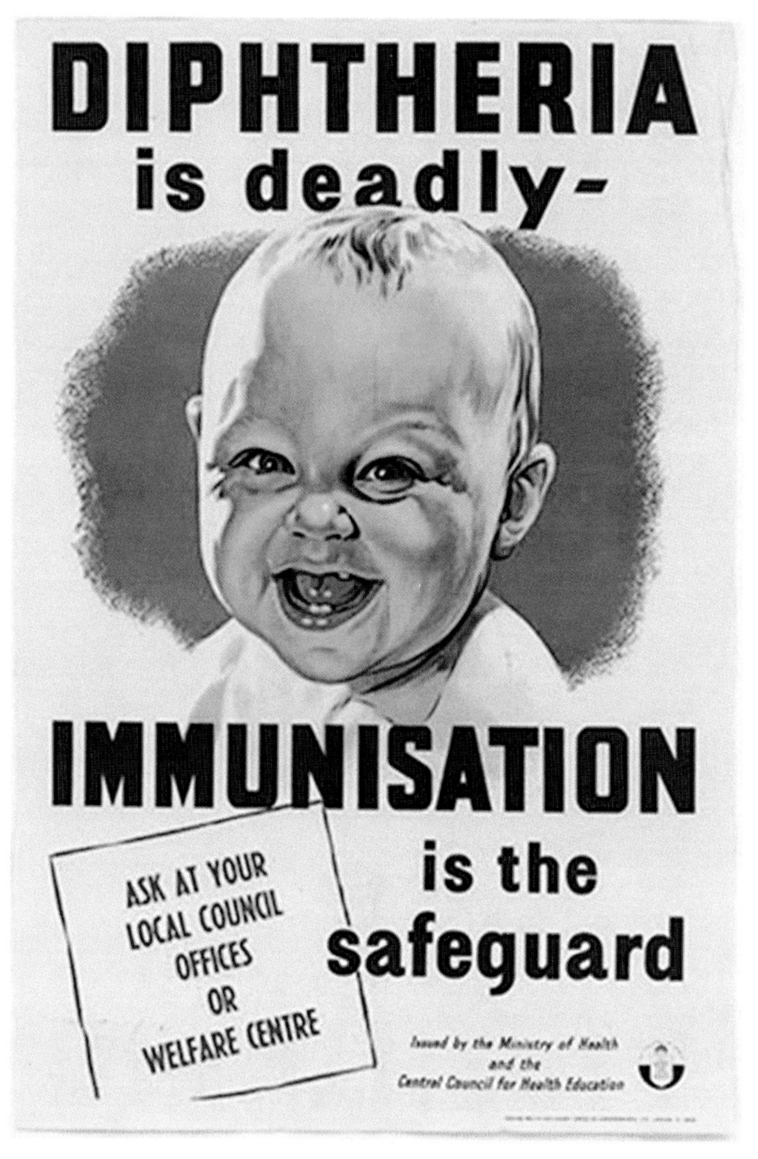

FIG. 3 "Diphtheria is deadly—Immunisation is the safeguard." UK Ministry of Health advertisement for diphtheria toxoid vaccine from the 1960s [80].

women as a priority group for influenza vaccination in 2012. Seasonal influenza vaccine is now recommended by the WHO and other national health authorities (see Fig. 4) [83–85]. The development of maternal influenza vaccination platforms is further discussed in the Influenza chapter.

FIG. 4 "Flu. I'm ready for you." National Health Service (NHS) Scotland advertisement for influenza vaccine in pregnancy [82].

Pertussis vaccine

Based on recent research demonstrating decreased morbidity and mortality in neonates following maternal immunization with acellular pertussis, the United States Centers for Disease Control and Prevention Advisory Committee on Immunization Practices (ACIP) recommended a single dose of acellular pertussis vaccine combined with tetanus and diphtheria toxoid (TdaP) for all pregnant women in 2011. The Joint Committee on Vaccination and Immunization in the United Kingdom (UK) introduced routine administration during pregnancy in 2012, although the vaccine utilized in the UK is the multivalent vaccine dTaP-IPV (low dose diphtheria toxoid, tetanus toxoid, acellular pertussis and inactivated polio antigen) [86]. That same year, the US ACIP changed their single dose recommendation to one recommending maternal acellular pertussis immunization during each pregnancy [87]. The WHO added pertussis vaccination during pregnancy to its pertussis vaccine position paper in 2015 [88]. Pertussis vaccine during each pregnancy is currently recommended by multiple international and national health authorities (see Pertussis chapter).

Current work

Routine vaccination during pregnancy as endorsed by the WHO and health authorities in the UK, the US and an increasing number of countries include tetanus toxoid and acellular pertussis-containing vaccines as well as inactivated influenza vaccine (see Table 1) [46,94–97]. Other vaccinations that may be recommended during pregnancy under certain circumstances, particularly in pregnant women who are unimmunized and at risk for endemic or epidemic exposure, include meningococcal conjugate vaccine, polio vaccine, yellow fever vaccine, and hepatitis A and B vaccines [47]. MenAfriVac, a meningococcus capsular group A vaccine that is conjugated with the tetanus toxoid vaccine, is currently administered to all demographic groups in sub-Saharan Africa including pregnant and lactating women and no adverse effects have been reported [89,90,98]. Given theoretical risks of primary viral infection to mother and fetus, live vaccines such as those for measles, mumps and rubella as well as influenza, varicella and zoster, are not recommended in pregnancy [47]. In cases of less commonly used vaccines, the benefit of administration of the vaccine is weighed against the risk of exposure to mother and fetus [99].

The impact of maternal vaccination on subsequent active vaccination in the infant has been conducted following maternal immunization with pertussis, *Haemophilus influenzae* type b, and influenza vaccines [100–102]. Further evaluation is needed, particularly when maternal immunization with vaccines utilized in primary infant immunization such as DTP (Diphtheria, tetanus and acellular pertussis) are used following maternal immunization with one or multiple similar or identical antigens (e.g., Tdap). This may become increasingly important if the WHO schedule of Expanded Programme on Immunization (EPI) is used in early infancy, as the EPI schedule begins immunization at 6 weeks of age and re-immunizes

TABLE 1 Vaccines currently recommended in pregnancy [28,47,48,51–53,88–93].

Vaccines—licensed	Year developed	Year of WHO recommendation for pregnancy	Type of vaccine	Pregnancy administration	Maternal/neonatal benefit	Safety concerns
Tetanus toxoid	1926	1999; 2006[a]	Protein or polysaccharide	Routine	✓/✓	None
Influenza-inactivated multivalent	1936	2012	Inactivated or recombinant	Routine	✓/✓	None
Polio—oral vaccine	1963	2014	Live attenuated	*If necessary*	✓/No data	None
Pertussis—acellular	1996	2014	Protein	Routine	✓/✓	None
Meningococcal A conjugate	2005	2011	Protein conjugate	*If necessary*	✓/No data	None

Vaccines—in development			Type of vaccine	Pregnancy administration	Maternal/neonatal benefit	Safety
CMV	Pending	–	Various	No data	Possibly/✓	No data
GBS	Pending	–	Multivalent conjugate	No data	✓/✓	No data
HSV	Pending	–	Various	No data	✓/✓	No data
RSV	Pending	–	Protein or polysaccharide	No data	Possibly/possibly	No data
ZiV	Pending	–	Various	No data	Possibly/possibly	No data
EBOV	Pending	–	Recombinant	No data	Possibly/possibly	No data

[a]*The WHO recommendation for elimination of both maternal and neonatal tetanus occurred in 1999. The first WHO position paper on tetanus vaccine in pregnancy was published in 2006.*
Abbreviations: *CMV*, cytomegalovirus; *EBOV*, Ebola virus; *GBS*, group B *Streptococcus*; *HSV*, herpes simplex virus; *RSV*, respiratory syncytial virus; *WHO*, World Health Organization; *ZiV*, Zika virus.

infants at 10 and 14 weeks of age. The first months of life represent a time period when maternal antibodies are likely to be at higher concentrations than later in infancy, such as in current schedules commonly utilized by countries not using EPI vaccine schedules (e.g., primary infant immunization given at 2, 4, 6 and 12 months in US or 3, 5, 12 months in Sweden or 2, 3, 4, and 12 months in the UK) [85,103,104].

Future work

Given the potential of far-reaching protective effect of maternal immunization, from mother to fetus to neonate, several new vaccines are currently in development for use in pregnancy (see Table 1). The RSV F nanoparticle vaccine by Novavax (Gaithersburg, MD) in phase III clinical trials in pregnant women (ClinicalTrials.gov NCT02624947) and smaller studies of other RSV vaccines such as the RSV-PFP2 (purified F protein) have shown immunogenicity and lack of reactogenicity in pregnant women, efficient transplacental antibody transfer, and potential protective effects in the infant [105–108]. Other vaccine candidates for RSV are also under development. Group B *Streptococcus* (GBS) is also a desirable target for vaccine production given its pathogenicity in pregnancy and the neonatal period. Novartis has recently conducted a clinical trial in South Africa with its multivalent conjugate GBS vaccine (ClinicalTrials.gov, NCT01412804). Other future targets for maternal immunization that are currently in the research pipeline and development stages are vaccines against cytomegalovirus (CMV) and herpes simplex virus (HSV). A vaccine against Zika virus is also currently in the research stages [91], refer to relevant chapter. In recent years a vaccine against Ebola infection has been developed and administered for the purposes of disease containment, however it is investigational and safety data for pregnant and lactating women is still pending [92].

Historical perspective on women's inclusion in research

Important factors in the development of maternal immunizations have been the changing perspectives and trends in women's involvement in research. Pharmaceutical development as well as medical research became more regulated in the early 1900s. Particularly "tonics" advertised for women that were found to be addictive played a role in introducing the Food and Drugs Act in 1906 in the US which helped establish the governmental regulatory agency, the Food and Drug Administration (FDA) [54]. The movement toward protection of the rights of human research subjects gained more traction during the Nuremburg trials in 1946 when

Nazi physicians were convicted of crimes against humanity after the end of World War II and ultimately led to the Universal Declaration of Human Rights [55].

In the 1960s and 1970s, several incidents occurred which shaped protectionist policy regarding participation of women in clinical research and use of pharmaceuticals in pregnancy for much of the next three to four decades. In 1961–1962, thalidomide, a drug specifically marketed to pregnant women for treatment of morning sickness and in widespread use in Europe, was found to be teratogenic resulting in severe abnormalities of fetal limb development and other organ systems [56]. In 1971 and 1974, severe adverse health effects were announced in female offspring of pregnant women who used diethylstilbestrol (DES) and the Dalkon Shield as a contraceptive device [57,58]. In addition, research studies conducted on reproductive age women in New Zealand and in the US raised serious ethical concerns [55,109,110]. In response to the increase in pharmaceutical and research scandals and activist outcry, in 1975, the US Department of Health and Human Services (DHHS) designated pregnant women as a "vulnerable population" in which research should only be conducted under special circumstances [55,59]. In 1977, due to concerns about liability, the FDA issued a policy to prohibit reproductive age women and pregnant women from participating in early phases of clinical trials on pharmaceuticals [54]. This policy was commonly interpreted to apply to all clinical phases for drug trials and effectively, within a short period of time, reproductive age women and pregnant women were essentially excluded from participation in clinical research [111].

The US Public Health Service Task Force's report on Women's health Issues in 1985 outlined lack of data specific to women's health and started the movement of policy regarding inclusion of women in research away from the protectionist and more toward an inclusionist approach [60,61]. In 1986, the US National Institutes of Health (NIH) introduced a policy that encouraged but did not mandate the inclusion of women in research trials [59]. Despite these changes, women were not represented in clinical trials at appropriate rates [61]. Following a public outcry, in 1990, the US Congress passed the Women's Health Equity Act that mandated research in areas of women's health and established an office for women's health at the NIH [62,63]. Nonetheless, in 1991 the Federal Policy for the Protection of Human Subjects, also known as the "Common Rule" was published by the US DHHS that contained additional protections for pregnant women, human fetuses, and neonates [76]. Under this "Common Rule," if research involving pregnant women is thought to benefit only the fetus, consent of not only the woman but also that of the biological father is required unless he is unavailable or incompetent or if the pregnancy was a result of rape [112].

In 1993, the Council for International Organizations of Medical Sciences (CIOMS) characterized the exclusion of pregnant women from research as a whole as unjust [64,65]. This was supported by the FDA and the European Medicines Agency, the agency of the European Union responsible for evaluation, regulation and monitoring of medicines which was established in 1995 [66–68]. In the same year, following the US General Accounting Office's report on the FDA's policy toward women titled, "Women's Health: FDA Needs to Ensure More Study of Gender Differences in Prescription Drug Testing," the FDA issued their "Guideline for the Study and Evaluation of Gender Differences in the Clinical Evaluation of Drugs" which outlined expectations that both men and women be involved in clinical drug trials [68,69]. The NIH also released its Revitalization Act of 1993 (P.L. 103-43) which encouraged the inclusion of women and minorities in research and clinical trials, and was followed by specific guidelines to this effect in 1994 [70,71]. In 2016, the US Congress' *21st Century Cures Act* in the established the Task Force on Research Specific to Pregnant Women and Lactating Women (PRGLAC) to advise the US DHHS on priorities for research and therapies for pregnant and lactating women [75,113]. Revisions of the Common Rule are projected to occur in early 2019 which may remove the requirement for paternal consent and the designation of pregnant women as a vulnerable population [113].

An example of inclusionist policy toward pregnant women in research is that of MenAfriVac mass immunization campaigns. In 2010, preparations were being made for mass immunizations in West Africa against endemic meningococcal disease. The discussion at the time was whether pregnant women should be excluded from this campaign given lack of long-term safety data [72]. The WHO's Global Advisory Committee on Vaccine Safety concluded that MenAfriVac should be offered to all women including pregnant women based on (1) clear benefit from immunization for pregnant women, (2) no alternative way to protect against meningococcal disease, (3) clear risk for disease in this geographical area and (4) lack of adverse safety data. At the same time, the committee also emphasized that data should be collected regarding safety and adverse events especially in pregnant and lactating women [72].

Despite advances, the discussion regarding representation of pregnant women in research continues. This was especially important during the Ebola treatment trials in West Africa in the outbreak in 2013–2016. Given extremely high maternal mortality and fetal and neonatal mortality nearing 100% with Ebola infection in pregnancy, the WHO Research Ethics Review Committee (WHO-ERC) emphasized inclusion of pregnant women and children in research protocols [73,114]. Nonetheless, most pregnant women were excluded from drug and vaccine trials during the Ebola outbreak [115]. Pregnant and lactating women were also excluded from receiving an investigational vaccine against Ebola during an outbreak in the Democratic Republic of Congo in 2018 due to lack of safety data [92]. In regards to other emerging diseases, in 2017 a WHO scoping meeting on vector-borne diseases encouraged inclusion of pregnant women and reproductive-age women in future vaccine trials against the Zika virus with ethical oversight [74]. In the 21st century, progress has been made due to policy changes by international and national regulatory and research agencies. However, more work is still needed [61,65].

Current challenges

Maternal immunizations continue to face several challenges including regulatory and legal issues, medical-legal liability and risk for vaccine manufacturers in the research and development of vaccines for use in pregnancy. Currently, none of the vaccines recommended for use in pregnancy has actually been licensed by regulatory authorities for use during pregnancy. In 2018 a proposal was submitted by the US Department of Health and Human Services to add the category of vaccines recommended for pregnant women to the Vaccine Injury Table within the National Vaccine Injury Compensation Program [116], a major potential advance for immunization during pregnancy. While there has been a paradigm shift in recent years from protection pregnant women *from* research to protecting them *through* research, there is still room for improvement [117].

Conclusion

Maternal immunization research and clinical application has a rich history and an even brighter future. Pregnancy and the neonatal period continue to come into focus as areas of medicine that necessitate more rigorous research and investment. The history of women's involvement in research is a cautionary tale of the need to avoid "protecting" women from research, while also recognizing the complex nature of pregnancy and emphasizing the need for research as a vehicle to making pregnancy and the postpartum period safer. It is also important to study and highlight maternal benefit from vaccines in pregnancy as an end in itself beyond emphasizing the woman serving solely as a vehicle for fetal and neonatal disease prevention. Maternal immunization is a complex field that must take into account multiple factors: maternal immunogenicity, transplacental antibody transfer, the immune system of the neonatal and subsequent neonatal immunogenicity, as well as potential adverse effects in mother and infant. The continued study of established vaccines as well as the development of new vaccines will continue to challenge and inspire efforts to improve maternal and neonatal health globally in the future.

Acknowledgments

We would like to express our gratitude to Jan Hamanishi for her help in designing Fig. 1 for this article.

Conflict of interest

AK and LOE have no financial conflicts to disclose. JAE has received research support to her institution from GlaxoSmithKline, Gilead, MedImmune, Novavax, and Chimerix. She has served as a consultant for Sanofi Pasteur and Gilead.

References

[1] Heath PT, Culley FJ, Jones CE, Kampmann B, Le Doare K, Nunes MC, et al. Group B streptococcus and respiratory syncytial virus immunisation during pregnancy: a landscape analysis. Lancet Infect Dis 2017;17(7):e223–34.
[2] Loudon I. Maternal mortality in the past and its relevance to developing countries today. Am J Clin Nutr 2000;72(1):241S–6S.
[3] World Health Organization. Trends in maternal mortality: 1990 to 2015: estimates by WHO, UNICEF, UNFPA, World Bank Group and the United Nations Population Division. Geneva: WHO; 2015.
[4] WHO, UNICEF, UNFPA, World Bank Group, United Nations Population Division Maternal Mortality Estimation Inter-Agency Group. Maternal mortality in 1990–2015. Democratic Republic of the Congo: World Health Organization; 2015. Available from: http://www.who.int/gho/maternal_health/countries/cod.pdf.
[5] Nishiura H. Smallpox during pregnancy and maternal outcomes. Emerg Infect Dis 2006;12(7):1119–21.
[6] Klotz H. Beitrage zur Pathologie der Schwangerschaft. Arch Gynakol 1887;29(3):448–75.
[7] Christensen PE, Schmidt H, Jensen O, Bang HO, Andersen V, Jordal B. An epidemic of measles in southern Greenland, 1951. I. Measles in virgin soil. Acta Med Scand 1953;144(4):313–22.
[8] Atmar RL, Janet AE, Hunter H. Complications of measles during pregnancy. Clin Infect Dis 1992;14(1):217–26.
[9] Panum P. Observations made during the epidemic of measles on the Faroe Islands in the year 1846. New York: Delta Omega Society; 1940.
[10] Rasmussen SA, Jamieson DJ, Bresee JS. Pandemic influenza and pregnant women. Emerg Infect Dis 2008;14(1):95–100.
[11] Nuzum JW, Pilot I, Stangl FH, Bonar BE. 1918 Pandemic influenza and pneumonia in a large civil hospital. IMJ Ill Med J 1976;150(6):612–6.
[12] Harris J. Influenza occurring in pregnant women. JAMA 1919;72:978–80.
[13] Freeman DW, Barno A. Deaths from Asian influenza associated with pregnancy. Am J Obstet Gynecol 1959;78:1172–5.
[14] Hardy JM, Azarowicz EN, Mannini A, Medearis Jr DN, Cooke RE. The effect of Asian influenza on the outcome of pregnancy, Baltimore, 1957-1958. Am J Public Health Nations Health 1961;51:1182–8.
[15] Acs N, Banhidy F, Puho E, Czeizel AE. Maternal influenza during pregnancy and risk of congenital abnormalities in offspring. Birth Defects Res A Clin Mol Teratol 2005;73(12):989–96.

[16] Coffey VP, Jessop WJ. Maternal influenza and congenital deformities. A follow-up study. Lancet 1963;1(7284):748–51.
[17] Wilson MG, Stein AM. Teratogenic effects of Asian influenza. An extended study. JAMA 1969;210(2):336–7.
[18] Saxen L, Hjelt L, Sjostedt JE, Hakosalo J, Hakosalo H. Asian influenza during pregnancy and congenital malformations. Acta Pathol Microbiol Scand 1960;49:114–26.
[19] Neuzil KM, Reed GW, Mitchel EF, Simonsen L, Griffin MR. Impact of influenza on acute cardiopulmonary hospitalizations in pregnant women. Am J Epidemiol 1998;148(11):1094–102.
[20] Mullooly JP, Barker WH, Nolan Jr TF. Risk of acute respiratory disease among pregnant women during influenza A epidemics. Public Health Rep 1986;101(2):205–11.
[21] Haberg SE, Trogstad L, Gunnes N, Wilcox AJ, Gjessing HK, Samuelsen SO, et al. Risk of fetal death after pandemic influenza virus infection or vaccination. N Engl J Med 2013;368(4):333–40.
[22] Siston AM, Rasmussen SA, Honein MA, Fry AM, Seib K, Callaghan WM, et al. Pandemic 2009 influenza A(H1N1) virus illness among pregnant women in the United States. JAMA 2010;303(15):1517–25.
[23] Omer SB, Goodman D, Steinhoff MC, Rochat R, Klugman KP, Stoll BJ, et al. Maternal influenza immunization and reduced likelihood of prematurity and small for gestational age births: a retrospective cohort study. PLoS Med 2011;8(5):e1000441.
[24] Centers for Disease Control and Prevention. Achievements in public health, 1900–1999: healthier mothers and babies. MMWR CDC Surveill Summ 1999;48(38):849–58.
[25] The World Bank. Mortality rate, infant (per 1,000 live births). World Bank Group; 2016. Available from: https://data.worldbank.org/indicator/SP.DYN.IMRT.IN.
[26] Liu L, Oza S, Hogan D, Chu Y, Perin J, Zhu J, et al. Global, regional, and national causes of under-5 mortality in 2000-15: an updated systematic analysis with implications for the sustainable development goals. Lancet 2016;388(10063):3027–35.
[27] All Timelines Overview. The College of Physicians of Philadelphia. Updated. Available from: https://www.historyofvaccines.org/timeline; 2018.
[28] Plotkin S, Plotkin S. A short history of vaccination. In: Plotkin S, Orenstein W, Offit P, editors. Vaccines. 6th ed. St. Louis, MO: Elsevier Saunders; 2013. p. 1–13.
[29] Burckhardt A. Zur intrauterinen vaccination. Deut Arch Klin Med 1879;24:506–9.
[30] Polano O. Der Antitoxinubergang von der Mutter auf das Kind. Ein Beitrag zur Physiologie der Placenta. Ztschr f Geburtsh u Gynak 1904;53:456–76.
[31] Lichty JA, Slavin B, Bradford WL. An attempt to increase resistance to pertussis in newborn infants by immunizing their mothers during pregnancy. J Clin Invest 1938;17(5):613–21.
[32] Cohen P, Scadron S. The placental transmission of protective antibodies against whooping cough by inoculation of the pregnant mother. JAMA 1943;121:656–62.
[33] Barr M, Glenny AT, Randall KJ. Diphtheria immunization in young babies; a study of some factors involved. Lancet 1950;1(6593):6–10.
[34] Barr M, Glenny AT, Randall KJ. Concentration of diphtheria antitoxin in cord blood and rate of loss in babies. Lancet 1949;2(6573):324–6.
[35] Brescia MA, Tartaglione EF. Prenatal diphtheria immunization. Arch Pediatr 1948;65(12):633–9.
[36] Heinonen OP, Shapiro S, Monson RR, Hartz SC, Rosenberg L, Slone D. Immunization during pregnancy against poliomyelitis and influenza in relation to childhood malignancy. Int J Epidemiol 1973;2(3):229–35.
[37] Schofield FD, Tucker VM, Westbrook GR. Neonatal tetanus in New Guinea. Effect of active immunization in pregnancy. Br Med J 1961;2(5255):785–9.
[38] Stahlie TD. The role of tetanus neonatorum in infant mortality in Thailand. J Trop Pediatr Afr Child Health 1960;6:15–8.
[39] Sanofi. 40 Years against meningococcal meningitis. Paris, France: Sanofi; 2014.

[40] Harjulehto-Mervaala T, Aro T, Hiilesmaa VK, Saxen H, Hovi T, Saxen L. Oral polio vaccination during pregnancy: no increase in the occurrence of congenital malformations. Am J Epidemiol 1993;138(6):407–14.
[41] Harjulehto T, Hovi T, Aro T, Saxén L. Congenital malformations and oral poliovirus vaccination during pregnancy. Lancet 1989;333(8641):771–2.
[42] Ornoy A, Ben IP. Congenital anomalies after oral poliovirus vaccination during pregnancy. Lancet 1993;341(8853):1162.
[43] Linder N, Handsher R, Fruman O, Shiff E, Ohel G, Reichman B, et al. Effect of maternal immunization with oral poliovirus vaccine on neonatal immunity. Pediatr Infect Dis J 1994;13(11):959–62.
[44] McCormick JB, Gusmao HH, Nakamura S, Freire JB, Veras J, Gorman G, et al. Antibody response to serogroup A and C meningococcal polysaccharide vaccines in infants born of mothers vaccinated during pregnancy. J Clin Invest 1980;65(5):1141–4.
[45] Amirthalingam G, Andrews N, Campbell H, Ribeiro S, Kara E, Donegan K, et al. Effectiveness of maternal pertussis vaccination in England: an observational study. Lancet 2014;384(9953):1521–8.
[46] Pertussis vaccines: WHO position paper. Wkly Epidemiol Rec 2010;85(40):385–400.
[47] WHO. Safety of immunization during pregnancy: a review of the evidence. Geneva: World Health Organization; 2014. Available from: http://www.who.int/vaccine_safety/publications/safety_pregnancy_nov2014.pdf.
[48] WHO. Meningococcal A conjugate vaccine: updated guidance, February 2015. Geneva: World Health Organization; 2015.
[49] WHO. Fact sheet: poliomyelitis. World Health Organization; 2017. Updated 2017 April. Available from: http://www.who.int/mediacentre/factsheets/fs114/en/.
[50] WHO. Maternal and neonatal tetanus elimination (MNTE). World Health Organization; 2017. Updated 2017 August 1. Available from: http://www.who.int/immunization/diseases/MNTE_initiative/en/.
[51] WHO. Tetanus vaccines: WHO position paper—February 2017. Geneva: World Health Organization; 2017.
[52] WHO. Vaccines against influenza WHO position paper—November 2012. Geneva: World Health Organization; 2012.
[53] WHO. Polio vaccines: WHO position paper—March, 2016. Geneva: World Health Organization; 2016.
[54] U.S. Food and Drug Administration. 100 Years of protecting and promoting women's health. U.S. Department of Health and Human Services; 2015. Available from: https://www.fda.gov/ForConsumers/ByAudience/ForWomen/ucm118458.htm#1906:_Fighting_Addictive__Medicines.
[55] Stevens PE, Pletsch PK. Informed consent and the history of inclusion of women in clinical research. Health Care Women Int 2002;23(8):809–19.
[56] Administration USFaD. About the Office of Scientific Investigations. U.S. Department of Health and Human Services; 2014. Available from: https://www.fda.gov/AboutFDA/CentersOffices/OfficeofMedicalProductsandTobacco/CDER/ucm091393.htm.
[57] Centers for Disease Control and Prevention. DES history. CDC; 2011. Available from: https://www.cdc.gov/des/consumers/about/history.html.
[58] Elevated risk of pelvic inflammatory disease among women using the Dalkon Shield. MMWR Morb Mortal Wkly Rep 1983;32(17):221–2.
[59] Johnson T, Fee E. Women's participation in clinical research: from protectionism to access. In: Mastroianni A, Faden R, Federman D, editors. Women and health research: ethical and legal issues of including women in clinical studies, volume 2 workship and commissioned papers. Washington, DC: National Academy Press (US); 1994.
[60] Women's health. Report of the public health service task force on women's health issues. Public Health Rep 1985;100(1):73–106.

[61] Denny C, Grady C. Research involving women. In: Emanuel E, Grady C, Crouch R, Lie R, Miller F, Wendler D, editors. The Oxford textbook of clinical research ethics. New York, NY: Oxford University Press; 2008. p. 407–22.
[62] U.S. Congress. Women's Health Equity Act of 1990. H. R. 5397, S.2961: 101st Congress, 2nd Session; 1990.
[63] National Institutes of Health. Office of research on women's health: legislative mandate. Available from: https://orwh.od.nih.gov/career/mentored/legislative-mandate.
[64] CIOMS. International ethical guidelines for biomedical research involving human subjects, Geneva, 1993 and 2002; and International ethical guidelines for health-related research involving humans. Geneva: Council for International Organizations of Medical Sciences (CIOMS); 2016.
[65] van der Graaf R, van der Zande ISE, den Ruijter HM, Oudijk MA, van Delden JJM, Oude Rengerink K, et al. Fair inclusion of pregnant women in clinical trials: an integrated scientific and ethical approach. Trials 2018;19:78.
[66] European Medicines Agency. 2016. About us: EMA, Updated March 29, 2019; Available from: http://www.ema.europa.eu/ema/index.jsp?curl=pages/about_us/document_listing/document_listing_000426.jsp&mid=; .
[67] EMA (European Medicines Agency). Guideline on the exposure to medicinal products during pregnancy: need for post-authorisation data. Available from: http://www.ema.europa.eu/docs/en_GB/document_library/Regulatory_and_procedural_guideline/2009/11/WC500011303.pdf; 2005.
[68] Guideline for the study and evaluation of gender differences in the clinical evaluation of drugs; notice. Fed Regist 1993;58(139):39406–16.
[69] General Accounting Office. Women's health: FDA needs to ensure more study of gender differences in prescription drug testing. GAO/HRD-93-17. Washington, DC: GAO; 1992.
[70] U.S. Congress. NIH Revitalization Act of 1993. Pub. L. 103-43, June 10,1993.
[71] National Institutes of Health, Department of Health and Human Services. NIH guidelines on the inclusion of women and minorities as subjects in clinical research; notice. Fed Regist 1994;59(59):14508–13.
[72] World Health Organization. Committee concludes that new meningitis vaccine is safe and should be offered to pregnant women. WHO; 2011. Available from: http://www.who.int/immunization/newsroom/newsstory_meningitis_vaccine_safe_pregnancy_jan2011/en/.
[73] World Health Organization. Managing ethical issues in infectious disease outbreaks. Geneva: WHO; 2016.
[74] World Health Organization. Ethical issues associated with vector-borne diseases. In: Report of a WHO scoping meeting, Geneva, 23–24 February 2017. Geneva: WHO; 2017.
[75] U.S. Congress. H.R. 34; 114th congress of the United States of America: congress.gov. Available from: https://www.congress.gov/114/bills/hr34/BILLS-114hr34enr.pdf; 2016.
[76] Office for Human Research Protection. Federal policy for the protection of human subjects ('common rule'). U.S. Department of Health & Human Services; 2016. Available from: https://www.hhs.gov/ohrp/regulations-and-policy/regulations/common-rule/index.html.
[77] Kindhauser MK, Allen T, Frank V, Santhana RS, Dye C. Zika: the origin and spread of a mosquito-borne virus. Bull World Health Organ 2016;94(9):675–686c.
[78] World Health Organization. 44 Countries eliminated MNT between 2000 & January 2018: unicef.org. Available from: https://www.unicef.org/health/index_43509.html; 2018.
[79] Ridpath AD, Scobie HM, Shibeshi ME, Yakubu A, Zulu F, Raza AA, et al. Progress towards achieving and maintaining maternal and neonatal tetanus elimination in the African region. Pan Afr Med J 2017;27(Suppl 3):24.
[80] UK Ministry of Health. Diphtheria warning poster from 1960s from UK Ministry of Health. NEWS; 2018. Available from: http://mobile.abc.net.au/news/2018-02-08/diptheria-warning-poster-from-1960s/9407962?pfm=sm.

[81] World Health Organization. Diphtheria vaccine: WHO position paper—August 2017. Geneva: WHO; 2017. p. 417–36.
[82] NHS Scotland. Flu. Pregnant Women 2017 Poster. Scottish Government; 2016. Available from: http://healthyhighlanders.co.uk/HPAC/MoreDetailsv4.jsp?id=1354&subjectId=45&referrer=http://healthyhighlanders.co.uk/HPAC/BrowseSearchv4.jsp?subjectId=45&typeId=P&sort=dater&page=1&submit=true&newSearch=true&dsn=hphighland.
[83] WHO. Vaccines against influenza WHO position paper. Wkly Epidemiol Rec 2012;87:461–76.
[84] CDC. Prevention and control of seasonal influenza with vaccines. Recommendations of the Advisory Committee on Immunization Practices—United States, 2013–2014. MMWR Recomm Rep 2013;62(Rr-07):1–43.
[85] The routine immunisation schedule; from Spring 2018. Public Health England; 2018. Available from: https://assets.publishing.service.gov.uk/government/uploads/system/uploads/attachment_data/file/699392/Complete_immunisation_schedule_april2018.pdf.
[86] Royal College of Obstetricians & Gynecologists. RCOG statement: pertussis (whooping cough) vaccination now offered from 20 weeks of pregnancy: RCOG; Updated 21 April 2016. Available from: https://www.rcog.org.uk/en/news/rcog-statement-pertussis-whooping-cough-vaccination-now-offered-from-20-weeks-of-pregnancy/.
[87] Matlow JN, Pupco A, Bozzo P, Koren G. Tdap vaccination during pregnancy to reduce pertussis infection in young infants. Can Fam Physician 2013;59(5):497–8.
[88] WHO. Pertussis vaccines: WHO position paper—August 2015. Geneva: World Health Organization; 2015. Updated 28 August 2015. Available from: http://www.who.int/wer/2015/wer9035.pdf?ua=1.
[89] Wak G, Williams J, Oduro A, Maure C, Zuber PL, Black S. The safety of PsA-TT in pregnancy: an assessment performed within the Navrongo Health and Demographic Surveillance Site in Ghana. Clin Infect Dis 2015;61(Suppl 5):S489–92.
[90] Ateudjieu J, Stoll B, Nguefack-Tsague G, Yakum MN, Mengouo MN, Genton B. Incidence and types of adverse events during mass vaccination campaign with the meningococcal a conjugate vaccine (MENAFRIVAC) in Cameroon. Pharmacoepidemiol Drug Saf 2016;25(10):1170–8.
[91] Wang R, Liao X, Fan D, Wang L, Song J, Feng K, et al. Maternal immunization with a DNA vaccine candidate elicits specific passive protection against post-natal Zika virus infection in immunocompetent BALB/c mice. Vaccine 2018;36(24):3522–32.
[92] World Health Organization. Frequently asked questions on compassionate use of investigational vaccine for the Ebola virus disease outbreak in Democratic Republic of the Congo. WHO; 2018. Available from: http://www.who.int/ebola/drc-2018/faq-vaccine/en/.
[93] Keller-Stanislawski B, Englund JA, Kang G, Mangtani P, Neuzil K, Nohynek H, et al. Safety of immunization during pregnancy: a review of the evidence of selected inactivated and live attenuated vaccines. Vaccine 2014;32(52):7057–64.
[94] CDC. Guidelines for vaccinating pregnant women. Centers for Disease Control and Prevention; 2012. Available from: https://www.cdc.gov/vaccines/pregnancy/hcp/guidelines.html.
[95] CDC. Recommended immunization schedule for adults aged 19 years or older, by vaccine and age group. Centers for Disease Control and Prevention; 2017. Updated 6 February 2017. Available from: https://www.cdc.gov/vaccines/schedules/hcp/imz/adult.html.
[96] The routine immunisation schedule, from Spring 2018: Public Health England. Available from: https://assets.publishing.service.gov.uk/government/uploads/system/uploads/attachment_data/file/699392/Complete_immunisation_schedule_april2018.pdf; 2018.
[97] WHO. Table 1: summary of WHO position papers—recommendations for routine immunization. World Health Organization; 2017. Updated March 2017. Available from: http://www.who.int/immunization/policy/Immunization_routine_table1.pdf?ua=1.

[98] WHO. Weekly epidemiological record: meningitis A conjugate vaccine. World Health Organization; 2011. Updated 28 January 2011. Available from: http://www.who.int/wer/2011/wer8605.pdf?ua=1.
[99] CDC. Guidelines for vaccinating pregnant women. Centers for Disease Control and Prevention; 2016. Last updated August 2016, Accessed July 6, 2017. Available from: https://www.cdc.gov/vaccines/pregnancy/hcp/guidelines.html.
[100] Walter EB, Englund JA, Blatter M, Nyberg J, Ruben FL, Decker MD. Trivalent inactivated influenza virus vaccine given to two-month-old children: an off-season pilot study. Pediatr Infect Dis J 2009;28(12):1099–104.
[101] Munoz FM, Bond NH, Maccato M, Pinell P, Hammill HA, Swamy GK, et al. Safety and immunogenicity of tetanus diphtheria and acellular pertussis (Tdap) immunization during pregnancy in mothers and infants: a randomized clinical trial. JAMA 2014;311(17):1760–9.
[102] Nahm MH, Glezen P, Englund J. The influence of maternal immunization on light chain response to Haemophilus influenzae type b vaccine. Vaccine 2003;21(24):3393–7.
[103] Centers for Disease Control and Prevention. Recommended immunization schedule for children and adolescents aged 18 years or younger, United States, 2018. CDC; 2018. Available from: https://www.cdc.gov/vaccines/schedules/hcp/imz/child-adolescent.html.
[104] Public Health Agency of Sweden. Vaccination programmes. Public Health Agency of Sweden; 2018. Available from: https://www.folkhalsomyndigheten.se/the-public-health-agency-of-sweden/communicable-disease-control/vaccinations/vaccination-programmes/.
[105] PATH. RSV vaccine and mAb snapshot. PATH; 2017. Updated March 3 2017. Available from: http://www.path.org/publications/files/CVIA_rsv_snapshot_final.pdf.
[106] Glenn GM, Smith G, Fries L, Raghunandan R, Lu H, Zhou B, et al. Safety and immunogenicity of a Sf9 insect cell-derived respiratory syncytial virus fusion protein nanoparticle vaccine. Vaccine 2013;31(3):524–32.
[107] Ochola R, Sande C, Fegan G, Scott PD, Medley GF, Cane PA, et al. The level and duration of RSV-specific maternal IgG in infants in Kilifi Kenya. PLoS One 2009;4(12):e8088.
[108] Munoz FM, Piedra PA, Glezen WP. Safety and immunogenicity of respiratory syncytial virus purified fusion protein-2 vaccine in pregnant women. Vaccine 2003;21(24):3465–7.
[109] Goldzieher JW, Moses LE, Averkin E, Scheel C, Taber BZ. Nervousness and depression attributed to oral contraceptives: a double-blind, placebo-controlled study. Am J Obstet Gynecol 1971;111(8):1013–20.
[110] Campbell AV. A report from New Zealand: an unfortunate experiment. Bioethics 1989;3(1):59–66.
[111] McCarthy CR. Historical background of clinical trials involving women and minorities. Acad Med 1994;69(9):695–8.
[112] Office for Human Research Protection. 45 CFR 46. U.S. Department of Health & Human Services; 2010. Available from: https://www.hhs.gov/ohrp/regulations-and-policy/regulations/45-cfr-46/index.html#subpartb.
[113] Rubin R. Addressing barriers to inclusion of pregnant women in clinical trials. JAMA 2018;320(8):742–4.
[114] Alirol E, Kuesel AC, Guraiib MM, de la Fuente-Núñez V, Saxena A, Gomes MF. Ethics review of studies during public health emergencies—the experience of the WHO ethics review committee during the Ebola virus disease epidemic. BMC Med Ethics 2017;18:43.
[115] Gomes MF, de la Fuente-Nunez V, Saxena A, Kuesel AC. Protected to death: systematic exclusion of pregnant women from Ebola virus disease trials. Reprod Health 2017;14(Suppl 3):172.
[116] Health Resources & Services Administration. National vaccine injury compensation program. HRSA; 2018. Available from: https://www.hrsa.gov/vaccine-compensation/index.html.
[117] Roberts JN, Graham BS, Karron RA, Munoz FM, Falsey AR, Anderson LJ, et al. Challenges and opportunities in RSV vaccine development: meeting report from FDA/NIH workshop. Vaccine 2016;34(41):4843–9.

CHAPTER

2

Vaccination of women in the pre-conception and post-partum periods

Nicholas Wood

National Centre for Immunisation Research and Surveillance, The Children's Hospital at Westmead, Sydney, NSW, Australia, The University of Sydney Children's Hospital Westmead Clinical School, Sydney, NSW, Australia

Introduction

Ideally pregnancies are planned, and pre-conception medical consultation occurs, however this is not often part of routine healthcare. Where a pre-conception medical consultation takes place, an important part of this consultation is the assessment of the need for vaccination, particularly for hepatitis B, varicella and rubella. The main reason is to identify women who are non-immune and therefore at risk of being infected with these diseases during pregnancy, where fetal outcomes can be severe. Identification of non-immune women will enable appropriate "catch up" vaccines to be administered before pregnancy. As it is contraindicated to vaccinate a pregnant woman with live viral vaccines it is important to identify those women needing vaccination with live viral vaccines prior to pregnancy. Serological screening is recommended to ascertain immunity to selected vaccine preventable diseases, including hepatitis B, measles, mumps and rubella as part of any pre-conception health check [1]. Routine serological testing to these viruses can indicate whether previous natural infection has occurred, however it does not provide a reliable measure of vaccine-induced immunity [2]. Influenza vaccine is recommended for any person who wishes to be protected against influenza and is strongly recommended for women planning pregnancy [3]. In some settings, women

https://doi.org/10.1016/B978-0-12-814582-1.00002-4

with additional pneumococcal disease risk factors, including smokers and Indigenous women, are recommended to be assessed for pneumococcal vaccination.

During any clinic consultation of women of child-bearing age who present for immunization it is recommended that they be questioned regarding the possibility of pregnancy as part of the routine pre-vaccination screening. This aims to prevent inadvertent administration of a vaccine(s), particularly as live viral vaccines are not recommended to be given in pregnancy. A non-pregnant woman who has been given a live viral vaccine is recommended to avoid becoming pregnant for approximately 1 month [2, 4, 5].

Overall, the goals of a pre-partum consultation are to identify women who may benefit from vaccination administration of live viral vaccines prior to becoming pregnant, ensure that if this does not occur that they be given these vaccines in the immediate postpartum period and inform and educate women about the need for both pertussis and influenza vaccines during pregnancy. The overall benefit is to improve protection against certain vaccine preventable diseases for both mother and infant.

The post-partum period is an opportune time to "catch up" on vaccines that women have not received during pregnancy by taking advantage of the woman being located in a medical facility or having contact with healthcare workers. Standing orders for the administration of specific vaccines may facilitate their administration in healthcare facilities. If a woman commenced the tetanus vaccine schedule during pregnancy, then it is important that she completes the remainder of the course to receive three doses in total of tetanus containing vaccine.

This chapter reviews the preconception vaccine assessment, including hepatitis B, rubella, varicella and administration of vaccines in the post-partum period and safety of vaccination whilst breastfeeding.

Assessment of vaccination status and vaccination in the pre-conceptual period

Hepatitis B

Pregnant women who are chronically infected with Hepatitis B Virus (HBV) can transmit the virus to the fetus either in utero or to the newborn at or around the time of delivery through exposure to blood or blood-contaminated fluids [6]. The risks of transmission to the fetus vary according to the gestation in which the woman becomes acutely infected. If a woman has an acute hepatitis B infection during the first trimester the risk of transmission is estimated at 10%. If acute infection occurs during the third trimester, the risk of transmission climbs significantly to 90% [1, 6].

Estimates are that transmission in the perinatal period is believed to account for 35–50% of hepatitis B carriers [7]. An important determinant of the risk of transmission is the hepatitis B envelope antigen (HBeAg) status of the mother. If a woman is both hepatitis surface antigen (HBsAg) and HBeAg positive, there is a risk of transmission of 70–90%, in the absence of preventative measures [8, 9]. If the mother is negative to HBsAg positive and HBeAg, the risk is significantly reduced [10–14]. In a cohort study of HBsAg-positive, hepatitis B DNA-positive women (n=313), where the infants received both hepatitis B immunoglobulin and vaccine, transmission rates were 3% among hepatitis B DNA-positive women overall, higher (7%) among HBeAg-positive mothers and highest (9%) among women with very high hepatitis B DNA levels when the infants were assessed at 9 months of age [15]. In addition to maternal-to-child HBV transmission, a large epidemiological study in China has shown that mothers with chronic HBV infection have an increased risk of preterm delivery [16]. This increased risk could potentially be reduced by early diagnosis of chronic HBV infection pre-pregnancy and initiation of appropriate medical intervention before a woman conceives [16, 17].

As part of preconception screening, HBV serology (including HBsAg and HBeAg) and HBV DNA viral load measurement (if found to be HBeAg/HBsAg positive), should be performed. Testing of all pregnant women for hepatitis B is recommended in many countries including the United Kingdom [18], Australia and the United States [19–21]. Universal testing of all women rather than selective testing based on risk of being infected is supported by the findings of several observational studies into selective testing. Several studies indicate that using risk factors as a basis to test and subsequently identify "high-risk" women for HBsAg may miss up to half of all pregnant women with hepatitis B infection [20, 22–24].

The rationale for testing prior to pregnancy is two-fold, first to identify women with chronic hepatitis B infection, as this then leads to further assessment with hepatitis B viral load measurement and allows the possibility of treatment with anti-viral medications prior to pregnancy – when treatment options are more limited. Second, detection in the antenatal period directs future recommendation for hepatitis B immunoglobulin to be given at birth (ideally within 12h of birth) to infants born to chronically infected women (both HBsAg and HBeAg positive). All non-immune or non-HBV vaccinated women are recommended to receive HBV vaccine. Vaccination is highly recommended for all high-risk women (this includes the following groups: household and sexual contacts of HBV chronically infected individuals, women with sexually transmitted diseases or other high-risk behaviors that include injecting drug use, multiple sex partners, travelers to regions were hepatitis B is endemic, workers in healthcare, public safety, and institutions) [1, 2, 19].

Varicella

In order to prevent congenital varicella infection, and its severe consequences, assessment of vaccine status or prior infection with varicella is important in the pre-conception period, as varicella vaccine is contraindicated during pregnancy, since it is a live vaccine. All women booking for antenatal care should be asked as to whether they have had previous varicella infection or varicella vaccine. Routine testing of women of child bearing age is not recommended in the majority of countries globally. This is in part due to variations in varicella epidemiology globally. Western Europe has endemicity of varicella, with over 98% of women being immune after disease in childhood [25]. Seroprevalence studies in the United Kingdom (UK), France and Spain have found seropositivity in individuals over 15 years of age to be >90% [26–28]. Therefore serological testing for immunity is not recommended as it is assumed that the majority of women are immune. In the UK the National Screening Committee following review of the evidence for antenatal susceptibility to varicella infection concluded that there was insufficient evidence to recommend antenatal screening [29]. Exceptions to this include women migrating from other regions to Europe, for example from the Mediterranean region, where varicella is not endemic. The UK guidelines also recommend that due to the lower likelihood of women born and raised in tropical climates being seropositive they should have prenatal varicella serology test performed [29]. In other countries for example the US and Australia, serological testing prior to or during pregnancy is recommended for women who have not had an age appropriate complete vaccine course [30–33]. It would be advisable to check local country specific guidelines relating to prenatal varicella screening. One of the reasons to consider screening and vaccination prior to pregnancy is that following exposure in pregnancy, varicella-zoster immunoglobulin can be administered to a non-immune woman however, in many countries this is expensive and can be logistically hard to access.

Evidence of immunity to varicella includes a confirmed history or laboratory diagnosis of varicella disease, records showing age-appropriate vaccination, or serological evidence of immunity. The US Advisory Committee on Immunization Practice recommends that "neither a self-reported dose nor a history of vaccination provided [...] is, by itself, considered adequate evidence of immunity" [31]. It is highly likely that a woman is protected if she has documented evidence of receipt of an age-appropriate dose(s) of a varicella-containing vaccine and in these situations serological testing is not required.

Non-immune household contacts of pregnant women are recommended to have the varicella vaccine. If varicella vaccine is given prior to pregnancy, women should be advised of the need to avoid conception for 1 month after each dose because the effects of the vaccine on fetal development are not definitively known [30, 31] (Table 1). However, importantly no adverse outcomes of pregnancy or in a fetus have been reported in several studies among women who inadvertently received varicella vaccine shortly before or during pregnancy [25, 30, 31, 34–37].

Testing to check for seroconversion after varicella vaccination is *not* routinely recommended. It is possible that commercially available laboratory tests may not be sufficiently sensitive to reliably detect antibody levels following vaccination [33, 38]. Several studies conducted in individuals over 13 years have described seroconversion rates after receipt of the single-antigen varicella vaccine. In these studies seropositivity estimates ranged from: 72–94% after 1 dose to 94–99% after a second dose administered 4–8 weeks later [39–43].

Rubella

Rubella (German measles) is usually a mild self-limiting disease and in most cases there are few complications. However, the main concern follows infection during the first trimester. In these cases rubella virus infection can lead to both complications in pregnancy and congenital rubella syndrome (CRS) at birth. CRS is rare in most developed countries, however cases still occur in the Americas, South East Asia and the European regions of WHO [44]. Maternal rubella infection can result in spontaneous miscarriage, fetal infection, stillbirth, or fetal growth restriction [45]. Congenital infection is most likely if the maternal infection occurs in the first trimester (12–16 weeks gestation) of pregnancy. CRS is very unlikely if infection occurs after 16 weeks of pregnancy [46]. Features of CRS include cardiac defects, hearing defects (deafness), ocular defects, haematologic abnormalities (thrombocytopenic purpura, hemolytic anemia), gastrointestinal abnormalities (enlarged liver and spleen), and inflammation of the meninges and brain [47]. Other late features of CRS may include pneumonitis, diabetes, thyroid dysfunction and progressive panencephalitis [48, 49].

One of the critical ways to prevent congenital infection is to ensure high levels of general population immunity to rubella [44, 50]. Overall it is estimated that approximately 10–20% of women of childbearing age do not have serologic evidence of rubella immunity, however this varies from region to region depending on local epidemiology and childhood rubella vaccination coverage [1, 51, 52]. Once rubella infection has been detected in pregnancy, at present there is no treatment to prevent or reduce mother-to-child transmission of rubella. Serological testing for rubella either pre-pregnancy or in pregnancy is recommended in many countries. The aim of this testing is to identify women who are non-immune, and therefore be vaccinated after birth and enable future pregnancies to be protected against rubella infection and its severe consequences, such as CRS.

TABLE 1 Summary recommendations of pre-conception screening and vaccination pre-conception of post-partum for selected diseases.[a]

Disease	Preconception Screening	Recommendation to vaccinate pre-pregnancy	Interval post vaccination before conception	Post pregnancy vaccination
Hepatitis B (HBV)	Hepatitis B surface and e antigen serology to identify chronic HBV infection	Women with no history of 3 doses of HBV vaccines should be vaccinated	none	Women with no history of 3 doses of HBV vaccine should be vaccinated
Varicella	Ascertain varicella vaccine status Serology may be performed	Recommended (2 doses) if no history of varicella vaccination or serology non-immune	1 month	Recommended (2 doses) if not administered pre-pregnancy
Rubella	Serology recommended	Recommended if serology non-immune	1 month	Recommended if not administered pre-pregnancy
Influenza	Ascertain vaccine status	Administer seasonal vaccine	none	Administer seasonal vaccine
Diphtheria, tetanus and pertussis	Ascertain vaccine status No serology testing for pertussis recommended	Preference for vaccination during pregnancy	none	Recommended if not administered pre or during pregnancy

[a] *Note: recommendations vary from country to country and therefore country-specific guidelines should be consulted.*

Rubella vaccination is contraindicated in pregnancy as it is a live vaccine [53, 54]. Women who are seronegative to rubella or have no record of receiving 2 doses of a rubella containing vaccine should be administered rubella vaccine, often in combination with measles and mumps vaccine, either prior to conception or immediately post-partum. If given prior to conception, similar to varicella vaccine, women should be counseled to avoid pregnancy for 1 month following vaccination (Table 1). However, several prospective cohort studies have reported no association between inadvertent vaccination in pregnancy and CRS and there is no recommendation for termination of pregnancy in cases where rubella vaccine has been inadvertently given [55–59]. In a large study from the Americas, women and their offspring who were inadvertently given MMR during pregnancy were followed up to determine newborn serological status and incidence of CRS. In this multi country study, over 30,000 women who were pregnant or became pregnant within 1 month after receiving vaccine were followed up. Nearly 3000 women (n=2894) were classified as non-immune (susceptible) at the time of vaccination – defined as having negative rubella IgM and IgG. Of these pregnancies, nearly three quarters (n=1980 [68%]) gave birth to live babies. Sera from 70 (3.5%) of these infants was rubella IgM antibody positive (indicating congenital infection), but none of the infants whose mothers had been given rubella vaccine during pregnancy had features of CRS [60].

In many countries administration of rubella (R) vaccine, requires receipt of measles (M) and mumps (M) as the vaccine is a combination MMR formulation. Antenatal screening for measles and mumps is not routinely recommended, however if a woman's measles infection or vaccination history is uncertain or not known she can have serology testing to determine her immune status. There are no adverse effects of administering a vaccine containing MMR to a non-pregnant woman who is non-immune to rubella however has pre-existing immunity to measles and mumps.

Influenza

It is recommended that all women who will be pregnant during the flu season are vaccinated either during pregnancy (preferred option, see chapter on influenza vaccination in pregnancy) or prior to pregnancy. This recommendation is particularly important for women who have co-morbid medical conditions, such as diabetes, obesity, heart or lung disease, as well as being pregnant [3, 61]. Influenza vaccine immediately prior to or during pregnancy has been shown to reduce influenza infections in the infant in the first few months of infancy [62–72]. If a woman has received the current season flu vaccine prior to pregnancy and then becomes pregnant there is no requirement to receive a second dose of the current flu season vaccine, however it's use is not contraindicated. If a woman has received the previous season influenza vaccine prior to pregnancy and then becomes pregnant in the following year influenza season, receipt of the current year influenza vaccine is recommended.

Vaccination in the post-partum period

Administration of inactivated vaccines

In general inactivated, recombinant, subunit, conjugate and polysaccharide and toxoid vaccines do not pose any risk for mothers who are breastfeeding or for their infants.

Pertussis vaccine

The most important vaccine to administer in the post-partum period if not given during pregnancy, is a pertussis-containing vaccine. Post-partum administration of a pertussis-containing vaccine to the mother may reduce the risk of the new mother becoming infected and subsequently transmission to her newborn and is recommended in many countries. In situations where a mother was not vaccinated during pregnancy, vaccination of mother, father and other close contacts in the peri-partum period can be beneficial to protect the newborn from pertussis infection. Vaccinating a mother and other persons caring for the newborn with a pertussis-containing vaccine in the post-partum period is known as the "cocoon" strategy. The effectiveness of the cocoon strategy in preventing early infant pertussis is limited [73–76]. In one Australian study, vaccination after delivery of both parents of the infant and more than 28 days prior to illness onset, was estimated to reduce pertussis infection by 64–77%, however the result was not statistically significant [75]. In another large linked Australian study there was no significant protection offered by "cocoon" vaccination. In this study there was no difference in the incidence of pertussis among infants whose mothers were vaccinated postpartum compared to those with unvaccinated mothers (adjusted HR 1.19; 95% CI 0.82–1.72) [76]. The Global pertussis Initiative recommends maternal vaccination as the primary strategy, given its superior effectiveness and logistical advantages, however vaccination of close contacts of the newborn is recommended where maternal vaccination does not occur [77, 78]. This recommendation is because in many cases the main source of infection for the newborn is via transmission from household contacts [79, 80]. In lower and middle income (LMIC) countries the barriers to the administration of the cocoon pertussis vaccine include: insufficient knowledge about the vaccine, cost, access issues including lack of transportation to the health center and fear of needles [78]. Given these additional factors in LMIC the administration of a pertussis vaccine during pregnancy is preferred. Even in high income countries the cost effectiveness of vaccinating all close contacts of the newborn (cocoon strategy) is not beneficial [81].

There are no contraindications to the administration of a pertussis-containing vaccine to a breastfeeding woman. Secretory immunoglobulin A (IgA) antibodies to *Bordetella pertussis* are found in the breastmilk of women vaccinated during pregnancy or in the post-partum period, however they are not considered sufficient to protect infants although there is limited data [82–84].

Hepatitis B

Hepatitis B vaccine can be administered to breastfeeding women. Infants born to mothers who are positive for hepatitis B surface antigen (HBsAg) can be breastfed. It is important that the infant born to a mother with chronic hepatitis B infection is appropriately immunized at birth with both hepatitis B immunoglobulin and hepatitis B vaccine. Hepatitis B virus has been detected in the breastmilk of mothers with hepatitis B virus infection, yet despite this there is no additional risk of virus transmission with breastfeeding, compared with formula feeding, as long as the infant is appropriately vaccinated, with the first dose immediately after birth and 2–3 more doses, depending on the country specific immunization schedule [85]. In a systematic review (including 32 studies), overall less than 5% (4.32% [244/5650]) of infants born to mothers with chronic hepatitis B infection and whom received hepatitis B vaccine in the newborn period developed chronic infection themselves and there was no difference in rates between breastfed and formula fed infants [85].

Administration of live viral vaccines

Following delivery, it is recommended that women found to be non-immune to rubella, measles, mumps or varicella prior to or during pregnancy receive these live viral vaccines in the post-partum period, unless there are other contraindications or they had been received pre-pregnancy. Live viruses in vaccines have been shown to replicate in the mother following vaccination, however the majority of live viruses in vaccines are not excreted in human milk or if found in milk do not appear to cause clinical infection in the breastfed infant, except for Yellow Fever [86–88]. Live attenuated vaccines should be administered to non-immune women in the post-partum period prior to discharge from the health-care facility. In some settings, the use of standing orders in health-care facilities will assist in the delivery of vaccines in the post-partum period.

Varicella vaccine

Varicella vaccine virus has not been found in human milk [88, 89]. In a small study by Bohlke et al. including 12 women, all of whom had demonstrated seroconversion after vaccination, varicella DNA was not detected by polymerase chain reaction (PCR) in more than 200 post-vaccination breastmilk specimens [89]. Samples from six infants were tested for varicella zoster virus DNA by PCR, and all were negative [89]. Varicella vaccine can be safely given to breastfeeding women if indicated [90]. It is recommended that 2 doses of varicella vaccine be administered, with an interval of 4–8 weeks between doses.

Rubella vaccine

Rubella vaccine virus has been found in human breastmilk samples. In a study by Losonsky et al. rubella virus or virus antigen was found in the breastmilk of 11 (68%) of 16 breastfeeding women vaccinated in the post-partum period. Rubella virus was subsequently isolated from the nasopharynx and throat of 56% of the breastfed infants, however none showed any evidence of clinical disease [91]. Infection, if it occurs, is thought to be well tolerated and of no clinical significance because the virus is attenuated [88, 92–94]. Based on this, administration of a rubella containing vaccine is not contraindicated in a breastfeeding woman unless other reasons not to vaccinate her exist [90]. Similar to varicella vaccine, it is recommended that 2 doses of a rubella-containing vaccine be administered, with an interval of 4–8 weeks between doses.

Measles vaccine

Measles virus has been detected in breastmilk samples [95], however is not associated with clinical disease in the infants. Administration of a measles containing vaccine is not contraindicated in a breastfeeding woman [90]. Similar to rubella, it is recommended that 2 doses of a measles containing vaccine be administered, with an interval of 4–8 weeks between doses.

Yellow fever vaccine

Yellow fever vaccine should be avoided in breastfeeding women if possible [86, 87, 96]. This is because three cases of yellow-fever vaccine associated acute neurotropic disease (YEL-AND) have been reported in exclusively breastfed infants whose mothers were vaccinated but the infants were not vaccinated themselves. A case in Brazil demonstrated laboratory confirmed evidence of yellow fever virus transmission. In this case yellow fever viral RNA was detected in a CSF specimen of the infant and the nucleotide sequence of the amplified PCR product was identical to 17DD yellow fever vaccine virus. However, in this case no breastmilk or maternal serum was collected for yellow fever virus testing [86, 87]. A second case has also been reported in Brazil [97]. In Canada, an infant who had been breastfed by a recently vaccinated mother had evidence of acute neurotropic disease. A sample of cerebrospinal fluid was positive for yellow fever antigen (as measured by IgM capture enzyme-linked immunosorbent assay). The sample was negative for yellow fever virus by polymerase chain reaction (PCR). Yellow fever virus was also not isolated in the breastmilk [98]. These infants were <1 month old at the time of exposure and in two infants, importantly it was not able to determine the precise mode of transmission because although vaccine virus was found in the cerebrospinal fluid of the infant, vaccine virus was not recovered from breastmilk [97, 98]. If nursing mothers cannot avoid or postpone travel to areas endemic for yellow fever and it is determined that the risk for acquisition is high and vaccination is recommended, these women should be cautioned about the potential risk of transmission to the infant following vaccination. There is no data on whether ceasing breastfeeding for a period post-vaccination will lessen the risk of transmission. It is known that viremia occurs between four and 10 days after primary vaccination and so avoidance of breastfeeding for a period after vaccination may be useful. Of note the Brazilian Ministry of Health recommends delaying vaccination of breastfeeding mothers until their infant is 6 months of age or older [97]. However, more data on the duration of vaccine virus excretion in breastmilk are needed to guide recommendations for temporary suspension of breastfeeding after vaccination [99].

Conclusion

An assessment of a woman's vaccination history and immunity to vaccine preventable diseases, should be an important part of pre-conception medical assessment. Identification of non-immunity to live viruses, such as rubella and varicella, will enable a woman to receive these vaccines prior to conception. An interval of 1 month after receipt of varicella, rubella and measles vaccines is recommended prior to conceiving. Identification of chronic hepatitis B infection is important, as potential treatment exists for the mother to reduce the viral load prior to and during pregnancy and ensures that the infant receives both hepatitis B immunoglobulin and vaccine immediately after birth.

The post-partum period is an important time to administer vaccines to a woman who has recently delivered. Although not the preferred strategy to prevent early infant pertussis, a "cocoon" dose of a pertussis-containing vaccine to both mother and any other close contacts may be beneficial in reducing early infant pertussis exposure. There is no contraindication to administering vaccines to a breastfeeding woman, except that caution is advised with yellow fever vaccine and a risk-based assessment needs to be undertaken prior to its administration. It is important to catch up live viral vaccines, such as measles, mumps and rubella in the post-partum period so that a woman has immunity prior to her next pregnancy.

References

[1] Coonrod D, Jack B, Boggess K, Long R, Conry J, Cox S, et al. The clinical content of preconception care: immunisations as part of preconception care. Am J Obstet Gynecol 2008;99(6 Suppl. 2):S290–5.
[2] Australian Technical Advisory Group on Immunisation (ATAGI). Australian immunisation handbook. Canberra: Australian Government Department of Health; 2018. [cited 2019 September 9]. Available from: immunisationhandbook.health.gov.au.
[3] World Health Organization (WHO). Vaccines against influenza WHO position paper—November 2012. Wkly Epidemiol Rec 2012;87(47):461–76.
[4] Marin M, Guris D, Chaves S, Schmid S, Seward J, Advisory Committee on Immunization Practices, Centres for Disease Control and Prevention (CDC). Prevention of Varicella: recommendations of the Advisory Committee on Immunization Practices (ACIP). MMWR Recomm Rep 2007;56(RR04):1–40.
[5] From the Centers for Disease Control and Prevention. Revised ACIP recommendation for avoiding pregnancy after receiving a rubella-containing vaccine. JAMA 2002;287(3):311–2.
[6] Lee C, Gong Y, Brok J, Boxall E, Gluud C. Hepatitis B immunisation for newborn infants of hepatitis B surface antigen-positive mothers. Cochrane Database Syst Rev 2006;(2).
[7] Yao J. Perinatal transmission of hepatitis B virus infection and vaccination in China. Gut 1996;38(Suppl. 2):S37–8.
[8] Stevens C, Beasley R, Tsui J, Lee W. Vertical transmission of hepatitis B antigen in Taiwan. N Engl J Med 1975;292(15):771–4.
[9] Akhter S, Talukder MQ, Bhuiyan N, Chowdhury T, Islam M, Beum S. Hepatitis B virus infection in pregnant mothers and its transmission to infants. Indian J Pediatr 1992;59(4):411–5.
[10] Okada K, Kamiyama I, Inomata M, Imai M, Miyakawa Y. E antigen and anti-e in the serum of asymptomatic carrier mothers as indicators of positive and negative transmission of hepatitis B virus to their infants. N Engl J Med 1976;294(14):746–9.
[11] Beasley R, Hwang L, Lee G, Lan C, Roan C, Huang F, et al. Prevention of perinatally transmitted hepatitis B virus infections with hepatitis B immune globulin and hepatitis B vaccine. Lancet 1983;2(8359):1099–102.
[12] Nayak N, Panda S, Zuckerman A, Bhan M, Guha D. Dynamics and impact of perinatal transmission of hepatitis B virus in north India. J Med Virol 1987;21(2):137–45.
[13] Aggarwal R, Ranjan P. Preventing and treating hepatitis B infection. BMJ 2004;329(7474):1080–6.
[14] Wong V, Ip H, Reesink H, Lelie P, Reerink-Brongers E, Yeung C, et al. Prevention of the HbsAg carrier state in newborn infants of mothers who are chronic carriers of HBsAg and HBeAg by administration of hepatitis-B vaccine and hepatitis-B immunoglobulin. Double-blind randomised placebo—controlled study. Lancet 1984;1(8383):921–6.
[15] Wiseman E, Fraser M, Holden S, Glass A, Kidson B, Heron L, et al. Perinatal transmission of hepatitis B virus: an Australian experience. Med J Aust 2009;190(9):489–92.
[16] Liu J, Zhang S, Liu M, Wang Q, Shen H, Zhang Y. Maternal pre-pregnancy infection with hepatitis B virus and the risk of preterm birth: a population-based cohort study. Lancet Glob Health 2017;5(6):e624–32.
[17] Hanson B, Dorais J. Reproductive considerations in the setting of chronic viral illness. Am J Obstet Gynecol 2017;217(1):4–10.
[18] National Institute for Health and Care Excellence (NICE). Antenatal care: Routine care for the healthy pregnant woman. National Collaborating Centre for Women's and Children's Health, London: RCOG Press; 2008.
[19] Mast E, Margolis H, Fiore A, Brink E, Goldstein S, Wang S, et al. A comprehensive immunization strategy to eliminate transmission of hepatitis B virus infection in the United States: recommendations of the Advisory Committee on Immunization Practices

(ACIP) part 1: immunization of infants, children, and adolescents. MMWR Recomm Rep 2005;54(RR-16):1–31.
[20] Lin K, Vickery J. Screening for hepatitis B virus infection in pregnant women: evidence for the U.S. Preventive Services Task Force reaffirmation recommendation statement. Ann Intern Med 2009;150(12):874–6.
[21] US Preventive Services Task Force, Owens D, Davidson K, Krist A, Barry M, Cabana M, et al. Screening for hepatitis B virus infection in pregnancy: United States Preventive Services Task Force reaffirmation recommendation statement. JAMA 2019;322(4):349–54.
[22] Summers P, Biswas M, Pastorek J, Pernoll M, Smith L, Bean B. The pregnant hepatitis B carrier: evidence favoring comprehensive antepartum screening. Obstet Gynecol 1987;69(5):701–4.
[23] Jensen L, Heilmann C, Smith E, Wantzin P, Peitersen B, Weber T, et al. Efficacy of selective antenatal screening for hepatitis B among pregnant women in Denmark: is selective screening still an acceptable strategy in a low-endemicity country? Scand J Infect Dis 2003;35(6–7):378–82.
[24] Cowan S, Bagdonaite J, Qureshi K. Universal hepatitis B screening of pregnant women in Denmark ascertains substantial additional infections: results from the first five months. Euro Surveill 2006;11(6):E060608.3.
[25] Royal College of Obstetricians and Gynaecologists (RCOG). Chickenpox in pregnancy (Green-top guideline number 13). London: Royal College of Obstetricians and Gynaecologists (RCOG); 2015. [cited 2019 June 10]. Available from: https://www.rcog.org.uk/en/guidelines-research-services/guidelines/gtg13/.
[26] Vyse A, Gay N, Hesketh L, Morgan-Capner P, Miller E. Seroprevalence of antibody to varicella zoster virus in England and Wales in children and young adults. Epidemiol Infect 2004;132(6):1129–34.
[27] Plans P, Costa J, Espuñes J, Plasència A, Salleras L. Prevalence of varicella-zoster antibodies in pregnant women in Catalonia (Spain). Rationale for varicella vaccination of women of childbearing age. BJOG 2007;114(9):1122–7.
[28] Saadatian-Elahi M, Mekki Y, Del Signore C, Lina B, Derrough T, Caulin E, et al. Seroprevalence of varicella antibodies among pregnant women in Lyon-France. Eur J Epidemiol 2007;22(6):405–9.
[29] Manikkavasagan G, Bedford H, Peckham C, Dezateux C. Antenatal screening for susceptibility to varicella zoster virus (VZV) in the United Kingdom: A review commissioned by the National Screening Committee. London: MRC Centre of Epidemiology for Child Health; 2009.
[30] Australian Technical Advisory Group on Immunisation (ATAGI). Varicella. In: Australian immunisation handbook. Canberra: Australian Government Department of Health; 2018. [cited 2019 September 9]. Available from: https://immunisationhandbook.health.gov.au/vaccine-preventable-diseases/varicella-chickenpox.
[31] The Royal Australasian College of General Practitioners. Preventive activities prior to pregnancy. In: Guidelines for preventive activities in general practice. The red book. East Melbourne, VIC: RACGP; 2016. [Chapter 1] [cited 2019 September 9]. Available from: https://www.racgp.org.au/clinical-resources/clinical-guidelines/key-racgp-guidelines/view-all-racgp-guidelines/red-book/preventive-activities-prior-to-pregnancy.
[32] National Center for Immunization and Respiratory Diseases. General recommendations on immunization: recommendations of the Advisory Committee on Immunization Practices (ACIP). MMWR Recomm Rep 2011;60(2):1–64.
[33] American Academy of Pediatrics, Committee on Infectious Diseases. Varicella vaccine update. Pediatrics 2000;105:136–41.
[34] Shields K, Galil K, Seward J, Sharrar R, Cordero J, Slater E. Varicella vaccine exposure during pregnancy: data from the first 5 years of the pregnancy registry. Obstet Gynecol 2001;98(1):14–9.

[35] Wilson E, Goss M, Marin M, Shields K, Seward J, Rasmussen S, et al. Varicella vaccine exposure during pregnancy: data from ten years of the pregnancy registry. J Infect Dis 2008;197(Suppl. 2):S178–84.
[36] Miller E, Lewis P, Shimabukuro T, Su J, Moro P, Woo E, et al. M. Post-licensure safety surveillance of zoster vaccine live (Zostavax®) in the United States, Vaccine Adverse Event Reporting System (VAERS), 2006-2015. Hum Vaccin Immunother 2018;14(8):1963–9.
[37] Swamy G, Heine R. Vaccinations for pregnant women. Obstet Gynecol 2015;125(1):212–26.
[38] Bogger-Goren S, Baba K, Hurley P, Yabuuchi H, Takahashi M, Ogra O. Antibody response to varicella-zoster virus after natural or vaccine-induced infection. J Infect Dis 1982;146(2):260–5.
[39] Gershon A, Steinberg S, LaRussa P, Ferrara A, Mammerschlag M, Gelb L. Immunization of healthy adults with live attenuated varicella vaccine. J Infect Dis 1988;158(1):132–7.
[40] Saiman L, LaRussa P, Steinberg S, Zhou J, Baron K, Whittier S, et al. Persistence of immunity to varicella-zoster virus after vaccination of health care workers. Infect Control Hosp Epidemiol 2001;22(5):279–83.
[41] Ampofo K, Saiman L, LaRussa P, Steinberg S, Annunziato P, Gershon A. Persistence of immunity to live attenuated varicella vaccine in healthy adults. Clin Infect Dis 2002;34(6):774–9.
[42] Zerboni L, Nader S, Aoki K, Arvin A. Analysis of the persistence of humoral and cellular immunity in children and adults immunized with varicella vaccine. J Infect Dis 1998;177(6):1701–4.
[43] Watson B, Boardman C, Laufer D, Piercy S, Tustin N, Olaleye D, et al. Humoral and cell-mediated immune responses in healthy children after one or two doses of varicella vaccine. Clin Infect Dis 1995;20(2):316–9.
[44] Grant G, Reef S, Patel M, Knapp J, Dabbagh A. Progress in rubella and congenital rubella syndrome control and elimination—Worldwide, 2000-2016. MMWR Morb Mortal Wkly Rep 2017;66(45):1256–60.
[45] Reef S, Plotkin S, Cordero J, Katz M, Cooper L, Zimmer-Swain L, et al. Preparing for elimination of congenital rubella syndrome (CRS): summary of a workshop on CRS elimination in the United States. Clin Infect Dis 2000;31(1):85–95.
[46] Miller E, Cradock-Watson J, Pollock T. Consequences of confirmed maternal rubella at successive stages of pregnancy. Lancet 1982;2(8302):781–4.
[47] Sanchez E, Atabani S, Kaplanova J, Griffiths P, Geretti A, Haque T. Forgotten, but not gone. BMJ 2010;341:c5246.
[48] Weil M, Itabashi H, Cremer N, Oshiro L, Lennette E, Carnay L. Chronic progressive panencephalitis due to rubella virus stimulating subacute sclerosing panencephalitis. N Engl J Med 1975;292(19):994–8.
[49] Cooper LZ, Preblub SR, Alford CA. Rubella. In: Remington JS, Klein JO, editors. Infectious diseases of the fetus and newborn. 4th ed. Philadelphia, PA: WB Saunders; 1995. p. 268.
[50] Martinez-Quintana E, Castillo-Solorzano C, Torner N, Rodriguez-Gonzalez F. Congenital rubella syndrome: a matter of concern. Rev Panam Salud Publica 2015;37(3):179–86.
[51] Ogundele M, Ghebrehewet S, Chawla A. Some factors affecting rubella seronegative prevalence among pregnant women in a North West England region between April 2011 and March 2013. J Public Health (Oxf) 2016;38(2):243–9.
[52] Skidmore S, Boxall E, Lord S. Is the MMR vaccination programme failing to protect women against rubella infection? Epidemiol Infect 2014;142(5):1114–7.
[53] McLean H, Fiebelkorn A, Temte J, Wallace G, Centers for Disease Control and Prevention. Prevention of measles, rubella, congenital rubella syndrome, and mumps, 2013: summary recommendations of the Advisory Committee on Immunization Practices (ACIP). [Erratum appears in MMWR Recomm Rep. 2015;13;64(9):259]. MMWR Recomm Rep 2013;62(RR-04):1–34.

[54] White S, Boldt K, Holditch S, Poland G, Jacobson R. Measles, mumps, and rubella. Clin Obstet Gynecol 2012;55(2):550–9.
[55] Bar-Oz B, Levichek Z, Moretti M, Mah C, Andreou S, Koren G. Pregnancy outcome following rubella vaccination: a prospective controlled study. Am J Med Genet 2004;130A(1):52–4.
[56] Centers for Disease Control and Prevention. Revised ACIP recommendation for avoiding pregnancy after receiving a rubella-containing vaccine. MMWR Morb Mortal Wkly Rep 2001;50(49):1117.
[57] Hamkar R, Jalilvand S, Abdolbaghi M, Esteghamati A, Hagh-Goo A, Jelyani K, et al. Inadvertent rubella vaccination of pregnant women: evaluation of possible transplacental infection with rubella vaccine. Vaccine 2006;24(17):3558–663.
[58] Badilla X, Morice A, Avila-Aguero M, Saenz E, Cerda L, Reef S, et al. Fetal risk associated with rubella vaccination during pregnancy. Pediatr Infect Dis J 2007;26(9):830–5.
[59] Sukumaran L, McNeil M, Moro P, Lewis P, Winiecki SK, Shimabukuro T. Adverse events following measles, mumps, and rubella vaccine in adults reported to the Vaccine Adverse Event Reporting System (VAERS), 2003–2013. Clin Infect Dis 2015;60(10):e58–65.
[60] Castillo-Soloranzo C, Reef S, Morice A, Vascones N, Chevez A, Castalia-Soares R, et al. Rubella vaccination of unknowingly pregnant women during mass campaigns for rubella and congenital rubella syndrome elimination, the Americas 2001–2008. J Infect Dis 2011;204(Suppl. 2):S713–7.
[61] Grohskopf L, Sokolow L, Broder K, Walter E, Fry A, Jernigan D. Prevention and control of seasonal influenza with vaccines: recommendations of the Advisory Committee on Immunization Practices—United States, 2018–19 Influenza Season. MMWR Recomm Rep 2018;67(3):1–20.
[62] Madhi S, Cutland C, Kuwanda L, Weinberg A, Hugo A, Jones S, et al. Influenza vaccination of pregnant women and protection of their infants. N Engl J Med 2014;371(10):918–31.
[63] Tapia M, Sow S, Tamboura B, Teguete I, Pasetti M, Kodio M, et al. Maternal immunisation with trivalent inactivated influenza vaccine for prevention of influenza in infants in Mali: a prospective, active-controlled, observer-blind, randomised phase 4 trial. Lancet Infect Dis 2016;16(9):1026–35.
[64] Zaman K, Roy E, Arifeen S, Rahman M, Raqib R, Wilson E, et al. Effectiveness of maternal influenza immunization in mothers and infants. N Engl J Med 2008;359(15):1555–64.
[65] Steinhoff M, Katz J, Englund J, Khatry S, Shrestha L, Kuypers J, et al. Year-round influenza immunization during pregnancy in Nepal: a phase 4, randomized, placebo-controlled Phase 4 trial. Lancet Infect Dis 2017;17(9):981–9.
[66] Dabrera G, Zhao H, Andrews N, Begum F, Green H, Ellis J, et al. Effectiveness of seasonal influenza vaccination during pregnancy in preventing influenza infection in infants, England, 2013/14. Euro Surveill 2014;19(45):20959.
[67] Shakib J, Korgenski K, Presson A, Sheng X, Varner M, Pavia A, et al. Influenza in infants born to women vaccinated during pregnancy. Pediatrics 2016;137(6):e20152360.
[68] Eick A, Uyeki T, Klimov A, Hall H, Reid R, Santosham M, et al. Maternal influenza vaccination and effect on influenza virus infection in young infants. Arch Pediatr Adolesc Med 2011;165(2):104–11.
[69] Benowitz I, Esposito D, Gracey K, Shapiro E, Vazquez M. Influenza vaccine given to pregnant women reduces hospitalization due to influenza in their infants. Clin Infect Dis 2010;51(12):1355–61.
[70] Poehling K, Szilagyi P, Staat M, Snively B, Payne D, Bridges C, et al. Impact of maternal immunization on influenza hospitalizations in infants. Am J Obstet Gynecol 2011;204(6 Suppl. 1):S141–8.
[71] Nunes M, Cutland C, Jones S, Downs S, Weinberg A, Ortiz J, et al. Efficacy of maternal influenza vaccination against all-cause lower respiratory tract infection hospitalizations in young infants: results from a randomized controlled trial. Clin Infect Dis 2017;65(7):1066–71.

[72] Omer S, Clark D, Aqil A, Tapia M, Nunes M, Kozuki N, et al. Maternal influenza immunization and prevention of severe clinical pneumonia in young infants: analysis of randomized controlled trials conducted in Nepal, Mali, and South Africa. Pediatr Infect Dis J 2018;37(5):436–40.
[73] Swamy G, Wheeler S. Neonatal pertussis, cocooning and maternal immunization. Expert Rev Vaccines 2014;13(9):1107–14.
[74] Healy C, Rench M, Wootton S, Castagnini L. Evaluation of the impact of a pertussis cocooning program on infant pertussis infection. Pediatr Infect Dis J 2015;34(1):22–6.
[75] Rowe S, Tay E, Franklin L, Stephens N, Ware R, Kaczmarek M, et al. Effectiveness of parental cocooning as a vaccination strategy to prevent pertussis infection in infants: a case-control study. Vaccine 2018;36(15):2012–9.
[76] Carcione D, Regan A, Tracey L, Mak D, Gibbs R, Dowse G, et al. The impact of parental postpartum pertussis vaccination on infection in infants: a population-based study of cocooning in Western Australia. Vaccine 2015;33(42):5654–61.
[77] Forsyth K, Plotkin S, Tan T, Wirsing Von Konig C. Strategies to decrease pertussis transmission to infants. Pediatrics 2015;135(6):e1475–82.
[78] Forsyth K, Tan T, von König C, Heininger U, Chitkara A, Plotkin S. Recommendations to control pertussis prioritized relative to economies: a Global Pertussis Initiative update. Vaccine 2018;36(48):7270–5.
[79] Bisgard K, Pascual F, Ehresmann K, Miller C, Cianfrini C, Jennings C, et al. Infant pertussis: who was the source? Pediatr Infect Dis J 2004;23(11):985–9.
[80] Mertsola J, Ruskanen O, Eerola E, Viljanen M. Intrafamilial spread of pertussis. J Pediatr 1983;103(3):359–63.
[81] Van Rie A, Hethcote H. Adolescent and adult pertussis vaccination: computer simulations of five new strategies. Vaccine 2004;22(23–24):3154–65.
[82] Pandolfi E, Gesualdo F, Carloni E, Villani A, Midulla F, Carsetti R, et al. Does breastfeeding protect young infants from pertussis? Case-control study and immunologic evaluation. Pediatr Infect Dis J 2017;36(3):e48–53.
[83] Abu Raya B, Srugo I, Kessel A, Peterman M, Bader D, Peri R, et al. The induction of breast milk pertussis specific antibodies following gestational tetanus-diphtheria-acellular pertussis vaccination. Vaccine 2014;32(43):5632–7.
[84] Maertens K, De Schutter S, Braeckman T, Baerts L, Van Damme P, De Meester I, et al. Breastfeeding after maternal immunisation during pregnancy: providing immunological protection to the newborn: a review. Vaccine 2014;32(16):1786–92.
[85] Zheng Y, Lu Y, Ye Q, Xia Y, Zhou Y, Yao Q, et al. Should chronic hepatitis B mothers breastfeed? A meta analysis. BMC Public Health 2011;11:502.
[86] Staples J, Gershman M, Fischer M, Centers for Disease Control and Prevention (CDC). Yellow fever vaccine: recommendations of the Advisory Committee on Immunization Practices (ACIP). MMWR Recomm Rep 2010;59(RR-7):1–27.
[87] Centers for Disease Control and Prevention (CDC). Transmission of yellow fever vaccine virus through breast-feeding—Brazil, 2009. MMWR Morb Mortal Wkly Rep 2010;59(5):130–2.
[88] Centers for Disease Control and Prevention (CDC). General best practice guidelines for immunization: special situations. In: Best practices guidance of the Advisory Committee on Immunization Practices (ACIP). Atlanta, GA: Centers for Disease Control and Prevention; 2017. [cited 2019; September 9]. Available from: https://www.cdc.gov/vaccines/hcp/acip-recs/general-recs/special-situations.html.
[89] Bohlke K, Galil K, Jackson L, Schmid D, Starkovich P, Loparev V, et al. Postpartum varicella vaccination: is the vaccine virus excreted in breast milk? Obstet Gynecol 2003;102(5 Pt 1):970–7.
[90] Munoz F, Jamieson D. Maternal Immunization. Obstet Gynecol 2019;133(4):739–53.
[91] Losonsky G, Fishaut J, Strussenberg J, Ogra P. Effect of immunization against rubella on lactation products. II. Maternal-neonatal interactions. J Infect Dis 1982;145(5):661–6.

[92] Krogh V, Duffy L, Wong D, Rosenband M, Riddlesberger K, Ogra P. Postpartum immunization with rubella virus vaccine and antibody response in breast-feeding infants. J Lab Clin Med 1989;113(6):695–9.
[93] Klein E, Byrne T, Cooper Z. Neonatal rubella in a breast-fed infant after postpartum maternal infection. J Pediatr 1980;97(5):774–5.
[94] Landes R, Bass J, Millunchick E, Oetgen W. Neonatal rubella following postpartum maternal immunization. J Pediatr 1980;97(3):465–7.
[95] Hisano M, Kato T, Inoue E, Sago H, Yamaguchi K. Evaluation of measles-rubella vaccination for mothers in early puerperal phase. Vaccine 2016;34(9):1208–14.
[96] Centers for Disease Control and Prevention (CDC). Travel-related infectious diseases. In: CDC yellow book. New York: Oxford University Press; 2017. [cited 2019 June 15]. Available from: https://wwwnc.cdc.gov/travel/yellowbook/2018/infectious-diseases-related-to-travel/yellow-fever.
[97] Traiber C, Coelho-Amaral P, Ritter V, Winge A. Infant meningoencephalitis caused by yellow fever vaccine virus transmitted via breastmilk. J Pediatr (Rio J) 2011;87(3):269–72.
[98] Kuhn S, Twele-Montecinos L, MacDonald J, Webster P, Law B. Case report: probable transmission of vaccine strain of yellow fever virus to an infant via breast milk. CMAJ Can Med Assoc J 2011;183:E243–5.
[99] Sachs H, Committee On Drugs. The transfer of drugs and therapeutics into human breast milk: an update on selected topics. Pediatrics 2013;132(3):e796–809.

CHAPTER

3

Immunobiological aspects of vaccines in pregnancy: Maternal perspective

Helen Y. Chu[a], *Arnaud Marchant*[b]

[a]Departments of Medicine and Epidemiology, University of Washington, Seattle, WA, United States, [b]Institute for Medical Immunology, Université libre de Bruxelles, Brussels, Belgium

Introduction

Pregnancy is associated with important changes in the maternal immune system that have profound consequences for the fetus and the newborn infant. Tolerance of the fetus and successful pregnancy require a highly regulated immunological environment at the materno-fetal interface [1]. Maternal immune components are transferred across the placenta and in breastmilk that provide protection against pathogens in early life and contribute to immune development. Although pregnancy cannot be considered a state of immunodeficiency, infections with some pathogens cause more severe diseases in pregnant as compared to non-pregnant women. Vaccination during pregnancy has the potential to provide protection against pathogens affecting the mother and the newborn infant [2,3]. However, the efficacy of this approach can be limited by diseases affecting maternal immune responses and transfer of maternal immune components to the offspring. Recognizing that the health of the mother during pregnancy is essential to the health of the newborn infant, maternal vaccination has a great potential to reduce morbidity and mortality in mothers and children.

https://doi.org/10.1016/B978-0-12-814582-1.00003-6

Susceptibility to infectious diseases in pregnancy

Infections with some pathogens, including influenza virus, varicella zoster, hepatitis E, listeria, and malaria, cause more severe diseases in pregnant as compared to non-pregnant women. The mechanisms underlying this increased susceptibility have not been fully defined. Pregnancy-induced changes in the maternal immune system may play a role. On the other hand, pregnant women also have hormonal and physiologic changes including increasing levels of estradiol and progesterone and decreased pulmonary reserve and increased cardiac output that may contribute to reduced pathogen control and more severe clinical symptoms. As discussed below, emerging evidence suggest a central role for pathogen-specific mechanisms rather than processes common to multiple pathogens (Fig. 1). Understanding these mechansims is critical for the development of interventions protecting pregnant women and their offspring.

Influenza and other respiratory viruses

Pregnant women have higher morbidity and, in some studies, mortality due to influenza infection as compared to non-pregnant adults [4]. In the 1918 pandemic, 27% mortality was observed in pregnant women [5,6].

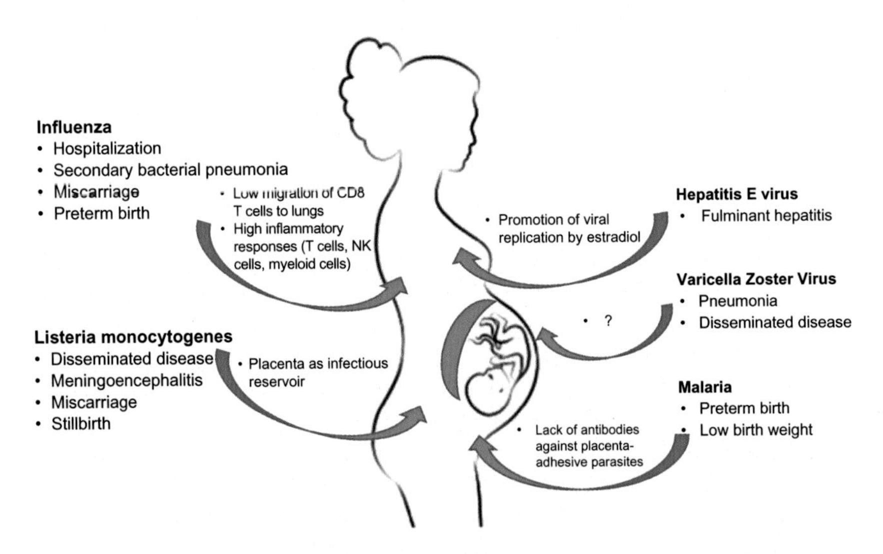

FIG. 1 Pathogens causing severe infections in pregnant women. Several pathogens cause more severe diseases in pregnant as compared to non-pregnant women. The pathogenesis underlying this increased susceptibility remains incompletely understood. Studies indicate that they involve pathogen-specific mechanisms rather than a global state of immunodeficiency *(arrows)*. Severe symptoms can compromise the survival of the mother and of the unborn fetus.

In the 2009 influenza A/H1N1 pandemic, pregnant women with influenza particularly towards the end of pregnancy in the third trimester had increased disease severity, higher rates of secondary bacterial pneumonia, and increased risk of both hospitalization and admission to the intensive care unit as compared to non-pregnant adults [7]. Higher rates of stillbirth, miscarriage, and preterm birth were additionally observed with A/H1N1 pandemic influenza infection during pregnancy [8]. Epidemiologic data do not clearly support an increased risk of acquisition of infection; rather the risk is in progression to severe disease once infected [9]. However, a recent systematic review emphasized the relatively low quality of available information on the incidence rate of severe influenza in pregnant women [10]. The mechanisms underlying severe influenza in pregnancy remain unclear and could involve both anatomic and physiologic changes increasing the risk of respiratory failure, reduced immune control and increased inflammatory responses [9]. Enhanced natural killer (NK) cell and T cell responses to influenza infection were observed in pregnant as compared to non-pregnant women, suggesting that suppression of cell-mediated immune responses is unlikely to be a major contributor to disease severity [11]. Mouse studies suggest a role for impaired migration of anti-viral CD8 T lymphocytes to the lungs [12]. Increased proinflammatory responses of monocytes and dendritic cells to *in vitro* influenza infection were observed in pregnant as compared to non-pregnant women, suggesting a role for immunopathologic responses [13]. Influenza vaccine is safe, immunogenic and efficacious in pregnancy, as has been demonstrated in several randomized clinical trials [14–17].

Few studies have evaluated the role of respiratory viruses other than influenza during pregnancy [18]. However, recently, two maternal influenza immunization trials have conducted secondary analyses of respiratory illness data during pregnancy to describe the incidence of maternal respiratory disease due to respiratory syncytial virus (RSV) during pregnancy [14,16]. In Nepal, where women with a fever and respiratory symptoms had a nasal swab collected, RSV prevalence was low at 0.2%, with an incidence of 3.9/1000 person-years overall [19]. In South Africa, RSV prevalence based on the presence of respiratory symptoms was much higher at 2%, or an incidence of 14.4–48.0 cases per 1000 person-years overall [20]. Rhinovirus, coronavirus, parainfluenza viruses 1–4, and human metapneumovirus have also recently been detected in cases of respiratory viral infections in pregnant women [21,22]. Rhinovirus is described as a cause of influenza-like illness in several studies in pregnant women, and Middle Eastern Respiratory Syndrome and Severe Acute Respiratory Syndrome coronaviruses have been shown in case reports to be associated with severe disease in pregnant women [23,24]. Other respiratory viruses were commonly detected in Nepal in women with fever and respiratory symptoms, with the most common being rhinovirus [22]. Health care-seeking was also common, ranging from 0% to 33% depending on the viral etiology. In a region of the world with limited access to medical care, this was a notable finding. Additionally, rhinovirus and human metapneumovirus infections with fever and respiratory symptoms during pregnancy both were found to be associated with increased risk of fetal growth restriction, manifested as low birth weight or small-for-gestational-age births [21]. Overall, these studies demonstrate that respiratory viral infections during pregnancy may adversely affect maternal and fetal outcomes.

Varicella zoster virus

Pregnant women may be at increased risk of progression of varicella to pneumonitis, particularly in the third trimester, with a mortality rate of 44% [25]. However, much of these data are based on case reports [26]. To our knowledge, no prospective studies have been conducted comparing the risk of varicella pneumonitis in pregnant versus non-pregnant women. Varicella infection during pregnancy is associated with disseminated disease and with adverse outcomes in the fetus, including congenital varicella, characterized by limb hypoplasia, optic nerve atrophy, microcephaly, seizures, cutaneous lesions [27]. Varicella zoster vaccination is now routine as part of the childhood immunization series in some countries; prior to vaccination, the majority of individuals were exposed during childhood with acquisition of immunity prior to pregnancy. Treatment for varicella infection in pregnancy is with acyclovir; use of live vaccines are contraindicated during pregnancy due to potential risk of fetal transmission.

Hepatitis E

Hepatitis E is acquired via fecal-oral transmission, and is a pathogen often associated with limited access to running water and basic sanitation. In a field study of viral hepatitis in pregnant versus non-pregnant adults in the 1980s, incidence of non-A non-B hepatitis was increased in pregnant women (17.3% vs. 2.1%, respectively). In this study, fulminant hepatitis developed in 22% of pregnant women as compared to no cases of fulminant hepatitis in non-pregnant women [28]. The incidence is increased in the second and third trimester as compared to the first trimester. Mortality has been estimated at 25–30% in pregnant women [29]. The mechanisms underlying the increased susceptibility of pregnant women to hepatitis E infection and not to other hepatitis viruses is not clear. A role for estradiol promoting hepatis E virus replication has been suggested [30]. Thus far, no treatment is available other than supportive care and no vaccine is available.

Listeria monocytogenes

Listeria is a bacterial infection that is associated with consumption of raw meats or vegetables, or drinking unpasteurized milk. Risk is higher in Hispanic populations, the elderly, those with malignancy, as well as pregnant women. Pregnant women are considered a high-risk group for development of disseminated disease, commonly manifested as bacteremia or meningoencephalitis [31]. Listeria has a predilection for invasion of the placenta, and is associated with adverse pregnancy outcomes, including miscarriage and stillbirth [32,33]. In a study in Britain between 1967 and 1985, 34% cases of listeria were during pregnancy, with an association with intrauterine fetal death and neonatal infection [32]. Although the pathogenesis of severe listeriosis during pregnancy is not well understood, it may involve establishment of an infectious reservoir at the level of placenta and uncontrolled dissemination to maternal organs [34].

Malaria (*Plasmodium falciparum*)

In malaria-endemic regions, pregnant women are estimated to have three-times increased risk of severe malaria as compared to non-pregnant adults [35]. *Plasmodium falciparum* has a tropism for the placenta and increases the risk of low birth weight and preterm birth. Women who are in their first pregnancy, as well as those who are younger, are at higher risk than multigravida women. This reflects acquisition of antibodies against placenta-adhesive parasites during consecutive pregnancies [36]. Chemoprevention against malaria is recommended during the second and third trimesters in endemic areas [37]. Therapy with artemesin was thought to be toxic to fetal development. However, the use of artemisin derivatives has now been demonstrated to be safe and effective in treatment of pregnant women early in pregnancy [38]. Maternal immunity transferred across the placenta provides protection against severe malaria in infants [39].

Other pathogens transmitted from the mother to the fetus and newborn infant

Several pathogens that are often asymptomatic in pregnant women cause a serious threat to the fetus and the newborn infant. These pathogens are therefore considered as targets for immunization of children, teenagers or women of childbearing age. This approach has been successfully applied to the prevention of congenital rubella syndrome in high income countries and, increasingly, in low and middle income countries [40].

Herpes simplex virus

Primary genital herpes infection during pregnancy may be associated with fulminant hepatitis and disseminated disease, as compared to primary herpes simplex virus (HSV) infections in non-pregnant adults, though the data for this is not conclusive [41]. The risk of neonatal herpes is substantially higher in women with primary disease with both HSV1 and HSV2 during pregnancy, rather than those who have disease reactivation during pregnancy [41]. Because many women with genital herpes are asymptomatic or have mild symptoms, diagnosis and treatment may be challenging, requiring the use of serology for evidence of preexisting antibody to HSV1 and HSV2. The target population for vaccination against herpes is likely prior to onset of sexual activity; however, currently no licensed vaccine exists. Multiple vaccine candidates have not shown efficacy in prevention of primary infection; however, multiple new vaccines are under development [42,43]. Mouse studies recently demonstrated the maternal HSV immunization confers protection against neonatal mortality and behavioral morbidity [44]. Neonatal herpes is associated with a 60% mortality rate if untreated. The route of acquisition is most commonly through exposure in the genital tract during vaginal delivery; therefore caesarean section is indicated in women with known active genital herpes at the time of delivery.

Cytomegalovirus

Women with cytomegalovirus (CMV) infection during pregnancy generally exhibit a range of disease severity similar to non-pregnant adults, ranging from asymptomatic infection to a mononucleosis-like syndrome with fevers, rash, and lymphadenophathy. Risk factors for primary CMV infection during pregnancy include exposure to infected young children excreting the virus for prolonged periods of time. Clinical manifestations of CMV infection in the fetus include sensorineural hearing loss, chorioretinitis, intrauterine growth restriction, preterm birth, hepatosplenomegaly, microcephaly, and fulminant disease [45]. Primary infection early in pregnancy is associated with a higher risk of transmission to the fetus, with estimates of 32% with primary infection as compared to 1% following recurrent infection. Recurrent infections include both reactivation of latent virus and acquisition of a new strain of CMV [46]. Similar to herpes simplex, the target of CMV vaccination would be young children or adolescents prior to pregnancy. A number of CMV vaccine candidates are currently in development [47,48], though none are licensed.

Zika virus

Zika virus is a re-emergent pathogen that has particular implications for pregnant women. Symptoms of Zika virus infection during pregnancy include maculopapular rash, arthralgia, conjunctivitis and low grade fever [49]. Like CMV and rubella, infection with zika earlier in pregnancy is associated with increased risk of congenital birth defects, including microcephaly and intrauterine growth retardation. Pregnant women are likely to have prolonged viremia due to zika virus, as compared to non-pregnant adults, and infection during pregnancy is associated with increased risk of congenital birth defects [50]. The United States Centers for Disease Control guidelines recommend that laboratory studies for zika virus by molecular testing or serology are indicated in women with clinical suspicion of zika virus infection, with history of exposure via travel or residence in a region where mosquito-borne transmission of zika is documented, or unprotected sexual contact with a partner who has traveled to these areas [51]. Viral infection during pregnancy is thought to lead to placental infection, and transmission to fetal neuronal cells, leading to cell injury and death. The risks of congenital infection appear to be high in women with documented zika virus infection during pregnancy; in a prospective study in Brazil, 42% of infants born to women with diagnosed infection during pregnancy had abnormal clinical or brain imaging findings [52]. These occurred in 55%, 52% and 29% of pregnancies where infection occurred in the first, second or third trimester of pregnancy, respectively. No specific treatment or vaccine is currently available, with current recommendations including avoidance of travel to endemic areas, use of barrier contraception to protect against sexual transmission, and use of standard precautions to prevent mosquito bites.

Innate and adaptive immune responses in pregnancy

The fetus expresses paternal antigens and is therefore a semi-allogeneic graft for the pregnant women. Multiple redundant mechanisms have been selected to promote immunological tolerance of the fetus and to allow successful pregnancy. At the materno-fetal interface, the placenta is a highly immunoregulated environment containing large numbers of innate and adaptive immune cells, including NK cells, T lymphocytes and myeloid cells, actively suppressing fetal tissue rejection [1].

Pregnancy is also associated with significant changes in the number and function of immune cells at the systemic level. Recent analyses of immune cell responsiveness to activation signals indicate dynamic and coordinated changes from the first to the third trimester of pregnancy, suggesting an "immune clock" regulating immune functions in pregnant women [53]. Relevant to antibody responses to vaccination are B cells, follicular helper T cells and antigen-presenting cells. Studies have examined the influence of pregnancy on B cells and antigen-presenting cells but its impact on follicular helper T cells remains to be assessed. Most studies showed that pregnancy is associated with decreased B cell numbers in peripheral blood [54–59]. Studies in mice indicated that estrogens produced during pregnancy reduce B cell lymphopoiesis [60]. Studies of the influence of hormones on B cell functions indicate that pregnancy may impact the production of immunoglobulins. Estrogen increases the production of immunoglobulin G (IgG) by human B cells and prolactin decreases the threshold of B cell activation [61,62]. However, total serum IgG levels are lower in pregnant than in non-pregnant women in both low and high income country settings, a phenomenon that probably involves hemodilution [63,64]. On the other hand, populations living in low income countries, including pregnant women, commonly have elevated serum levels of IgG, probably as a result of high and chronic exposure to microbial antigens, such as malaria [63,65,66]. A recent study indicated that estrogen stimulates the production of natural antibodies against bacterial oligosaccharides [67]. These antibodies were transferred from the mother to the offspring and protected mice against entheropathogenic *Escherichia coli* infection. Together, these studies indicate that sex hormones and pregnancy modulate the number of B lymphocytes and their production of immunoglobulins.

Significant changes in the quality of IgG are also observed during pregnancy. IgG are glycoproteins carrying N-glycans at both the Fc and Fab segments [68,69]. The composition of the N-glycans influences the three-dimensional structure of the Fc segment of IgG and the interaction of IgG with Fcγ receptors and complement. In total, 36 glycovariants can be attached to the Fc segment of IgG. In combination with the four IgG subclasses, this diversity offers the potential to fine tune IgG effector functions [69]. Pregnancy is associated with increased galactosylation and sialylation and with decreased fucosylation of total IgG Fc [70]. These changes have the potential to modulate the capacity of IgG to activate NK cells and complement and to reduce their inflammatory properties [71,72]. The clinical relevance of these modifications is supported by the association of IgG galactosylation and remission of rheumatoid arthritis in pregnant women [73]. Mouse studies suggest that estrogen upregulates the expression of the activation-induced deaminase, the enzyme that initiates class switch recombination and somatic hypermutation of immunoglobulins [74]. The impact of estrogen and pregnancy on the subclass and the avidity of antigen-specific IgG remains poorly characterized.

Changes in antigen-presenting cells are also observed in pregnant women. The absolute number of myeloid dendritic cells (mDC) was shown to increase in the first trimester and decrease as pregnancy progressed to reach similar counts in the third trimester as in non-pregnant women [75,76]. In contrast, the numbers of plasmacytoid (p)DCs were shown to be reduced during the third trimester of pregnancy, resulting in a higher mDC: pDC ratio [75,77,78]. Pregnancy is also associated with changes in Toll-like receptor (TLR) expression by circulating DC subsets. Increased expression of TLR-1 by mDC and of TLR-7 and TLR-9 by pDC was detected [79]. The role of these changes in the modulation of immune responses during pregnancy have not yet been explored.

Potentially relevant to the modulation of immune responses during pregnancy are the changes in the composition of the maternal microbiome. The composition and the diversity of the microbiome is influenced by pregnancy at several body sites, including the oral cavity, the gut and the genital tract [80]. Mouse studies suggest that changes in the gut microbiome may promote inflammatory responses and participate in the physiological changes in metabolism associated with pregnancy [81].

Effect of pregnancy on vaccine responses

The impact of pregnancy and sex hormones on B cells, antigen-presenting cells and microbiome suggests a possible influence on the magnitude and quality of antibody responses to vaccines. This notion is supported by the observation that the magnitude of antibody responses to vaccines is often higher in women than in men [82]. Overall, vaccines are immunogenic in pregnant women [2,3]. However, relatively few controlled studies have compared vaccine responses in pregnant and non-pregnant women. Available data come from studies of influenza or pertussis immunization and are not consistent across studies. Four studies described similar antibody responses to seasonal influenza vaccines in pregnant and non-pregnant women [83–86]. On the other hand, other studies reported lower responses in pregnant women following immunization with seasonal and pandemic influenza vaccines [87–89]. Whether the gestational stage of pregnancy affects responses to vaccines is uncertain. Similar antibody responses to seasonal and pandemic influenza vaccination were observed throughout pregnancy in two studies whereas a recent study reported a decline in the antibody response to seasonal influenza immunization and a relative increase in the proportion of the IgG4 subclass [84,89,90]. Fewer studies have examined the impact of pregnancy on the antibody response to pertussis immunization. Two studies reported similar antibody responses to pertussis immunization in pregnant and non-pregnant women [91,92]. However, one of the two studies reported lower T cell responses to pertussis antigens in pregnant as compared to non-pregnant women [92]. A recent study involving a larger sample size observed lower antibody responses to pertussis toxin and filamentous hemagglutinin in pregnant as compared to non-pregnant women [93]. Factors responsible for the discordant results obtained in different studies are unclear but are possibly attributable to the relatively small size of the population included in most studies and to differences in tested vaccines and in participant characteristics, including pre-vaccination antibody titers that are an important determinant of vaccine responses [94]. The persistence of antibodies following maternal immunization will influence the optimal timing of immunization and the requirement to repeat immunization during consecutive pregnancies. Pertussis antibodies decay rapidly supporting the need to immunize during each pregnancy, as is recommended in an increasing number of high income countries [95,96]. Antibody decay following immunization with adjuvanted pandemic influenza vaccine was similar in pregnant and non-pregnant women [88]. The impact of pregnancy on the quality of the antibody response to vaccines remains largely uncharacterized. A study showed that the avidity of cord blood antibodies is higher following pertussis immunization at 27–30 weeks as compared to after 31 weeks of gestation, a difference that is likely related to the time needed to increase antibody avidity following booster immunization rather than to an impact of gestational stage [97]. The impact of pregnancy on the glycosylation profile and on the functional properties of vaccine-induced antibodies remains to be investigated.

The evidence that pregnancy induces changes in circulating antigen-presenting cells and in the composition of the microbiome suggests that innate immune responses and inflammatory reactions to vaccines may also be influenced. In one study, pregnant women given seasonal influenza vaccine had increased plasma levels of inflammatory cytokines during the first days after vaccination but these responses were similar to those in non-pregnant women [85]. In a recent study, pregnant women were more likely to report moderate and severe pain at the injection site following pertussis immunization as compared to non-pregnant women but the occurrence of other local and systemic reactions following vaccination was similar in both groups [93].

With the exception of maternal human immunodeficiency virus (HIV) infection, the impact of maternal diseases on responses to vaccination remains largely uncharacterized. In South Africa, maternal HIV infection was associated with lower seroconversion rates after seasonal influenza vaccination but vaccine efficacy was comparable to that observed in HIV-uninfected pregnant women [16]. HIV infection was also associated with lower immunogenicity of a candidate glycoconjugate Group B streptococcus vaccine in pregnant women [98]. Immunogenicity of pandemic A/H1N1 influenza vaccination in HIV-infected pregnant women correlated positively with pre-vaccination immunity and negatively with HIV replication [99]. Hypergammaglobulinemia is commonly observed in pregnant women living in low income countries but its impact on antibody responses to vaccines has not been characterized.

Transfer of maternal antibodies to the newborn infant

Transfer of maternal antibodies to the fetus and newborn infant across the placenta and breastmilk provides protection against infectious pathogens to which the mother has been exposed. In addition, recent studies indicate that maternal antibodies also contribute to infant gut homeostasis by shielding commensal bacteria, impacting the development of the gut immune system and to the prevention of allergic responses in early life [100–102].

Of the immunoglobulins in the maternal circulation, only IgG is transferred transplacentally [68,103]. Studies suggest that maternal IgE could be transported complexed with IgG [104]. Transfer of antibodies is an active process beginning at the end of the first trimester of pregnancy. By the end of the second trimester, transplacental antibody ratios are about 50% and are greater than 100% at birth in full-term newborns [105,106]. As a result, infants born prematurely have lower levels of maternal antibodies, increasing risk for vaccine-preventable diseases [107]. IgG subclass is an important determinant of transfer across the placenta. Antibody transfer ratio is highest for IgG1 and lowest for IgG2 [68,103,106]. Transfer of IgG is also influenced by their antigen specificity [107]. Transfer of polysaccharide antigen-specific IgG is lower than protein antigen-specific IgG, a difference that is likely related to the lower transfer of IgG2. However, there is considerable variation in transfer ratios among protein antigen-specific and polysaccharide antigen-specific antibodies, suggesting that other factors than IgG subclass may play a role. Among them, IgG glycosylation could be involved. Indeed, IgG of different antigen-specificities have different glycosylation profiles and some studies suggest selective transfer of IgG expressing specific glycovariants [108,109].

Transport of maternal IgG to the fetus involves crossing several placental layers, including syncytiotrophoblasts, the fetal endothelium and the stroma that separates the two cell layers [103,106]. The neonatal Fc receptor (FcRn) is the main receptor transporting maternal IgG across syncytiotrophoblasts. Following endocytosis from the maternal circulation, the Fc segment of IgG binds the FcRn at acidic pH. The endosome is then transported to the basal surface of the syncytiotrophoblast where IgG dissociate from the FcRn at physiologic pH. As the FcRn is not expressed by fetal endothelial cells, other receptors have to participate in the transport of IgG to the fetal circulation. Studies suggest that the FcγRIIb2 expressed by fetal endothelial cells could be involved but its role in maternal antibody transfer is still debated [110,111]. A better understanding of the molecular and cellular basis of maternal antibody transfer across the placenta would help designing vaccines inducing antibodies with optimal transferability to the fetus.

Timing of maternal immunization

For maternal vaccine delivery, the decision of when to vaccinate during pregnancy needs to consider several variables, including limited antenatal care for many women in developing countries leading to a need for an increased vaccination window, the need to protect preterm infants who would not benefit from vaccination late in pregnancy, and the optimal timing for antibody affinity maturation and transplacental transfer. In the United States, current recommendations are to administer pertussis-containing vaccines during pregnancy from 27 to 36 weeks gestation, which then requires about 2 weeks to have a serologic response, followed by transplacental antibody transfer [112]. This recommendation leads to an at-risk group of infants who are born earlier than 30 weeks gestation, who are also at highest risk for severe disease due to pertussis. It also potentially limits protection of infants born closer to term when maternal immunization is performed shortly before delivery. A recent prospective observational study examined concentrations of anti-pertussis toxin and anti-filamentous hemagglutinin IgG in newborns of women who received second versus third trimester vaccination [113]. Overall, higher concentrations of cord blood antibodies to both pertussis toxin and filamentous hemagglutinin were detected following second trimester versus third trimester vaccination. Although maternal vaccine responses were not measured, the data were interpreted as the consequence of the longer duration of antibody transfer following vaccination during the second trimester of pregnancy. Immunization against pertussis is recommended from the second trimester of pregnancy in several countries, including the United Kingdom and Belgium, but currently not in the United States [112,114].

For influenza vaccination in pregnancy, the recommendation in industrialized countries is to vaccinate any time during pregnancy during influenza season. There are many countries where influenza circulates many months during the year, including subtropical Asia and sub-Saharan Africa. In these settings, influenza vaccination may be considered as a year-round vaccination strategy and has been shown to be efficacious in a randomized clinical trial of maternal influenza vaccination in Nepal [14]. In this study, women were additionally randomized to second or third trimester vaccination to evaluate the effect on infant influenza vaccine efficacy [115]. The results from this study showed that there was no significant difference in influenza vaccine efficacy by second or third trimester vaccination, and there was a nonsignificant trend towards improved birth weight with earlier vaccination. Cord blood anti-hemagluttination inhibition titers against A/H3N2, A/H1N1 and B antigens was not different between the two groups, though the numbers of samples tested were small. Therefore, available data regarding pertussis and influenza vaccine timing in pregnancy suggest that earlier vaccination is not associated with decreased efficacy against disease in mothers and infants. As vaccination from the second trimester of pregnancy offers the advantage to protect preterm infants and to increase the potential window for vaccination in settings where it may be challenging to access pregnant women, further studies should be conducted to consolidate these observations in different populations. Limited information is available regarding the impact of timing of vaccination during pregnancy for other vaccines. The principle that optimal transfer of maternal antibodies requires sufficient time between maternal immunization and delivery is likely to apply. In a study of *Haemophilus influenzae* type B conjugate vaccine, transmission of antibodies was greatest in mothers vaccinated more than 4weeks before delivery [116].

Impact of chronic maternal infections on antibody transfer

Chronic maternal infections, such as HIV and malaria, reduce the transfer of maternal IgG [117–120]. Studies conducted across different populations reported an association between reduced transfer of maternal IgG and placental malaria [117,121–124]. Trophozoites cause direct invasion of the placenta and could thereby alter IgG transfer. However, a recent study in Papua New Guinea indicated that this decreased transfer may not be directly related to placental malaria but rather due to hypergammaglobulinemia [125]. Hypergammaglobulinemia results from chronic inflammation and polyclonal B cell activation. It is generally considered that hypergammaglobulinemia impairs maternal antibody transfer by saturating the FcRn [106]. However, studies reported a variable impact of maternal hypergammaglobulinemia on transplacental transfer according to IgG subclass and antigen-specificity, suggesting that other factors involving placental receptors and cells or biophysical characteristics of IgG may be involved [117,121]. Little is know regarding the impact of hypergammaglobulinemia on transfer of IgG induced by vaccination during pregnancy.

Studies from low and high income countries reported decreased transfer of maternal IgG in children born to HIV-infected mothers across multiple vaccine and pathogen-specific antigens [119,126,127]. This is particularly significant given the increased infectious morbidity and mortality observed in HIV-exposed uninfected as compared to HIV-unexposed infants [120,128]. This notion is supported by a recent study conducted in Belgium, showing an association between reduced transfer of maternal antibodies, as well as immune activation in the newborn, and risk of hospitalization for infection during the first months of life in HIV-exposed uninfected infants [127]. Few studies have examined the impact of maternal HIV infection on the transfer of antibodies induced by vaccination during pregnancy. Following pandemic A/H1N1 influenza vaccination of HIV-infected pregnant women, cord blood antibody levels were higher than maternal antibody levels in some studies and lower in others [129]. This variability could be related to differences in the activity of maternal HIV infection across studies. Indeed, higher transfer ratios of maternal antibodies were observed in mothers who initiated anti-retroviral therapy before as compared to during pregnancy [127]. Immunization of HIV-infected pregnant women is an important strategy to protect their vulnerable newborn infants but its efficacy can be limited by reduced vaccine immunogenicity and reduced transfer of maternal antibodies. The impact of HIV infection on the quality of maternal antibodies has not been explored. A better understanding of the impact of HIV in maternal immunity and its transfer to the newborn should help the design of optimal vaccination strategies.

Maternal antibody transfer through breastmilk

Breastfeeding is associated with improvement in a multitude of infant outcomes, particularly in low and middle income countries [130]. Exclusive breastfeeding is recommended by multiple professional groups for the first 6 months of life. Specifically, in relation to infectious risk, breastmilk decreases risk of sepsis and necrotizing enterocolitis in preterm infants, and is associated with decreased risk of respiratory viral infections, acute otitis media, and gastroenteritis [131]. Breastmilk contains secretory IgA and IgG. Other components of breastmilk include lactoferrin, lysozyme, white blood cells (predominantly neutrophils and macrophages), hormones, growth factors, and cytokines which could contribute to antiviral and antibacterial properties [3]. The synthesis of IgA is via plasma cells in the enteromammary and bronchomammary immune system. After exposure to antigens, maternal plasma cells synthesize secretory IgA antibody in the mammary gland and secrete this into milk. The polymeric Ig receptor (pIgR) transports secretory IgA and IgM into breastmilk [132]. Secretory IgA represents the major immunoglobulin in breastmilk, followed by secretory IgM and then IgG. In humans, ingested breastmilk antibodies do not enter the neonatal circulation, but secretory antibodies may prevent microbial colonization and invasion by coating mucosal surfaces [100]. Studies have shown that secretory IgA in colostrum can prevent HIV transcytosis across epithelium by direct neutralization [133]. Breastmilk IgG has also been shown to protect against HIV infection through antibody-dependent cytotoxicity [134]. A correlation between breastmilk IgG against respiratory syncytial virus with protection against acute respiratory infection has recently been reported [135]. Although the role of breastmilk antibodies in protection against disease is not well-studied, pathogen-specific IgA and IgG are induced in breastmilk by maternal immunization and could contribute to protection [3,136,137].

Conclusion

Immunization during pregnancy is an efficient strategy to protect both the mother and the newborn infant against infectious pathogens. As the momentum for maternal immunization is growing, we have an opportunity to gain fundamental insights in the determinants of maternal immunity against pathogens and of its transfer to the newborn infant. Pregnancy is associated with many dynamic changes in maternal immune cells and molecules that are critical for tolerance of the fetus. These changes could have also been selected to modulate maternal immune components that are transferred to the newborn infant and thereby provide optimal protection against infectious pathogens after birth. Pregnancy modifies the glycosylation profile of maternal IgG and could also influence the production of individual IgG subclasses. Transfer of maternal IgG through the placenta, and potentially through breastmilk, could select antibodies with optimal biophysical and functional profiles. Unraveling these fundamental processes and identifying determinants of the magnitude and quality of vaccine responses in pregnant women offer the potential to optimize maternal immunization strategies. Most of the burden of infectious diseases affecting young infants is in low income countries where prevalence of maternal diseases, including chronic infections, is also highest. A more systematic evaluation and a better understanding of the impact of maternal diseases on vaccine-induced immunity and its transfer to the newborn are required to achieve the broadest impact of maternal immunization worldwide.

References

[1] Arck PC, Hecher K. Fetomaternal immune cross-talk and its consequences for maternal and offspring's health. Nat Med 2013;19(5):548–56.
[2] Edwards KM. Maternal immunisation in pregnancy to protect newborn infants. Arch Dis Child 2019;104(4):316–9. https://doi.org/10.1136/archdischild-2017-313530. PMID: 29909381.
[3] Marchant A, Sadarangani M, Garand M, Dauby N, Verhasselt V, Pereira L, et al. Maternal immunisation: collaborating with mother nature. Lancet Infect Dis 2017;17(7):e197–208.
[4] Ortiz JR, Englund JA, Neuzil KM. Influenza vaccine for pregnant women in resource-constrained countries: a review of the evidence to inform policy decisions. Vaccine 2011;29(27):4439–52.
[5] Rasmussen SA, Jamieson DJ, Bresee JS. Pandemic influenza and pregnant women. Emerg Infect Dis 2008;14(1):95–100.
[6] Van Kerkhove MD, Vandemaele KAH, Shinde V, Jaramillo-Gutierrez G, Koukounari A, Donnelly CA, et al. Risk factors for severe outcomes following 2009 influenza A (H1N1) infection: a global pooled analysis. PLoS Med 2011;8(7):e1001053.
[7] Siston AM, Rasmussen SA, Honein MA, Fry AM, Seib K, Callaghan WM, et al. Pandemic 2009 influenza A(H1N1) virus illness among pregnant women in the United States. JAMA 2010;303(15):1517–25.
[8] Centers for Disease Control and Prevention (CDC). Maternal and infant outcomes among severely ill pregnant and postpartum women with 2009 pandemic influenza A (H1N1)—United States, April 2009–August 2010. MMWR Morb Mortal Wkly Rep 2011;60(35):1193–6.
[9] Memoli MJ, Harvey H, Morens DM, Taubenberger JK. Influenza in pregnancy. Influenza Other Respi Viruses 2013;7(6):1033–9.

[10] Katz MA, Gessner BD, Johnson J, Skidmore B, Knight M, Bhat N, et al. Incidence of influenza virus infection among pregnant women: a systematic review. BMC Pregnancy Childbirth 2017;17(1):155.
[11] Kay AW, Fukuyama J, Aziz N, Dekker CL, Mackey S, Swan GE, et al. Enhanced natural killer-cell and T-cell responses to influenza A virus during pregnancy. Proc Natl Acad Sci U S A 2014;111(40):14506–11.
[12] Engels G, Hierweger AM, Hoffmann J, Thieme R, Thiele S, Bertram S, et al. Pregnancy-related immune adaptation promotes the emergence of highly virulent H1N1 influenza virus strains in allogenically pregnant mice. Cell Host Microbe 2017;21(3):321–33.
[13] Le Gars M, Kay AW, Bayless NL, Aziz N, Dekker CL, Swan GE, et al. Increased proinflammatory responses of monocytes and plasmacytoid dendritic cells to influenza A virus infection during pregnancy. J Infect Dis 2016;214(11):1666–71.
[14] Steinhoff MC, Katz J, Englund JA, Khatry SK, Shrestha L, Kuypers J, et al. Year-round influenza immunisation during pregnancy in Nepal: a phase 4, randomised, placebo-controlled trial. Lancet Infect Dis 2017;17(9):981–9.
[15] Tapia MD, Sow SO, Tamboura B, Téguété I, Pasetti MF, Kodio M, et al. Maternal immunisation with trivalent inactivated influenza vaccine for prevention of influenza in infants in Mali: a prospective, active-controlled, observer-blind, randomised phase 4 trial. Lancet Infect Dis 2016;16(9):1026–35.
[16] Madhi SA, Cutland CL, Kuwanda L, Weinberg A, Hugo A, Jones S, et al. Influenza vaccination of pregnant women and protection of their infants. N Engl J Med 2014;371(10):918–31.
[17] Zaman K, Roy E, Arifeen SE, Rahman M, Raqib R, Wilson E, et al. Effectiveness of maternal influenza immunization in mothers and infants. N Engl J Med 2008;359(15):1555–64.
[18] Datta S, Walsh EE, Peterson DR, Falsey AR. Can analysis of routine viral testing provide accurate estimates of respiratory syncytial virus disease burden in adults? J Infect Dis 2017;215(11):1706–10.
[19] Chu HY, Katz J, Tielsch J, Khatry SK, Shrestha L, LeClerq SC, et al. Clinical presentation and birth outcomes associated with respiratory syncytial virus infection in pregnancy. PLoS One 2016;11(3):e0152015.
[20] Madhi SA, Cutland CL, Downs S, Jones S, van Niekerk N, Simoes EAF, et al. Burden of respiratory syncytial virus infection in South African human immunodeficiency virus (HIV)-infected and HIV-uninfected pregnant and postpartum women: a longitudinal cohort study. Clin Infect Dis 2018;66(11):1658–65.
[21] Lenahan JL, Englund JA, Katz J, Kuypers J, Wald A, Magaret A, et al. Human metapneumovirus and other respiratory viral infections during pregnancy and birth, Nepal. Emerg Infect Dis 2017;23(8).
[22] Philpott EK, Englund JA, Katz J, Tielsch J, Khatry S, LeClerq SC, et al. Febrile rhinovirus illness during pregnancy is associated with low birth weight in Nepal. Open Forum Infect Dis 2017;4(2):ofx073.
[23] Assiri A, Abedi GR, Al Masri M, Bin Saeed A, Gerber SI, Watson JT. Middle east respiratory syndrome coronavirus infection during pregnancy: a report of 5 cases from Saudi Arabia. Clin Infect Dis 2016;63(7):951–3.
[24] Pilorgé L, Chartier M, Méritet J-F, Cervantes M, Tsatsaris V, Launay O, et al. Rhinoviruses as an underestimated cause of influenza-like illness in pregnancy during the 2009-2010 influenza pandemic. J Med Virol 2013;85(8):1473–7.
[25] Esmonde TF, Herdman G, Anderson G. Chickenpox pneumonia: an association with pregnancy. Thorax 1989;44(10):812–5.
[26] Triebwasser JH, Harris RE, Bryant RE, Rhoades ER. Varicella pneumonia in adults. Report of seven cases and a review of literature. Medicine (Baltimore) 1967;46(5):409–23.
[27] Paryani SG, Arvin AM. Intrauterine infection with varicella-zoster virus after maternal varicella. N Engl J Med 1986;314(24):1542–6.

[28] Khuroo MS, Teli MR, Skidmore S, Sofi MA, Khuroo MI. Incidence and severity of viral hepatitis in pregnancy. Am J Med 1981;70(2):252–5.
[29] Pérez-Gracia MT, Suay-García B, Mateos-Lindemann ML. Hepatitis E and pregnancy: current state. Rev Med Virol 2017;27(3):e1929. https://doi.org/10.1002/rmv.1929.
[30] Yang C, Yu W, Bi Y, Long F, Li Y, Wei D, et al. Increased oestradiol in hepatitis E virus-infected pregnant women promotes viral replication. J Viral Hepat 2018;25(6):742–51.
[31] Lamont RF, Sobel J, Mazaki-Tovi S, Kusanovic JP, Vaisbuch E, Kim SK, et al. Listeriosis in human pregnancy: a systematic review. J Perinat Med 2011;39(3):227–36.
[32] McLauchlin J. Human listeriosis in Britain, 1967-85, a summary of 722 cases. 1. Listeriosis during pregnancy and in the newborn. Epidemiol Infect 1990;104(2):181–9.
[33] Robbins JR, Skrzypczynska KM, Zeldovich VB, Kapidzic M, Bakardjiev AI. Placental syncytiotrophoblast constitutes a major barrier to vertical transmission of Listeria monocytogenes. PLoS Pathog 2010;6(1):e1000732.
[34] Bakardjiev AI, Theriot JA, Portnoy DA. Listeria monocytogenes traffics from maternal organs to the placenta and back. PLoS Pathog 2006;2(6):e66.
[35] Luxemburger C, Ricci F, Nosten F, Raimond D, Bathet S, White NJ. The epidemiology of severe malaria in an area of low transmission in Thailand. Trans R Soc Trop Med Hyg 1997;91(3):256–62.
[36] Ataíde R, Mayor A, Rogerson SJ. Malaria, primigravidae, and antibodies: knowledge gained and future perspectives. Trends Parasitol 2014;30(2):85–94.
[37] Radeva-Petrova D, Kayentao K, ter Kuile FO, Sinclair D, Garner P. Drugs for preventing malaria in pregnant women in endemic areas: any drug regimen versus placebo or no treatment. Cochrane Database Syst Rev 2014;10:CD000169.
[38] McGready R, Lee SJ, Wiladphaingern J, Ashley EA, Rijken MJ, Boel M, et al. Adverse effects of falciparum and vivax malaria and the safety of antimalarial treatment in early pregnancy: a population-based study. Lancet Infect Dis 2012;12(5):388–96.
[39] Kurtis JD, Raj DK, Michelow IC, Park S, Nixon CE, McDonald EA, Nixon CP, Pond-Tor S, Jha A, Taliano RJ, Kabyemela ER, Friedman JF, Duffy PE, Fried M. Maternally-derived antibodies to Schizont Egress Antigen-1 and protection of infants from severe malaria. Clin Infect Dis 2019;68(10):1718–24. https://doi.org/10.1093/cid/ciy728. PMID: 30165569.
[40] Vynnycky E, Papadopoulos T, Angelis K. The impact of measles-rubella vaccination on the morbidity and mortality from congenital rubella syndrome in 92 countries. Hum Vaccin Immunother 2019;15(2):309–16.
[41] Corey L, Wald A. Maternal and neonatal herpes simplex virus infections. N Engl J Med 2009;361(14):1376–85.
[42] Roth K, Ferreira VH, Kaushic C. HSV-2 vaccine: current state and insights into development of a vaccine that targets genital mucosal protection. Microb Pathog 2013;58:45–54.
[43] Belshe RB, Leone PA, Bernstein DI, Wald A, Levin MJ, Stapleton JT, et al. Efficacy results of a trial of a herpes simplex vaccine. N Engl J Med 2012;366(1):34–43.
[44] Patel CD, Backes IM, Taylor SA, Jiang Y, Marchant A, Pesola JM, et al. Maternal immunization confers protection against neonatal herpes simplex mortality and behavioral morbidity. Sci Transl Med 2019;11(487).
[45] Adler SP, Nigro G. Prevention of maternal-fetal transmission of cytomegalovirus. Clin Infect Dis 2013;57(Suppl 4):S189–92.
[46] Ross SA, Arora N, Novak Z, Fowler KB, Britt WJ, Boppana SB. Cytomegalovirus reinfections in healthy seroimmune women. J Infect Dis 2010;201(3):386–9.
[47] Pass RF, Zhang C, Evans A, Simpson T, Andrews W, Huang M-L, et al. Vaccine prevention of maternal cytomegalovirus infection. N Engl J Med 2009;360(12):1191–9.
[48] Plotkin SA, Boppana SB. Vaccination against the human cytomegalovirus. Vaccine 2018; https://doi.org/10.1016/j.vaccine.2018.02.089. (e-pub ahead of print). PMID: 29622379.

[49] Driggers RW, Ho C-Y, Korhonen EM, Kuivanen S, Jääskeläinen AJ, Smura T, et al. Zika virus infection with prolonged maternal viremia and fetal brain abnormalities. N Engl J Med 2016;374(22):2142–51.
[50] Suy A, Sulleiro E, Rodó C, Vázquez É, Bocanegra C, Molina I, et al. Prolonged Zika virus viremia during pregnancy. N Engl J Med 2016;375(26):2611–3.
[51] Oduyebo T, Polen KD, Walke HT, Reagan-Steiner S, Lathrop E, Rabe IB, et al. Update: interim guidance for health care providers caring for pregnant women with possible Zika virus exposure—United States (including U.S. territories), July 2017. MMWR Morb Mortal Wkly Rep 2017;66(29):781–93.
[52] Brasil P, Pereira JP, Moreira ME, Ribeiro Nogueira RM, Damasceno L, Wakimoto M, et al. Zika virus infection in pregnant women in Rio de Janeiro. N Engl J Med 2016;375(24):2321–34.
[53] Aghaeepour N, Ganio EA, Mcilwain D, Tsai AS, Tingle M, Van Gassen S, et al. An immune clock of human pregnancy. Sci Immunol 2017;2(15).
[54] Mahmoud F, Abul H, Omu A, Al-Rayes S, Haines D, Whaley K. Pregnancy-associated changes in peripheral blood lymphocyte subpopulations in normal Kuwaiti women. Gynecol Obstet Invest 2001;52(4):232–6.
[55] Watanabe M, Iwatani Y, Kaneda T, Hidaka Y, Mitsuda N, Morimoto Y, et al. Changes in T, B, and NK lymphocyte subsets during and after normal pregnancy. Am J Reprod Immunol 1997;37(5):368–77.
[56] Zimmer JP, Garza C, Butte NF, Goldman AS. Maternal blood B-cell (CD19+) percentages and serum immunoglobulin concentrations correlate with breast-feeding behavior and serum prolactin concentration. Am J Reprod Immunol 1998;40(1):57–62.
[57] Matthiesen L, Berg G, Ernerudh J, Håkansson L. Lymphocyte subsets and mitogen stimulation of blood lymphocytes in normal pregnancy. Am J Reprod Immunol 1996;35(2):70–9.
[58] Valdimarsson H, Mulholland C, Fridriksdottir V, Coleman DV. A longitudinal study of leucocyte blood counts and lymphocyte responses in pregnancy: a marked early increase of monocyte-lymphocyte ratio. Clin Exp Immunol 1983;53(2):437–43.
[59] Moore MP, Carter NP, Redman CW. Lymphocyte subsets defined by monoclonal antibodies in human pregnancy. Am J Reprod Immunol 1983;3(4):161–4.
[60] Kincade PW, Medina KL, Smithson G, Scott DC. Pregnancy: a clue to normal regulation of B lymphopoiesis. Immunol Today 1994;15(11):539 44.
[61] Kanda N, Tamaki K. Estrogen enhances immunoglobulin production by human PBMCs. J Allergy Clin Immunol 1999;103(2 Pt 1):282–8.
[62] Correale J, Farez MF, Ysrraelit MC. Role of prolactin in B cell regulation in multiple sclerosis. J Neuroimmunol 2014;269(1–2):76–86.
[63] McGregor IA, Rowe DS, Wilson ME, Billewicz WZ. Plasma immunoglobulin concentrations in an African (Gambian) community in relation to season, malaria and other infections and pregnancy. Clin Exp Immunol 1970;7(1):51–74.
[64] Amino N, Tanizawa O, Miyai K, Tanaka F, Hayashi C, Kawashima M, et al. Changes of serum immunoglobulins IgG, IgA, IgM, and IgE during pregnancy. Obstet Gynecol 1978;52(4):415–20.
[65] Rowe DS, McGregor IA, Smith SJ, Hall P, Williams K. Plasma immunoglobulin concentrations in a West African (Gambian) community and in a group of healthy British adults. Clin Exp Immunol 1968;3(1):63–79.
[66] Logie DE, McGregor IA, Rowe DS, Billewicz WZ. Plasma immunoglobulin concentrations in mothers and newborn children with special reference to placental malaria: studies in the Gambia, Nigeria, and Switzerland. Bull World Health Organ 1973;49(6):547–54.
[67] Zeng Z, Surewaard BGJ, Wong CHY, Guettler C, Petri B, Burkhard R, et al. Sex-hormone-driven innate antibodies protect females and infants against EPEC infection. Nat Immunol 2018;19(10):1100–11.

[68] Vidarsson G, Dekkers G, Rispens T. IgG subclasses and allotypes: from structure to effector functions. Front Immunol 2014;5:520.
[69] Jennewein MF, Alter G. The immunoregulatory roles of antibody glycosylation. Trends Immunol 2017;38(5):358–72.
[70] Bondt A, Rombouts Y, Selman MHJ, Hensbergen PJ, Reiding KR, Hazes JMW, et al. Immunoglobulin G (IgG) Fab glycosylation analysis using a new mass spectrometric high-throughput profiling method reveals pregnancy-associated changes. Mol Cell Proteomics 2014;13(11):3029–39.
[71] Dekkers G, Treffers L, Plomp R, Bentlage AEH, de Boer M, Koeleman CAM, et al. Decoding the human immunoglobulin G-glycan repertoire reveals a spectrum of Fc-receptor- and complement-mediated-effector activities. Front Immunol 2017;8:877.
[72] Kaneko Y, Nimmerjahn F, Ravetch JV. Anti-inflammatory activity of immunoglobulin G resulting from Fc sialylation. Science 2006;313(5787):670–3.
[73] Bondt A, Selman MHJ, Deelder AM, Hazes JMW, Willemsen SP, Wuhrer M, et al. Association between galactosylation of immunoglobulin G and improvement of rheumatoid arthritis during pregnancy is independent of sialylation. J Proteome Res 2013;12(10):4522–31.
[74] Pauklin S, Sernández IV, Bachmann G, Ramiro AR, Petersen-Mahrt SK. Estrogen directly activates AID transcription and function. J Exp Med 2009;206(1):99–111.
[75] Della Bella S, Giannelli S, Cozzi V, Signorelli V, Cappelletti M, Cetin I, et al. Incomplete activation of peripheral blood dendritic cells during healthy human pregnancy. Clin Exp Immunol 2011;164(2):180–92.
[76] Yoshimura T, Inaba M, Sugiura K, Nakajima T, Ito T, Nakamura K, et al. Analyses of dendritic cell subsets in pregnancy. Am J Reprod Immunol 2003;50(2):137–45.
[77] Shin S, Jang JY, Roh EY, Yoon JH, Kim JS, Han KS, et al. Differences in circulating dendritic cell subtypes in pregnant women, cord blood and healthy adult women. J Korean Med Sci 2009;24(5):853–9.
[78] Ueda Y, Hagihara M, Okamoto A, Higuchi A, Tanabe A, Hirabayashi K, et al. Frequencies of dendritic cells (myeloid DC and plasmacytoid DC) and their ratio reduced in pregnant women: comparison with umbilical cord blood and normal healthy adults. Hum Immunol 2003;64(12):1144–51.
[79] Young BC, Stanic AK, Panda B, Rueda BR, Panda A. Longitudinal expression of Toll-like receptors on dendritic cells in uncomplicated pregnancy and postpartum. Am J Obstet Gynecol 2014;210(5):445. e1–6.
[80] Nuriel-Ohayon M, Neuman H, Koren O. Microbial changes during pregnancy, birth, and infancy. Front Microbiol 2016;7:1031.
[81] Koren O, Goodrich JK, Cullender TC, Spor A, Laitinen K, Bäckhed HK, et al. Host remodeling of the gut microbiome and metabolic changes during pregnancy. Cell 2012;150(3):470–80.
[82] Klein SL, Jedlicka A, Pekosz A. The Xs and Y of immune responses to viral vaccines. Lancet Infect Dis 2010;10(5):338–49.
[83] Hulka JF. Effectiveness of polyvalent influenza vaccine in pregnancy. Report of a controlled study during an outbreak of Asian influenza. Obstet Gynecol 1964;23:830–7.
[84] Murray DL, Imagawa DT, Okada DM, St Geme JW. Antibody response to monovalent A/New Jersey/8/76 influenza vaccine in pregnant women. J Clin Microbiol 1979;10(2):184–7.
[85] Christian LM, Porter K, Karlsson E, Schultz-Cherry S, Iams JD. Serum proinflammatory cytokine responses to influenza virus vaccine among women during pregnancy versus non-pregnancy. Am J Reprod Immunol 2013;70(1):45–53.
[86] Kay AW, Bayless NL, Fukuyama J, Aziz N, Dekker CL, Mackey S, et al. Pregnancy does not attenuate the antibody or plasmablast response to inactivated influenza vaccine. J Infect Dis 2015;212(6):861–70.

[87] Schlaudecker EP, McNeal MM, Dodd CN, Ranz JB, Steinhoff MC. Pregnancy modifies the antibody response to trivalent influenza immunization. J Infect Dis 2012;206(11):1670–3.
[88] Bischoff AL, Følsgaard NV, Carson CG, Stokholm J, Pedersen L, Holmberg M, et al. Altered response to A(H1N1)pnd09 vaccination in pregnant women: a single blinded randomized controlled trial. PLoS One 2013;8(4):e56700.
[89] Schlaudecker EP, Ambroggio L, McNeal MM, Finkelman FD, Way SS. Declining responsiveness to influenza vaccination with progression of human pregnancy. Vaccine 2018;36(31):4734–41.
[90] Ohfuji S, Fukushima W, Deguchi M, Kawabata K, Yoshida H, Hatayama H, et al. Immunogenicity of a monovalent 2009 influenza A (H1N1) vaccine among pregnant women: lowered antibody response by prior seasonal vaccination. J Infect Dis 2011;203(9):1301–8.
[91] Munoz FM, Bond NH, Maccato M, Pinell P, Hammill HA, Swamy GK, et al. Safety and immunogenicity of tetanus diphtheria and acellular pertussis (Tdap) immunization during pregnancy in mothers and infants: a randomized clinical trial. JAMA 2014;311(17):1760–9.
[92] Huygen K, Caboré RN, Maertens K, Van Damme P, Leuridan E. Humoral and cell mediated immune responses to a pertussis containing vaccine in pregnant and nonpregnant women. Vaccine 2015;33(33):4117–23.
[93] Fortner KB, Swamy GK, Broder KR, Jimenez-Truque N, Zhu Y, Moro PL, et al. Reactogenicity and immunogenicity of tetanus toxoid, reduced diphtheria toxoid, and acellular pertussis vaccine (Tdap) in pregnant and nonpregnant women. Vaccine 2018;36(42):6354–60.
[94] Tsang JS. Utilizing population variation, vaccination, and systems biology to study human immunology. Trends Immunol 2015;36(8):479–93.
[95] Healy CM, Rench MA, Baker CJ. Importance of timing of maternal combined tetanus, diphtheria, and acellular pertussis (Tdap) immunization and protection of young infants. Clin Infect Dis 2013;56(4):539–44.
[96] Abu Raya B, Edwards KM, Scheifele DW, Halperin SA. Pertussis and influenza immunisation during pregnancy: a landscape review. Lancet Infect Dis 2017;17(7):e209–22.
[97] Abu Raya B, Bamberger E, Almog M, Peri R, Srugo I, Kessel A. Immunization of pregnant women against pertussis: the effect of timing on antibody avidity. Vaccine 2015;33(16):1948–52.
[98] Heyderman RS, Madhi SA, French N, Cutland C, Ngwira B, Kayambo D, et al. Group B streptococcus vaccination in pregnant women with or without HIV in Africa: a non-randomised phase 2, open-label, multicentre trial. Lancet Infect Dis 2016;16(5):546–55.
[99] Weinberg A, Muresan P, Richardson KM, Fenton T, Dominguez T, Bloom A, et al. Determinants of vaccine immunogenicity in HIV-infected pregnant women: analysis of B and T cell responses to pandemic H1N1 monovalent vaccine. PLoS One 2015;10(4):e0122431.
[100] Koch MA, Reiner GL, Lugo KA, Kreuk LSM, Stanbery AG, Ansaldo E, et al. Maternal IgG and IgA antibodies dampen mucosal T helper cell responses in early life. Cell 2016;165(4):827–41.
[101] Gomez de Agüero M, Ganal-Vonarburg SC, Fuhrer T, Rupp S, Uchimura Y, Li H, et al. The maternal microbiota drives early postnatal innate immune development. Science 2016;351(6279):1296–302.
[102] Ohsaki A, Venturelli N, Buccigrosso TM, Osganian SK, Lee J, Blumberg RS, et al. Maternal IgG immune complexes induce food allergen-specific tolerance in offspring. J Exp Med 2018;215(1):91–113.
[103] Roopenian DC, Akilesh S. FcRn: the neonatal Fc receptor comes of age. Nat Rev Immunol 2007;7(9):715–25.

[104] Bundhoo A, Paveglio S, Rafti E, Dhongade A, Blumberg RS, Matson AP. Evidence that FcRn mediates the transplacental passage of maternal IgE in the form of IgG anti-IgE/IgE immune complexes. Clin Exp Allergy 2015;45(6):1085–98.
[105] Linder N, Ohel G. In utero vaccination. Clin Perinatol 1994;21(3):663–74.
[106] Palmeira P, Quinello C, Silveira-Lessa AL, Zago CA, Carneiro-Sampaio M. IgG placental transfer in healthy and pathological pregnancies. Clin Dev Immunol 2012;2012:985646.
[107] van den Berg JP, Westerbeek EA, van der Klis FR, Berbers GA, van Elburg RM. Transplacental transport of IgG antibodies to preterm infants: a review of the literature. Early Hum Dev 2011;87(2):67–72.
[108] Einarsdottir HK, Selman MHJ, Kapur R, Scherjon S, Koeleman CAM, Deelder AM, et al. Comparison of the Fc glycosylation of fetal and maternal immunoglobulin G. Glycoconj J 2013;30(2):147–57.
[109] Jansen BC, Bondt A, Reiding KR, Scherjon SA, Vidarsson G, Wuhrer M. MALDI-TOF-MS reveals differential N-linked plasma- and IgG-glycosylation profiles between mothers and their newborns. Sci Rep 2016;6:34001.
[110] Ishikawa T, Takizawa T, Iwaki J, Mishima T, Ui-Tei K, Takeshita T, et al. Fc gamma receptor IIb participates in maternal IgG trafficking of human placental endothelial cells. Int J Mol Med 2015;35(5):1273–89.
[111] Jennewein MF, Abu-Raya B, Jiang Y, Alter G, Marchant A. Transfer of maternal immunity and programming of the newborn immune system. Semin Immunopathol 2017;39(6):605–13.
[112] Liang JL, Tiwari T, Moro P, Messonnier NE, Reingold A, Sawyer M, et al. Prevention of pertussis, tetanus, and diphtheria with vaccines in the United States: recommendations of the advisory committee on immunization practices (ACIP). MMWR Recomm Rep Morb Mortal Wkly Rep Recomm Rep 2018;67(2):1–44.
[113] Eberhardt CS, Blanchard-Rohner G, Lemaître B, Boukrid M, Combescure C, Othenin-Girard V, et al. Maternal immunization earlier in pregnancy maximizes antibody transfer and expected infant seropositivity against pertussis. Clin Infect Dis 2016;62(7):829–36.
[114] Amirthalingam G, Andrews N, Campbell H, Ribeiro S, Kara E, Donegan K, et al. Effectiveness of maternal pertussis vaccination in England: an observational study. Lancet Lond Engl 2014;384(9953):1521–8.
[115] Katz J, Englund JA, Steinhoff MC, Khatry SK, Shrestha L, Kuypers J, et al. Impact of timing of influenza vaccination in pregnancy on transplacental antibody transfer, influenza incidence, and birth outcomes: a randomized trial in rural Nepal. Clin Infect Dis 2018;67(3):334–40.
[116] Englund JA, Glezen WP, Turner C, Harvey J, Thompson C, Siber GR. Transplacental antibody transfer following maternal immunization with polysaccharide and conjugate Haemophilus influenzae type b vaccines. J Infect Dis 1995;171(1):99–105.
[117] de Moraes-Pinto MI, Verhoeff F, Chimsuku L, Milligan PJ, Wesumperuma L, Broadhead RL, et al. Placental antibody transfer: influence of maternal HIV infection and placental malaria. Arch Dis Child Fetal Neonatal Ed 1998;79(3):F202–5.
[118] Dauby N, Goetghebuer T, Kollmann TR, Levy J, Marchant A. Uninfected but not unaffected: chronic maternal infections during pregnancy, fetal immunity, and susceptibility to postnatal infections. Lancet Infect Dis 2012;12(4):330–40.
[119] Abu-Raya B, Smolen KK, Willems F, Kollmann TR, Marchant A. Transfer of maternal antimicrobial immunity to HIV-exposed uninfected newborns. Front Immunol 2016;7:338.
[120] Evans C, Jones CE, Prendergast AJ. HIV-exposed, uninfected infants: new global challenges in the era of paediatric HIV elimination. Lancet Infect Dis 2016;16(6):e92–107.

[121] Okoko BJ, Wesumperuma LH, Ota MO, Pinder M, Banya W, Gomez SF, et al. The influence of placental malaria infection and maternal hypergammaglobulinemia on transplacental transfer of antibodies and IgG subclasses in a rural west African population. J Infect Dis 2001;184(5):627–32.
[122] Brair ME, Brabin BJ, Milligan P, Maxwell S, Hart CA. Reduced transfer of tetanus antibodies with placental malaria. Lancet Lond Engl 1994;343(8891):208–9.
[123] Cumberland P, Shulman CE, Maple PAC, Bulmer JN, Dorman EK, Kawuondo K, et al. Maternal HIV infection and placental malaria reduce transplacental antibody transfer and tetanus antibody levels in newborns in Kenya. J Infect Dis 2007;196(4):550–7.
[124] Ogolla S, Daud II, Asito AS, Sumba OP, Ouma C, Vulule J, et al. Reduced transplacental transfer of a subset of Epstein-Barr virus-specific antibodies to neonates of mothers infected with Plasmodium falciparum malaria during pregnancy. Clin Vaccine Immunol 2015;22(11):1197–205.
[125] Atwell JE, Thumar B, Robinson LJ, Tobby R, Yambo P, Ome-Kaius M, et al. Impact of placental malaria and hypergammaglobulinemia on transplacental transfer of respiratory syncytial virus antibody in Papua New Guinea. J Infect Dis 2016;213(3):423–31.
[126] Jones CE, Naidoo S, De Beer C, Esser M, Kampmann B, Hesseling AC. Maternal HIV infection and antibody responses against vaccine-preventable diseases in uninfected infants. JAMA 2011;305(6):576–84.
[127] Goetghebuer T, Smolen KK, Adler C, Das J, McBride T, Smits G, Lecomte S, Haelterman E, Barlow P, Piedra PA, van der Klis F, Kollmann TR, Lauffenburger DA, Alter G, Levy J, Marchant A. Initiation of antiretroviral therapy before pregnancy reduces the risk of infection-related hospitalization in human immunodeficiency virus-exposed uninfected infants born in a high-income country. Clin Infect Dis 68 (7), 1193–1203. https://doi.org/10.1093/cid/ciy673. PMID: 30215689.
[128] Ruck C, Reikie BA, Marchant A, Kollmann TR, Kakkar F. Linking susceptibility to infectious diseases to immune system abnormalities among HIV-exposed uninfected infants. Front Immunol 2016;7:310.
[129] Abzug MJ, Nachman SA, Muresan P, Handelsman E, Watts DH, Fenton T, et al. Safety and immunogenicity of 2009 pH1N1 vaccination in HIV-infected pregnant women. Clin Infect Dis 2013;56(10):1488–97.
[130] Victora CG, Bahl R, Barros AJD, França GVA, Horton S, Krasevec J, et al. Breastfeeding in the 21st century: epidemiology, mechanisms, and lifelong effect. Lancet Lond Engl 2016;387(10017):475–90.
[131] Sankar MJ, Sinha B, Chowdhury R, Bhandari N, Taneja S, Martines J, et al. Optimal breastfeeding practices and infant and child mortality: a systematic review and meta-analysis. Acta Paediatr 2015;104(467):3–13.
[132] Johansen FE, Braathen R, Brandtzaeg P. The J chain is essential for polymeric Ig receptor-mediated epithelial transport of IgA. J Immunol 2001;167(9):5185–92.
[133] Hocini H, Bomsel M. Infectious human immunodeficiency virus can rapidly penetrate a tight human epithelial barrier by transcytosis in a process impaired by mucosal immunoglobulins. J Infect Dis 1999;179(Suppl 3):S448–53.
[134] Mabuka J, Nduati R, Odem-Davis K, Peterson D, Overbaugh J. HIV-specific antibodies capable of ADCC are common in breastmilk and are associated with reduced risk of transmission in women with high viral loads. PLoS Pathog 2012;8(6):e1002739.
[135] Mazur NI, Horsley NM, Englund JA, Nederend M, Magaret A, Kumar A, Jacobino SR, de Haan CAM, Khatry SK, LeClerq SC, Steinhoff MC, Tielsch JM, Katz J, Graham BS, Bont LJ, Leusen JHW, Chu HY. Breastmilk prefusion F immunoglobulin G as a correlate of protection against respiratory syncytial virus acute respiratory illness. J Infect Dis 2019;219(1):59–67. https://doi.org/10.1093/infdis/jiy477. PMID: 30107412; PMCID: PMC6284547.

[136] Shao H-Y, Chen Y-C, Chung N-H, Lu Y-J, Chang C-K, Yu S-L, et al. Maternal immunization with a recombinant adenovirus-expressing fusion protein protects neonatal cotton rats from respiratory syncytia virus infection by transferring antibodies via breastmilk and placenta. Virology 2018;521:181–9.
[137] Caballero-Flores G, Sakamoto K, Zeng MY, Wang Y, Hakim J, Matus-Acuña V, et al. Maternal immunization confers protection to the offspring against an attaching and effacing pathogen through delivery of IgG in breastmilk. Cell Host Microbe 2019;25(2):313–323.e4.

CHAPTER

4

Immunobiological aspects of vaccines in pregnancy: Infant perspective

Christopher R. Wilcox[a], *Christine E. Jones*[b]

[a]NIHR Clinical Research Facility, University Hospital Southampton NHS Foundation Trust, Southampton, United Kingdom, [b]Faculty of Medicine and Institute for Life Sciences, University of Southampton and University Hospital Southampton NHS Foundation Trust, Southampton, United Kingdom

Introduction

At birth, neonates encounter a wide range of new pathogens and are immunologically inexperienced, making them particularly vulnerable to infection. Antenatal vaccination is an intervention that works predominantly by boosting the concentration of maternal vaccine-induced immunoglobulin G (IgG), and thereby the quantity transferred across the placenta from mother to fetus. There is growing evidence showing that antenatal vaccination is a highly effective means of protecting infants until the time of primary infant vaccination, or until the window-period of greatest susceptibility to severe disease has passed.

Following birth, the concentration of maternally-derived antibodies declines over a period of weeks to months [1], with the half-life of specific IgG estimated to be from 30 to 47 days [2–7], depending on antigen specificity, and therefore the benefit of maternal vaccination is likely to be limited to the first months of life. Antenatal vaccination has been demonstrated to protect infants from pertussis and influenza up to 3 and 6 months-of-age, with an effectiveness of up to 95% and 68%, respectively [8–10]. Whether the duration of maternally-derived antibody is sufficient to cover the highest risk-period of other diseases, currently progressing through the

https://doi.org/10.1016/B978-0-12-814582-1.00004-8

vaccine development pipeline [such as late-onset Group B *Streptococcus* (GBS) and Respiratory Syncytial Virus (RSV)], remains to be seen.

Much of the immunobiological research in this field has been focused on better understanding the process of transplacental transfer and the factors which may affect this (details of which are covered in other chapters). That said, it is becoming increasingly recognized that vaccination in pregnancy may have additional implications for the infant other than passive protection via placental antibody transfer. In this chapter, we will therefore consider two aspects of recent debate with regards to infant health following vaccination in pregnancy: (1) the potential for interference with infant responses to vaccination, and (2) the potential for additional protection to be conferred to the newborn via alteration of breastmilk composition.

Interference with infant responses to vaccination

It is well acknowledged that maternally-derived immunoglobulin (naturally induced following previous maternal infection) can have significant inhibitory effects on infant IgG responses to vaccines administered in infancy. Such effects were first observed in studies of measles vaccination [11], and were key to influencing decisions regarding the optimum timing of measles vaccination in infancy [12]. Since this time, inhibitory effects have been observed for a range of live and non-live vaccines, including tetanus, pneumococcus, influenza, *Haemophilus influenzae* type b (Hib), mumps, hepatitis A & B, rotavirus and poliovirus [13].

A matter of recent debate, however, has been the observation that the presence of maternal antibodies induced through antenatal vaccination can also interfere with humoral responses to vaccination in infancy—an effect referred to as "interference" or "blunting." Vaccination in pregnancy provides protection early in life as a result of transplacental transfer of vaccine-specific antibody from mother to infant; however, the concentration of maternally derived antibody is likely to still be high at the time of the first doses of primary infant vaccination, given the half-life of maternally-derived antibody is around 4–6 weeks [2–7]. The majority of this research has concerned pertussis-containing vaccines. A number of trials based in the United Kingdom (UK), United States of America (US), Belgium, Vietnam and Canada, have assessed the immune response to primary vaccination (as well as post-booster) amongst infants born to mothers vaccinated with pertussis-containing vaccines (tetanus-diphtheria-acellular pertussis, Tdap) during pregnancy. Many of these studies have demonstrated statistically significant differences in antibody (IgG) concentrations to important pertussis antigens (as well as diphtheria, tetanus, pneumococcus and Hib antigens in some studies) compared with infants born to unvaccinated mothers; however, the clinical relevance of these differences is yet to be fully understood [14–19]. These studies are summarized in Table 1.

Interference with infant responses to pertussis antigens

Hardy-Fairbanks et al. [19], were the first to investigate the impact of maternal Tdap vaccination on infant immune responses to routine vaccination in infancy. This was a small US-based prospective cohort study in which infant blood samples were taken at birth, as well as before and after completion of primary and booster vaccinations. Infant serum was tested for antibodies against pertussis toxoid (PT) and important virulence factors which allow pertussis to bind ciliated epithelial cells in the upper respiratory tract, including: filamentous hemagglutinin (FHA), pertactin (PRN), and fimbriae (FIM). The study demonstrated significantly lower antibody concentrations to PT, FHA and PRN (but not FIM) amongst 16 infants born to mothers vaccinated in pregnancy with Adacel (tetanus toxoid, diphtheria toxoid, and five acellular pertussis antigens; $Tdap_5$) after primary vaccination; however, the levels were comparable with the 54 control group infants (whose mothers did not receive Tdap) following booster vaccination at 12–18 months.

Following this, Munoz et al. [18] conducted a randomized controlled trial (RCT) in which antibody concentrations were compared between 33 infants born to Adacel-vaccinated mothers and 15 control group infants whose mothers received placebo. One month following completion of the primary vaccination, significantly lower IgG concentrations to FHA (but not to PT, PRN and FIM) were observed in the intervention group. However, similarly to Hardy-Fairbanks, no difference was observed 1 month following booster vaccination at 12 months. Another cohort study based in Vietnam (Hoang [22] and Maertens [14]) demonstrated that infant PRN IgG concentrations were significantly lower amongst infants born to Adacel-vaccinated mothers, compared to infants of unvaccinated women, but similar anti-PT and anti-FHA concentrations were observed in the two groups of infants. Following booster vaccination in the second year of life however, this difference disappeared [14]. Interestingly, a small UK cohort study, published very recently, found that, for all pertussis antigens studied, the antibody concentrations were comparable between infants whose mothers received antenatal vaccination (Repevax or Boostrix) and infants of non-vaccinated mothers [23].

In contrast, two recent trials have demonstrated a blunting effect post-primary vaccination, which also persisted beyond the booster dose in the second year of life [15,17,20].

TABLE 1 Summary of clinical trials of pertussis vaccination in pregnancy and the effect on the infant responses to primary and booster immunization.

	Country	Period	Study design	Number of participants	Maternal vaccine; gestational age at vaccination
Hardy-Fairbanks et al. [19]	United States of America	2006–2009	Prospective cohort study	Tdap group: 16 mother-infant pairs Control group: 54 mother-infant pairs	Adacel ($Tdap_5$; Sanofi Pasteur); any trimester
Munoz et al. [18]	United States of America	2008–2012	Randomized, double-blind, placebo controlled trial with cross-over design	Tdap group: 33 mother-infant pairs Control group: 15 mother-infant pairs	Intervention: Adacel ($Tdap_5$; Sanofi Pasteur); 30–32 weeks of gestation with placebo post-delivery Control: Placebo at 30–32 weeks gestation and Adacel post-delivery
Ladhani et al. [16]	United Kingdom	2012–2014	Prospective cohort with historical control group	Tdap group: 141 mother-infant pairs Historical control group: 246 mother-infant pairs	REPEVAX-IPV ($Tdap_5$-IPV; Sanofi Pasteur); Median 9 weeks prior to delivery

Infant vaccine; infant vaccination schedule	Blood sampling time points	Infant's outcome measure	Results
Different vaccine products from different brands; 2, 4, 6 months and 12–18 months	Birth, before and 1 month after primary vaccination; before and 1 month after booster vaccination	IgG concentration to PT, FHA, PRN, FIM 2&3, TT, DT, HBV and polio 1/2/3	– Significantly higher IgG levels to all antigens in infants born to vaccinated mothers at birth and prior to 1st infant vaccination – Post-primary immunization, significantly lower IgG levels to PT, FHA and PRN (0.76, 0.73 and 0.67-fold, respectively) but not to FIM – IgG levels comparable pre- and post-booster at 12–18 months. *Blunting post-primary immunization*
Pentacel (DTaP$_5$-IPV-Hib; Sanofi Pasteur); 2, 4, 6 months and 12 months	Birth, before and 1 month after primary vaccination; 1 month after booster vaccination	IgG concentrations to PT, FHA, PRN, FIM 2&3, TT and DT	– Significantly higher IgG levels to all antigens at birth and at 2 months in infants born to vaccinated mothers – Post-primary vaccinations, significantly lower IgG levels to FHA (0.52-fold), but not to PT, PRN or FIM – IgG levels comparable following booster at 12 months *Blunting post-primary immunization*
Pediacel (DTaP$_5$-IPV-Hib; Sanofi Pasteur); 2, 3, 4 months PCV13 at 2–4 months, and MenC at 3 and/or 4 months	Before and 1 month after primary vaccination (for historical control group, only 1 month after primary vaccination)	IgG concentrations to PT, FHA, FIM 2&3, DT, TT, Hib, MenC and 13 pneumococcal serotypes	– Significantly lower IgG levels to PT, FHA and FIM post-primary vaccination (0.67, 0.62 and 0.51-fold, respectively) in infants born to vaccinated mothers – Significantly lower IgG levels to diphtheria (0.55-fold) and some CRM-conjugated MCC – Significantly higher IgG levels to TT (1.24-fold) and Hib (2.30-fold) – IgG levels comparable for PCV13 serotypes, except 3, 5 and 9 V (0.34, 0.59 and 0.78-fold, respectively) *Blunting post-primary immunization*

Continued

TABLE 1 Summary of clinical trials of pertussis vaccination in pregnancy and the effect on the infant responses to primary and booster immunization—cont'd

	Country	Period	Study design	Number of participants	Maternal vaccine; gestational age at vaccination
Maertens et al. [15] Maertens et al. [20] Maertens et al. [21]	Belgium	2012–2015	Prospective controlled cohort study	Tdap group: 55 mother-infant pairs Control group: 26 mother-infant pairs	Intervention: BOOSTRIX ($Tdap_3$; GlaxoSmithKline); 22–33 weeks of gestation Control Group: Td in pregnancy
Hoang et al. [22] Maertens et al. [14]	Vietnam	2013–2015	Randomized controlled trial	Tdap group: 51 mother-infant pairs. Control group: 48 mother-infant pairs	Intervention group: Adacel (Sanofi Pasteur); 18–36 weeks of gestation Control: TT in pregnancy
Halperin et al. [17]	Canada	2012–2015	Multi-center, observer-blind, randomized controlled trial	Tdap group: 135 mother-infant pairs Control group: 138 mother-infant pairs	Adacel ($Tdap_5$; Sanofi Pasteur); 32–36 weeks of gestation Control: Td in pregnancy

Infant vaccine; infant vaccination schedule	Blood sampling time points	Infant's outcome measure	Results
Infanrix hexa ($DTaP_3$-HBV-IPV-Hib GlaxoSmithKline); 2, 3, 4 months and 15 months	Birth, before and 1 month after primary vaccination and before and 1 month after booster vaccination	IgG concentrations to PT, FHA, PRN, TT and DT	– Significantly higher IgG levels to all antigens at birth and at 2 months in infants born to vaccinated mothers – Post-primary vaccination, significantly lower IgG levels to PT (0.54-fold), but not to FHA or PRN; and significantly lower levels to DT (0.81-fold), 9 pneumococcal serotypes but not TT – Significantly lower IgG levels to PT (0.64-fold) and 2 pneumococcal serotypes post-booster *Blunting post-primary immunization* *Blunting post-booster immunization*
Infanrix hexa ($DTaP_3$-HBV-IPV-Hib; GlaxoSmithKline); 2, 3, 4 months and second year of life	Birth, before and 1 month after primary vaccination and 1 month after booster vaccination	IgG concentrations to PT, FHA, PRN, TT and DT	– Significantly higher IgG levels to all antigens at birth and at 2 months in infants born to vaccinated mothers – Post-primary vaccination, significantly lower IgG levels to PRN (0.63-fold), but not to PT or FHA, and significantly lower levels to DT and TT (0.7 and 0.67-fold, respectively) – Post-booster, significantly lower IgG levels to TT only (0.64-fold) *Blunting post-primary immunization* *Blunting post-booster immunization*
Pediacel ($DTaP_5$-IPV-Hib; Sanofi Pasteur); 2, 4, 6 months and 12 months	Birth, 2, 4, 6 months of age and 1 month after primary vaccination and before and 1 month after booster vaccination	IgG concentrations to PT, FHA, PRN, FIM2&3, TT, DT and Hib	– Significantly higher IgG concentrations to PT at birth and at 2 months, and to FHA, PRN and FIM at birth, 2 months and 4 months in infants born to vaccinated mothers – Post-primary vaccination, significantly lower IgG concentrations to PT and FHA at 6 months (0.58 and 0.55-fold, respectively) and at 7 months (0.74 and 0.60-fold), and to PRN and FIM at 7 months (0.59 and 0.39)

Continued

TABLE 1 Summary of clinical trials of pertussis vaccination in pregnancy and the effect on the infant responses to primary and booster immunization—cont'd

	Country	Period	Study design	Number of participants	Maternal vaccine; gestational age at vaccination
Rice et al. [23]	United Kingdom	2014–2016	Prospective controlled cohort study	Tdap group: 16 mother-infant pairs Control group: 15 mother-infant pairs	REPEVAX_IPV ($DTaP_5$-IPV; Sanofi Pasteur) or BOOSTRIX-IPV ($DTaP_3$-IPV; GlaxoSmithKline); 24–37 weeks of gestation

N, number of infants; *PT*, pertussis toxoid; *FHA*, filamentous hemagglutinin; *PRN*, pertactin; *FIM*, fimbriae; *DP*, diphtheria toxoid; *TT*, tetanus toxoid; *RCT*, randomized controlled trial; *$DTaP_3$/Tdap3*, tetanus, diphtheria, acellular pertussis (containing three pertussis antigen components); *$DTaP_5$/Tdap5*, tetanus, diphtheria, acellular pertussis (containing five pertussis antigen components); *Td*, tetanus, diphtheria; *Hib*, hemophilus influenzae type b vaccine; *IPV*, inactivated poliomyelitis vaccine; *HBV*, hepatitis B vaccine; *PCV13*, 13-valent pneumococcal vaccine; *MenC*, meningococcus C vaccine; *IPV*, inactivated polio vaccine.

Halperin et al. [17] conducted a multi-center, observer-blind, RCT based in Canada involving 261 infants, whose mothers were randomized (1:1) to receive either vaccination with Adacel or Td (tetanus and diphtheria) during pregnancy. Infants of Adacel recipients showed significantly lower concentrations of antibodies to PT, FHA, PRN and FIM antigens at 7 months of age, following completion of primary vaccinations. These differences also persisted 1 month following booster for antibodies to PT, FHA and FIM. A smaller study in Belgium (Maertens [15], Maertens [20] and Maertens [21]) showed lower concentrations of IgG to PT, but similar anti-FHA and -PRN, in infants born to Boostrix ($Tdap_3$) vaccinated mothers ($n=57$), compared to infants of unvaccinated mothers ($n=42$) at 5 months of age, 1 month after completion of primary vaccinations. Furthermore, in this study population, anti-PT IgG remained lower at 1 month post-booster amongst infants born to vaccinated mothers; however, anti-FHA and -PRN levels were comparable between groups [20].

Only one trial [Immunizing Mums Against Pertussis 2 (iMAP2); not yet published], to our knowledge, has directly compared two pertussis-

Infant vaccine; infant vaccination schedule	Blood sampling time points	Infant's outcome measure	Results
			– A significant difference persisted pre-booster for all antigens, and post-booster for PT, FHA and FIM *Blunting post-primary immunization* *Blunting post-booster immunization*
Pediacel ($DTaP_5$-IPV-Hib Sanofi Pasteur) or Infanrix Hexa ($DTaP_3$-HBV-IPV-Hib; GlaxoSmithKline); 2, 3, 4 months 8 and 16 weeks with Prevenar13 (PCV13; Pfizer)	Birth, before and 1 month after primary vaccination	IgG concentrations to PT, FHA, PRN, TT, DT, Hib and 10 pneumococcal serotypes	– Significantly higher IgG concentrations to all antigens at birth and at 7 weeks in infants born to vaccinated mothers – Levels of vaccine antigens comparable post-primary vaccinations, except lower anti-pneumococcal 7F and higher anti-14 *Blunting post-primary immunization*

containing vaccines recommended for use in pregnancy: Repevax ($Tdap_5$) and Boostrix ($Tdap_3$). Antenatal vaccination with Boostrix ($n=77$) was associated with higher concentrations of anti-PT and anti-FHA IgG in infants at birth and at 2 months, compared with Repevax vaccinated mothers ($n=77$), yet there were no differences in anti-PT concentrations following primary vaccination at 5 and 13 months of age [24].

The question therefore remains over what the underlying reasons are for the differences observed between these trials. It has been argued that earlier studies, which demonstrate a blunting effect that disappeared after the administration of a booster in the second year of life, were not sufficiently powered to detect significant differences between groups at this time point. On the other hand, more recent trials showing persistent blunting following booster vaccination had larger sample sizes. Other possible explanations for these differences include variation in maternal/infant vaccine formulations, differing laboratory measurement techniques, timing of administration of maternal/infant vaccines, differences in maternal priming status, and different epidemiological backgrounds (including natural exposure to *Bordetella pertussis*) which may vary between lower- and higher-income countries. Studies involving historical control groups, in which mothers and infants were not vaccinated over the same time-period, may also introduce further confounding factors, including differences in pertussis exposure (particularly as it is seasonal).

Interference with infant responses to other vaccine antigens

Conjugate vaccines can use different carrier proteins such as tetanus toxoid (TT) or a naturally occurring diphtheria toxin variant, CRM_{197}. It has therefore been proposed that high maternally-derived tetanus and diphtheria antibody concentrations could potentially interfere with infant immune responses to these conjugate vaccines. This might be particularly important in countries (such as the UK) where reduced priming schedules are used for meningococcal C (single dose given) and PCV13 (two doses given).

To investigate this, a UK study (Ladhani [16]) assessed infant responses to primary vaccination amongst infants born to 141 women who received Repevax-IPV ($DTaP_5$-IPV) during pregnancy, and compared their responses to a historical cohort of infants born to women who did not receive a pertussis-containing vaccine in pregnancy ($n=246$). Following primary vaccination, antibody responses were found to be significantly lower to diphtheria and CRM-conjugated meningococcal C vaccines, amongst infants born to vaccinated mothers, compared to those born to unvaccinated mothers. Whilst these differences were statistically significant, most infants (92%) still achieved the protective threshold irrespective of meningococcal C vaccine schedule; however, the proportion was lowest (84%) amongst those who received a single dose of CRM-conjugated meningococcal vaccine at 3 months of age. For 13-valent pneumococcal vaccine (PCV13) serotypes, most infants achieved protective antibody concentrations, although the proportion achieving the protective threshold for serotypes 3, 5 and 9 V was significantly lower in the antenatal vaccination group. Interestingly, infant responses to tetanus, meningococcal and Hib vaccines conjugated to tetanus were actually enhanced following maternal TdaP vaccination; however, given the lack of association between pre-vaccination tetanus antibody concentrations and Hib-TT responses, the mechanisms underlying this difference remain uncertain. This finding is also supported by recent studies of antenatal vaccination with TT in developing countries, which have shown no association between transplacentally-acquired TT antibody and HiB-TT responses [25].

What are the underlying mechanisms for this interference?

The mechanisms of blunting of the infant immune responses after maternal immunization remain incompletely understood, and a few possible mechanisms have been proposed. One theory is termed "epitope masking," which proposes that maternally-derived antibodies may bind to vaccine epitopes and mask them from infant B-lymphocytes, thereby reducing the infant's subsequent responses to those antigens [13,26]. Consequently, a prediction of this model is that the blunting effect is dependent on the concentration of antibody present in the circulation, and therefore an IgG antibody lacking its Fc (constant)-region [termed an $F(ab')_2$ fragment] should suppress the generation of neutralizing antibodies to the same extent as complete IgG. However, it has been demonstrated experimentally, both *in vitro* and *in vivo* (in cotton rats), that the Fc-region is required for inhibition of antibody, and that inhibition was not epitope-specific [27]. This implies that the Fc region of IgG is essential for the blunting effect, and therefore does not support this model.

Another suggested explanation is that macrophages may remove antibody-antigen complexes from the circulation and thereby eliminate immune responses [26]; however, there is currently no experimental evidence to support this, and this mechanism would not explain why B cell responses are preferentially inhibited over T cell responses [13]. It has also been proposed that, in the presence of maternal antibodies, neutralization of vaccine virus may occur, thereby reducing the quantity of viral antigen below a certain threshold and preventing immune recognition [13,28]. However, this mechanism wouldn't explain why non-replicating protein vaccines are also suppressed, why non-neutralizing antibodies inhibit vaccination with live-attenuated vaccines [27], and why vaccination with vector systems that express measles proteins (which are insensitive to neutralizing antibodies) are also inhibited [29].

Finally, another possible explanation is that B cell inhibition occurs through the Fcγ-receptor IIB (FcγRIIB) [13]. FcγRIIB is an inhibitory Fc receptor, and is the only Fcγ-receptor expressed on the surface of B cells. This theory proposes that inhibition is based on maternally-derived IgG-vaccine antigen complexes, which cross-link the B cell receptor to the inhibitory FcγRIIB. Vaccine antigen is recognized by the B cell receptor on the surface of infant B cells, and if maternally-derived IgG binds to this vaccine antigen, then it may bind to the FcγRIIB, resulting in a negative signal which would inhibit both the proliferation of B cells and the secretion of antibodies [27,30]. This model is supported by *in vitro* data demonstrating that Fc region glycosylation (which is required for IgG-FcγRIIB binding) is necessary for inhibition [31], as well as recent *in vivo* experiments in rodents showing that a monoclonal antibody, with an Fc region that does not bind to FcγRIIB, cannot inhibit B cell responses [27]. These data therefore seem to confirm the importance of the interaction between FcγRIIB and the Fc region of IgG, and support the idea that the blunting effect depends on vaccine-IgG complex formation. This explanation remains contentious, however, as other *in vivo* data do not unequivocally support this mechanism, such as experiments in mice with a genetic deletion of FcγRIIB (as well as FcγRI and FcγRIII) showing that inhibition could still be induced by IgG [32]. Yet, it is worth noting the mice displayed an array of immunological abnormalities as a result of these deletions that were not restricted to B cells, making these results more difficult to interpret and translate to humans.

What is the clinical significance of these findings?

As highlighted above, maternal vaccination is highly effective at protecting infants up to 3 months of age against pertussis, with a vaccine effectiveness of up to 95% [8]. This is especially important given that the vast majority of deaths from pertussis occur in infants aged less than 3 months [33], particularly in those who have not had a first dose of the pertussis vaccine in infancy [34]. However, if antibody concentrations are lower after primary infant vaccination amongst those infants born to mothers who received a pertussis-containing vaccine during pregnancy, there is potential for an increased burden of disease from the second half of the first year of life onwards. Given that cases of disease occurring in later infancy are rarely associated with mortality, an increased incidence might actually be deemed acceptable if deaths in early infancy are prevented; however, recent surveillance data from England and Wales have not identified such trends [8].

It is also important to note that, whilst higher levels of pertussis-specific antibodies are associated with protection from disease (in particular anti-PT and anti-PRN IgG) [35,36], an actual protective antibody threshold (serocorrelate of protection) is not currently known [37]. This makes it more challenging to determine what the clinical significance of blunting is likely to be, and longer-term surveillance of pertussis rates in all children up until school-age is essential. The advantages and disadvantages of different vaccination schedules and designs also needs to be carefully evaluated. The development of combination vaccines for use in pregnancy containing only acellular pertussis antigens with tetanus, for example, may be preferable to the current TdaP design, due to the level of interference observed from diphtheria antibodies [16]. With regards to vaccine schedules, the US currently employs a 2-4-6-month schedule, with a second-year booster. Evaluation of the clinical significance of blunting is particularly important for countries which currently recommend a three-dose schedule with no TdaP booster until school age, such as the UK (2-3-4 months) and countries using the Expanded Program on Immunization schedule (6-10-14 weeks).

Given that maternally-derived antibodies decline in the infant circulation over 3–6 months following birth, it has also been suggested that this interference is unlikely to impact on subsequent boosting vaccinations, and that individual protection and disease control in the population is unlikely to be affected. In fact, responses to booster vaccines at 1 year of age may even be greater if the IgG concentration pre-booster is lower. As outlined above, early trials demonstrated that the blunting effect on the infants' immune response to pertussis vaccination disappeared after the administration of a booster dose in the second year of life [18,19], yet these studies may have been insufficiently powered to detect significant differences at this time point. In contrast, more recent larger trials have shown blunting effects that persisted beyond the booster dose in the second year of life [15,17]. Other possible confounding factors (discussed above) also make comparison of these trials difficult, and it therefore remains unclear at what age in childhood the blunting effect of maternally derived antibody disappears, and whether this might vary between different vaccines, vaccine schedules and epidemiological backgrounds.

Finally, it is also worth acknowledging that blunting of infant pertussis vaccination may differ between acellular vaccines (in which much of the research in this field has been based) and whole-cell vaccines [38]. Due to concerns over reactogenicity and a corresponding drop in vaccine coverage, the use of inactivated whole-cell vaccines has been largely replaced by acellular vaccines in developed countries since the early 1990s [39]. However, whole-cell vaccines are still used in many parts of the world, and we are aware that trials assessing the blunting effect of these vaccines are currently ongoing in South-East Asia and South America (Clincialtrials.gov identifiers: NCT02301702, NCT0168346 and NCT02408926).

Looking forwards, it should be noted that knowledge is lacking with regards to the functionality of infant antibodies induced in the presence of maternal antibodies, as well as the effect of maternal antibodies on the cellular immune response of infants. To our knowledge, these issues have not been directly described in any literature to-date; however, trials are currently being undertaken to specifically answer these questions. Development of vaccines with a specific pregnancy indication needs to take into consideration the potential for maternal vaccines to have an effect on the infant's response to vaccination, particularly if these vaccines are conjugated to CRM_{197} or TT.

Breastfeeding after antenatal vaccination: Potential for additional protection for the newborn?

Breastfeeding is well-acknowledged to have a role in protection against gastrointestinal and respiratory tract infections after birth [40], as well as allergic disease [41] and necrotizing enterocolitis [42]. The World Health Organization (WHO) currently recommends exclusive breastfeeding up to 6 months of age, followed by a combination of breastmilk and supplementary food up to 2 years of age or older [43].

Secretory immunoglobulin A (sIgA)

In recent years, it has been recognized that in addition to the protection conferred to the newborn via transplacental transfer of IgG following vaccination in pregnancy, there is potential for additional protection to be provided in early life by vaccine-specific antibody in breastmilk. However, unlike placental antibody transfer, little is currently known about the beneficial effects of vaccination in pregnancy on the composition of breastmilk and the transfer of vaccine-specific antibody to the newborn via breastfeeding [44].

Immunoglobulin in breastmilk is predominantly secretory IgA (sIgA), which is particularly high in concentration early in lactation (up to 1–2 g/L). SIgA is considered to be the primary immunologically protective component of breastmilk and is thought to act through a number of possible mechanisms in order to prevent colonization and invasion of the infant's mucus membrane in the gut and respiratory tract [45]. A major mode of action is thought to be "immune exclusion," whereby access of pathogens to the child's epithelium is inhibited via a step-wise process of agglutination, mucus entrapment and mucociliary clearance [46]. Alternatively, sIgA may bind to lectin-like bacterial adhesins, thereby preventing bacterial adhesions to epithelial cell receptors [47]. SIgA may also provide a barrier underneath the mucosal lining, meaning pathogens are intercepted within epithelial cell vesicular compartments, translocated through the epithelium and subsequently excreted. SIgA has also been shown to act directly on pathogens via neutralization [48,49] and via modification of virulence factors, such as toxins of *Escherichia coli* [50] and the surface protein A of *Streptococcus pneumoniae* [51].

Several trials have demonstrated significantly higher vaccine-specific sIgA in the breastmilk of women vaccinated against influenza [52], pertussis [53], meningococcus [54], and pneumococcus [55] compared to unvaccinated women, and high levels of sIgA have been maintained for up to 7 months post-partum [44,56]. Determining the actual protection that such antibodies provide has proved difficult however, as many of these studies lack clear correlates of protection. The strongest evidence comes from studies of influenza vaccination, in which a lower incidence of respiratory tract infection has been observed amongst infants aged up to 6 months born to vaccinated mothers who were exclusively breastfeeding [52,57]. On the other hand, whilst pneumococcal vaccination also results in a significant rise in breastmilk sIgA, no effect has been observed on incidence of respiratory tract infection [52,58]. Further work is therefore required to determine the cellular and immunological mechanisms of breastmilk-mediated protection after vaccination in pregnancy, and establish the clinical benefits.

Interestingly, it has been shown that many of the maternal and fetal conditions, that can negatively affect transplacental IgG transfer, do not affect IgA transfer through breastmilk. Maternal human immunodeficiency virus (HIV) infection [59] and malnutrition [60] have not been shown to reduce the concentration of total or antigen-specific IgA in breastmilk, and prematurity has actually been shown to increase the transfer of immune factors, including IgA, in breastmilk [61].

A limitation of the research to-date is that there is no standardized method of detecting sIgA in breastmilk samples. No validated commercial assay is available, and studies are therefore limited to "in house" enzyme-linked immunosorbent assay (ELISA) techniques, which may be subject to variation in sensitivities and specificities [44]. Furthermore, variation in infants' ages, the time-point of collection during breastfeeding, and storage/handling of milk samples between studies may all have an influence on the concentration of antibodies found [62]. Unfortunately, the use of animal studies to probe this area is also very limited, as the immune system and role of breastfeeding are very different from humans. In mice, for example, foster feeding studies have revealed that maternal antibodies are predominantly transferred via breastfeeding, rather than via placental transfer, and also provide longer lasting protection if transferred via breastfeeding [63].

Finally, there is potential for blunting of infant immune responses to live oral vaccines due to high breastmilk IgA concentration following vaccination. Support for this comes from trials (involving a number of breastfeeding women recruited from both developed and developing countries) demonstrating that higher rotavirus-specific IgA in breastmilk was associated with inhibitory effects against rotavirus vaccination in infancy [64]. However, these findings need further confirmation and the clinical implications remain unclear, as recent trials directly assessing the impact of withholding breastfeeding at the time of vaccination on the immunogenicity of oral rotavirus vaccines, have shown no significant differences [65,66].

Other immunologically active components of breastmilk

Finally, it is briefly worth considering two other components of breastmilk, that some evidence suggests may be altered following vaccination in pregnancy. The first of these is IgG, which originates from maternal serum and from breastmilk-derived B cells [67]. The concentration of IgG in breastmilk is approximately 10% of the IgA concentration; however, this is shown to increase with duration of breastfeeding, and higher concentrations of vaccine-specific IgG have been detected in breastmilk following vaccination against RSV and pneumococcus [68]. Evidence of a protective role for breastmilk IgG has come from studies demonstrating that breastmilk IgG is inversely correlated with transmission of human cytomegalovirus [69] and HIV [70]; however, other pathogens have not been studied and this area remains poorly understood.

Secondly, the potential role of cellular immunity should also be considered. It has been observed that colostrum contains a significantly higher proportion of antigen-primed T and B cell subsets possessing effector and memory functions in comparison to peripheral blood, suggesting their selective migration to breastmilk [71]. There is some evidence to suggest that maternal vaccination may lead to in utero vaccine-induced priming of the fetal immune system, thereby providing protection independent of antibody-mediated passive immunity [72]. However, to our knowledge, the role of the cellular immune response and its effect on breastmilk after vaccination has not been directly investigated.

Conclusion

Vaccination in pregnancy is a safe and highly effective strategy for reducing early infant morbidity and mortality; however, this may have an effect on the infant's immune response to vaccines delivered in infancy. A number of trials have demonstrated that blunting of infant immune responses to primary childhood vaccination can occur amongst those born to recipients of vaccines in pregnancy, yet the underlying mechanisms for this phenomenon remain a matter of significant debate. Furthermore, it is still unclear at which point this blunting effect disappears, and further work is required to understand the clinical implications—particularly the potential for an increased burden of disease from the second half of the first year of life onwards, and with regards to optimum vaccine design and schedules (of both vaccines in pregnancy and in infancy). Epidemiological data has not shown significantly increased rates of pertussis in older infants, however it is important to continue to monitor this. With regards to breastfeeding, several trials have demonstrated significantly higher vaccine-specific sIgA in the breastmilk of women who have been vaccinated in pregnancy, and there is indeed potential for additional protection from vaccination to be provided through breastmilk. However, determining the actual protection that such antibodies provide has proved difficult, and further work is required in order to determine the influence of vaccination on other immunologically active components of breastmilk. Looking forwards, it is vital that we address these knowledge gaps. The field of maternal vaccination is rapidly advancing and, as such, determining the potential implications for infants (both positive and negative) will be of increasing importance.

References

[1] Niewiesk S. Maternal antibodies: clinical significance, mechanism of interference with immune responses, and possible vaccination strategies. Front Immunol 2014;5:446.

[2] Dixon FJ, Talmage DW, Maurer PH, Deichmiller M. The half-life on homologous gamma globulin (antibody) in several species. J Exp Med 1952;95:313–8.

[3] Barr M, Glenny AT, Randall KJ. Concentration of diphtheria antitoxin in cord blood and rate of loss in babies. Lancet 1949;254:324–6.

[4] Vilajeliu A, Ferrer L, Munros J, Gonce A, Lopez M, Costa J, et al. Pertussis vaccination during pregnancy: antibody persistence in infants. Vaccine 2016;34:3719–22.

[5] Van Savage J, Decker MD, Edwards KM, Sell SH, Karzon DT. Natural history of pertussis antibody in the infant and effect on vaccine response. J Infect Dis 1990;161:487–92.

[6] Luffer-Atlas D, Reddy VR, Hilbish KG, Grace CE, Breslin WJ. PDGFRα monoclonal antibody: assessment of embryo-fetal toxicity and time-dependent placental transfer of a murine surrogate antibody of olaratumab in mice. Birth Defects Res 2018;110:1358–71.

[7] Sarvas H, Seppälä I, Kurikka S, Siegberg R, Mäkelä O. Half-life of the maternal IgG1 allotype in infants. J Clin Immunol 1993;13:145–51.

[8] Amirthalingam G, Campbell H, Ribeiro S, Fry NK, Ramsay M, Miller E, et al. Sustained effectiveness of the maternal pertussis immunization program in England 3 years following introduction. Clin Infect Dis 2016;63:S236–43.

[9] Zaman K, Roy E, Arifeen SE, Rahman M, Raqib R, Wilson E, et al. Effectiveness of maternal influenza immunization in mothers and infants. N Engl J Med 2008;359:1555–64.

[10] Tapia MD, Sow SO, Tamboura B, Teguete I, Pasetti MF, Kodio M, et al. Maternal immunisation with trivalent inactivated influenza vaccine for prevention of influenza in infants in Mali: a prospective, active-controlled, observer-blind, randomised phase 4 trial. Lancet Infect Dis 2016;16:1026–35.

[11] Albrecht P, Ennis FA, Saltzman EJ, Krugman S. Persistence of maternal antibody in infants beyond 12 months: mechanism of measles vaccine failure. J Pediatr 1977;91:715–8.

[12] Leuridan E, Hens N, Hutse V, Ieven M, Aerts M, Van Damme P. Early waning of maternal measles antibodies in era of measles elimination: longitudinal study. BMJ 2010;340:c1626.

[13] Siegrist CA. Mechanisms by which maternal antibodies influence infant vaccine responses: review of hypotheses and definition of main determinants. Vaccine 2003;21:3406–12.

[14] Maertens K, Hoang TT, Nguyen TD, Cabore RN, Duong TH, Huygen K, et al. The effect of maternal pertussis immunization on infant vaccine responses to a booster pertussis-containing vaccine in Vietnam. Clin Infect Dis 2016;63:S197–204.

[15] Maertens K, Cabore RN, Huygen K, Hens N, Van Damme P, Leuridan E. Pertussis vaccination during pregnancy in Belgium: results of a prospective controlled cohort study. Vaccine 2016;34:142–50.

[16] Ladhani SN, Andews NJ, Southern J, Jones CE, Amirthalingam G, Waight PA. Antibody responses after primary immunization in infants born to women receiving a pertussis-containing vaccine during pregnancy: single arm observational study with a historical comparator. Clin Infect Dis 2015;61:1637–44.

[17] Halperin SA, Langley JM, Ye L, MacKinnon-Cameron D, Elsherif M, Allen VM, et al. A randomized controlled trial of the safety and immunogenicity of tetanus, diphtheria, and acellular pertussis vaccine immunization during pregnancy and subsequent infant immune response. Clin Infect Dis 2018;67:1063–71.

[18] Munoz FM, Bond NH, Maccato M, Pinell P, Hammill HA, Swamy GK, et al. Safety and immunogenicity of tetanus diphtheria and acellular pertussis (Tdap) immunization during pregnancy in mothers and infants: a randomized clinical trial. JAMA 2014;311:1760–9.

[19] Hardy-Fairbanks AJ, Pan SJ, Decker MD, Johnson DR, Greenberg DP, Kirkland KB, et al. Immune responses in infants whose mothers received Tdap vaccine during pregnancy. Pediatr Infect Dis J 2013;32:1257–60.
[20] Maertens K, Caboré R, Huygen K, Vermeiren S, Hens N, Damme P, Leuridan E. Pertussis vaccination during pregnancy in Belgium: follow-up of infants until 1 month after the fourth infant pertussis vaccination at 15 months of age. Vaccine 2016;34:3613–9.
[21] Maertens K, Burbidge P, Van Damme P, Goldblatt D, Leuridan E. Pneumococcal immune response in infants whose mothers received tetanus, diphtheria and acellular pertussis vaccination during pregnancy. Pediatr Infect Dis J 2017;36:1186–92.
[22] Hoang HT, Leuridan E, Maertens K, Nguyen TD, Hens N, Vu NH, et al. Pertussis vaccination during pregnancy in Vietnam: results of a randomized controlled trial pertussis vaccination during pregnancy. Vaccine 2016;34:151–9.
[23] Rice T, Diavatopoulos D, Smits G, Van Gageldonk P, Berbers G, Van Der Klis F, et al. Antibody responses to Bordetella pertussis and other childhood vaccines in infants born to mothers who received pertussis vaccine in pregnancy—a prospective, observational cohort study from the UK. Clin Exp Immunol 2019;197:1–10.
[24] Jones CE, Southern J, Calvert A, Matheson M, Andrews NJ, Amirthalingam G, Hallis B, England A, Heath PT, Miller E. A randomised controlled trial comparing two pertussis-containing vaccines in pregnancy and vaccine responses in UK mothers and their infants (Immunising Mums Against Pertussis, IMAP2). In: Presentation at the Annual Meeting of the European Society of Pediatric Infectious Diseases; 2018.
[25] Nohynek H, Gustafsson L, Capeding MR, Kayhty H, Olander RM, et al. Effect of transplacentally acquired tetanus antibodies on the antibody responses to haemophilus influenzae type b-tetanus toxoid conjugate and tetanus toxoid vaccines in Filipino infants. Pediatr Infect Dis J 1999;18:25–30.
[26] Getahun A, Heyman B. Studies on the mechanism by which antigen-specific IgG suppresses primary antibody responses: evidence for epitope masking and decreased localization of antigen in the spleen. Scand J Immunol 2009;70:277–87.
[27] Kim D, Huey D, Oglesbee M, Niewiesk S. Insights into the regulatory mechanism controlling the inhibition of vaccine-induced seroconversion by maternal antibodies. Blood 2011;117:6143–51.
[28] Naniche D. Human immunology of measles virus infection. Curr Top Microbiol Immunol 2009;330:151–71.
[29] Schlereth B, Buonocore L, Tietz A, Meulen Vt V, Rose JK, Niewiesk S. Successful mucosal immunization of cotton rats in the presence of measles virus-specific antibodies depends on degree of attenuation of vaccine vector and virus dose. J Gen Virol 2003;84:2145–51.
[30] Kim D, Niewiesk S. Synergistic induction of interferon α through TLR-3 and TLR-9 agonists identifies CD21 as interferon α receptor for the B cell response. PLoS Pathog 2013;9:e1003233.
[31] Heyman B, Nose M, Weigle WO. Carbohydrate chains on IgG2b: a requirement for efficient feedback immunosuppression. J Immunol 1985;134:4018–23.
[32] Karlsson MC, Getahun A, Heyman B. FcgammaRIIB in IgG-mediated suppression of antibody responses: different impact in vivo and in vitro. J Immunol 2001;167:5558–64.
[33] Chow MY, Khandaker G, McIntyre P. Global childhood deaths from pertussis: a historical review. Clin Infect Dis 2016;63:S134–41.
[34] Tiwari TS, Baughman AL, Clark TA. First pertussis vaccine dose and prevention of infant mortality. Pediatrics 2015;135:990–9.
[35] Mooi F, De Greeff S. The case for maternal vaccination against pertussis. Lancet Infect Dis 2007;7:614–24.
[36] Heininger U, Riffelmann M, Bär G, Rudin C, von König CH. The protective role of maternally derived antibodies against Bordetella pertussis in young infants. Pediatr Infect Dis J 2013;32:695–8. https://doi.org/10.1097/INF.0b013e318288b610.

[37] Plotkin SA. Correlates of protection induced by vaccination. Clin Vaccine Immunol 2010;17:1055–65.
[38] Englund JA, Anderson EL, Reed GF, Decker MD, Edwards KM, Pichichero ME, et al. The effect of maternal antibody on the serologic response and the incidence of adverse reactions after primary immunization with acellular and whole-cell pertussis vaccines combined with diphtheria and tetanus toxoids. Pediatrics 1995;96:580–4.
[39] Ausiello CM, Cassone A. Acellular pertussis vaccines and pertussis resurgence: revise or replace? MBio 2014;5.
[40] Tromp I, Kiefte-de Jong J, Raat H, Jaddoe V, Franco O, Hofman A, et al. Breastfeeding and the risk of respiratory tract infections after infancy: The Generation R Study. Plos One 2017;12:e0172763.
[41] Kull I, Wickman M, Lilja G, Nordvall SL, Pershagen G. Breast feeding and allergic diseases in infants—a prospective birth cohort study. Arch Dis Child 2002;87:478.
[42] Meinzen-Derr J, Poindexter B, Wrage L, Morrow AL, Stoll B, Donovan EF. Role of human milk in extremely low birth weight infants' risk of necrotizing enterocolitis or death. J Perinatol 2009;29:57.
[43] World Health Organisation. Exclusive breastfeeding for six months best for babies everywhere. World Health Organisation; 2011.
[44] Maertens K, De Schutter S, Braeckman T, Baerts L, Van Damme P, De Meester I, et al. Breastfeeding after maternal immunisation during pregnancy: providing immunological protection to the newborn: a review. Vaccine 2014;32:1786–92.
[45] Goldman AS. The immune system of human milk: antimicrobial, antiinflammatory and immunomodulating properties. Pediatr Infect Dis J 1993;12:664–71.
[46] Mantis NJ, Rol N, Corthésy B. Secretory IgA's complex roles in immunity and mucosal homeostasis in the gut. Mucosal Immunol 2011;4:603–11.
[47] Mantis NJ, Forbes SJ. Secretory IgA: arresting microbial pathogens at epithelial borders. Immunol Invest 2010;39:383–406.
[48] Trang NV, Braeckman T, Lernout T, Hau VT, Anh le TK, et al. Prevalence of rotavirus antibodies in breastmilk and inhibitory effects to rotavirus vaccines. Hum Vaccin Immunother 2014;10:3681–7. https://doi.org/10.4161/21645515.2014.980204.
[49] Sadeharju K, Knip M, Virtanen SM, Savilahti E, Tauriainen S, et al. Maternal antibodies in breastmilk protect the child from enterovirus infections. Pediatrics 2007;119:941–6.
[50] Giugliano LG, Ribeiro ST, Vainstein MH, Ulhoa CJ. Free secretory component and lactoferrin of human milk inhibit the adhesion of enterotoxigenic Escherichia coli. J Med Microbiol 1995;42:3–9.
[51] Hammerschmidt S, Talay SR, Brandtzaeg P, Chhatwal GS. SpsA, a novel pneumococcal surface protein with specific binding to secretory immunoglobulin A and secretory component. Mol Microbiol 1997;25:1113–24.
[52] Schlaudecker EP, Steinhoff MC, Omer SB, McNeal MM, Roy E, Arifeen SE, et al. IgA and neutralizing antibodies to influenza a virus in human milk: a randomized trial of antenatal influenza immunization. PLoS One 2013;8:e70867.
[53] De Schutter S, Maertens K, Baerts L, De Meester I, Van Damme P, Leuridan E. Quantification of vaccine-induced antipertussis toxin secretory IgA antibodies in breastmilk: comparison of different vaccination strategies in women. Pediatr Infect Dis J 2015;34:e149–52.
[54] Shahid NS, Steinhoff MC, Roy E, Begum T, Thompson CM, Siber GR. Placental and breast transfer of antibodies after maternal immunization with polysaccharide meningococcal vaccine: a randomized, controlled evaluation. Vaccine 2002;20:2404–9.
[55] Shahid NS, Steinhoff MC, Hoque SS, Begum T, Thompson C, Siber GR. Serum, breastmilk, and infant antibody after maternal immunisation with pneumococcal vaccine. Lancet 1995;346:1252–7.
[56] Munoz F, Englund J, Cheesman C. Maternal immunization with pneumococcal polysaccharide vaccine in the third trimester of gestation. Vaccine 2001;20:826–37.

[57] Henkle E, Steinhoff MC, Omer SB, Roy E, Arifeen SE, et al. The effect of exclusive breast-feeding on respiratory illness in young infants in a maternal immunization trial in Bangladesh. Pediatr Infect Dis J 2013;32:431–5.
[58] Lopes CR, Berezin EN, Ching TH, Canuto Jde S, Costa VO, Klering EM. Ineffectiveness for infants of immunization of mothers with pneumococcal capsular polysaccharide vaccine during pregnancy. Braz J Infect Dis 2009;13:104–6.
[59] Shapiro RL, Lockman S, Kim S, Smeaton L, Rahkola JT, Thior I, et al. Infant morbidity, mortality, and breastmilk immunologic profiles among breast-feeding HIV-infected and HIV-uninfected women in Botswana. J Infect Dis 2007;196:562–9.
[60] Brüssow H, Barclay D, Sidoti J, Rey S, Blondel A, Dirren H, et al. Effect of malnutrition on serum and milk antibodies in Zairian women. Clin Diagn Lab Immunol 1996;3:37–41.
[61] Castellote C, Casillas R, Ramirez-Santana C, Perez-Cano FJ, Castell M, Moretones MG, et al. Premature delivery influences the immunological composition of colostrum and transitional and mature human milk. J Nutr 2011;141:1181–7.
[62] Ramírez-Santana C, Perez-Cano FJ, Audi C, Castell M, Moretones MG, Lopez-Sabater MC, et al. Effects of cooling and freezing storage on the stability of bioactive factors in human colostrum. J Dairy Sci 2012;95:2319–25.
[63] Zhang F, Fang F, Chang H, Peng B, Wu J, et al. Comparison of protection against H5N1 influenza virus in mouse offspring provided by maternal vaccination with HA DNA and inactivated vaccine. Arch Virol 2013;158:1253–65.
[64] Moon SS, Wang Y, Shane AL, Nguyen T, Ray P, Dennehy P, et al. Inhibitory effect of breastmilk on infectivity of live oral rotavirus vaccines. Pediatr Infect Dis J 2010;29:919–23.
[65] Ali A, Kazi AM, Cortese MM, Fleming JA, Moon S, Parashar UD, et al. Impact of withholding breastfeeding at the time of vaccination on the immunogenicity of oral rotavirus vaccine—a randomized trial. PLoS One 2015;10:e0127622.
[66] Groome MJ, Moon SS, Velasquez D, Jones S, Koen A, Van Niekerk N, et al. Effect of breastfeeding on immunogenicity of oral live-attenuated human rotavirus vaccine: a randomized trial in HIV-uninfected infants in Soweto, South Africa. Bull World Health Organ 2014;92:238–45.
[67] Tuaillon E, Valea D, Becquart P, Al Tabaa Y, Meda N, Bolore K, et al. Human milk-derived B cells: a highly activated switched memory cell population primed to secrete antibodies. J Immunol 2009;182:7155–62.
[68] Marchant A, Sadarangani M, Garand M, Dauby N, Verthesselt V, Pereira L, et al. Maternal immunisation: collaborating with mother nature. Lancet Infect Dis 2017;17:e197–208.
[69] Ehlinger EP, Webster EM, Kang HH, Cangialose A, Simmons AC, Barbas KH, et al. Maternal cytomegalovirus-specific immune responses and symptomatic postnatal cytomegalovirus transmission in very low-birth-weight preterm infants. J Infect Dis 2011;204:1672–82.
[70] Mabuka J, Nduati R, Odem-Davis K, Peterson D, Overbaugh J. HIV-specific antibodies capable of ADCC are common in breastmilk and are associated with reduced risk of transmission in women with high viral loads. PLoS Pathog 2012;8:e1002739.
[71] Peroni D, Chirumbolo S, Veneri D, Piacentini GL, Tenero L, Vella A, et al. Colostrum-derived B and T cells as an extra-lymphoid compartment of effector cell populations in humans. J Matern Fetal Neonatal Med 2012;26:137–42.
[72] Wilcox CR, Jones CE. Beyond passive immunity: is there priming of the fetal immune system following vaccination in pregnancy and what are the potential clinical implications? Front Immunol 2018;9:1548.

CHAPTER

5

Global considerations on maternal vaccine introduction and implementation

Michelle L. Giles[a], *Pauline Paterson*[b], *Flor M. Munoz*[c], *Heidi Larson*[b], *Philipp Lambach*[d]

[a]Department of Obstetrics and Gynaecology, Monash University, Melbourne, VIC, Australia, [b]Department of Infectious Disease Epidemiology, The London School of Hygiene & Tropical Medicine, London, United Kingdom, [c]Department of Pediatrics, Section of Infectious Diseases, Baylor College of Medicine, Houston, TX, United States, [d]World Health Organization, Geneva, Switzerland

Introduction

Infections among infants and young children are common and can result in serious untoward outcomes [1]. In particular, newborns' increased susceptibility to infection has made improving child survival a key priority for global development efforts. To address this, the United Nations (UN) Sustainable Development Goals aim to end preventable deaths of newborns and children younger than 5-years-old by 2030, and to reduce neonatal mortality from the current rate of 19 per 1000 live births to a maximum of 12 per 1000 live births [2]. There has been substantial progress over the past several decades, with the total number of under-five deaths reduced by more than half from 12.6 million in 1990 to 5.4 million in 2017 [3]. Progress in protecting newborns however, has been slower. According to the World Health Organization (WHO), 2.5 million newborns died globally in 2017—a staggering 46% of deaths in children under 5 years of age, with pneumonia and sepsis/meningitis responsible for 30% of these neonatal deaths [3]. This lack of progress led to the United

https://doi.org/10.1016/B978-0-12-814582-1.00006-1

Nations Inter-Agency Group for Child Mortality Estimation (UN IGME) call for expanding effective preventive interventions targeting the main causes of mortality in this vulnerable group [4]. Each year, vaccines prevent more than 2.5 million child deaths globally [5]. Vaccines have an expansive reach and rapid impact, while at the same time saving lives and costs, making them an attractive intervention to target pregnant women against vaccine preventable disease.

Maternal immunization is a preventive strategy that provides protection to the newborn and young infant through the antenatal transfer of maternally-derived pathogen specific immunoglobulin G (IgG) antibodies via the placenta [6] and is considered safe for the mother and infant, as assessed by the WHO Global Advisory Committee on Vaccine Safety (GACVS) [7]. Vaccines such as tetanus toxoid, have been recommended for pregnant women worldwide since the 1960s, and influenza vaccine, which is recommended in certain settings and risk groups. Other vaccines such as pertussis-containing vaccines have been recommended in some middle and high-income countries since 2011. In general, inactivated vaccines that are safe and available for the general population, can be offered to pregnant women when they or their fetus are at risk from exposure to a vaccine preventable disease. Furthermore, promising vaccine candidates against Respiratory Syncytial Virus (RSV) and Group B *Streptococcus* (GBS) are currently under development, offering the potential to reduce a substantial burden of respiratory and bloodstream infections in the neonatal period (see Chapters 10 and 11). These vaccines hold promise to protect the mother, the newborn, or both, but their benefit will only be maximized if optimally implemented. To do this there are some key issues common to all vaccines, which need to be considered. These include: knowledge of disease burden, existing delivery platforms, education, training, communication, cost effectiveness and establishing the capacity for administration, including maintaining supply and cold chain requirements and following guidance on vaccination during routine antenatal care. There are also important differences in the vaccines that can be administered during pregnancy, which also need consideration, such as timing of administration, seasonality, number of doses and proportional benefit to the mother and/or newborn.

Despite existing recommendations for vaccination during pregnancy, the implementation of maternal immunization programs is often challenging. Elimination goals for maternal and neonatal tetanus as of March 2019 still have not been reached in 13 countries [8], and despite global policy recommendations since 2012, only 81 of all 194 WHO Member States target pregnant women in their influenza immunization policy (most recent data from 2014) [9]. The factors that affect this decision-making, and the successful implementation of maternal immunization programs, are not fully understood. To fill this information gap in relation to influenza,

POLICY MAKERS .... IMPLEMENTERS ....

Decision to introduce vaccine	Planning the introduction	Implementation aspects	Monitoring and evaluation
• Disease and economic burden • Vaccine effectiveness, safety and acceptance • System capacity • Choice of optimal delivery strategy	• Annual and multiyear planning • Phased vs full scale introduction • Vaccination timing and scheduling • Procurement, logistics, and cold chain management	• Training of health staff • Knowledge building among health professionals • Communication and acceptance strategies	• Coverage monitoring • Disease surveillance • AEFI surveillance • Post introduction evaluation

FIG. 1 Process of introducing a maternal vaccine into the national immunization program. *Adapted from World Health Organization. How to implement influenza vaccination of pregnant women: an introduction manual for national immunization programme managers and policy makers; 2017. http://apps.who.int/iris/bitstream/10665/250084/1/WHO-IVB-16.06-eng.pdf.*

the WHO engaged in the Maternal Influenza Vaccine Introduction project (2013–2016) [10]. Results from this project confirm that introduction of maternal influenza vaccines requires not only an assessment of the evidence available on disease and economic burden, introduction costs, and vaccine safety and efficacy; but also addressing information needs of implementation planners related to the operationalization of service delivery in different country settings (Fig. 1).

Based on the information needs identified during the Influenza Vaccine Introduction project [11], WHO subsequently initiated the Maternal Immunization and Antenatal Care Situation Analysis (MIACSA) project, which is assessing Tetanus and other existing maternal vaccine service delivery strategies in low and middle-income countries (LMICs) [12]. Ultimately the results of this project will inform discussions on barriers, facilitators, and gaps to the use of current and additional future maternal vaccines such as GBS or RSV vaccine in low resource settings [13,14].

Implementation

The importance of careful planning and national strategies/ leadership

In 1950, in the context of limited options for the prevention of neonatal tetanus in home deliveries, the WHO suggested a study that was conducted in high risk villages in Papua and New Guinea, which demonstrated a direct correlation between the number of Tetanus Toxoid

Containing Vaccine (TTCV) doses administered during pregnancy and a reduction of neonatal tetanus mortality [15]. Given the high mortality and characteristic clinical presentation of neonatal tetanus, disease incidence was estimated based on verbal history obtained from caregivers asked to report neonatal deaths that met five pre-specified criteria considered typical of neonatal tetanus. There are two important lessons learned from this experience. The first is that demonstrating the impact of an intervention targeting a disease that results in high mortality is more likely to be successful when the outcomes of the disease are readily measurable. The second is that disease ascertainment using clinical characteristics instead of laboratory tests, which may be potentially costly and inaccessible, is also likely to ensure success. Diseases that are difficult to specifically diagnose clinically and for which the actual burden and impact are unknown, pose challenges for the introduction of vaccine prevention strategies. Other important observations from this early study included the need to ensure that the intervention—administration of TTCV vaccine to pregnant women—was practical and feasible. In this setting, using a low-cost vaccine that was given in already existing antenatal clinics where pregnant women could be seen several times to facilitate the administration of multiple vaccine doses with adequate interval prior to delivery to ensure maternal immune responses and antibody passage through the placenta, was critical.

Currently, maternal and neonatal tetanus (MNT) has yet to be eliminated from 13 countries: Afghanistan, Angola, Central African Republic, Chad, Congo DR, Guinea, Mali, Nigeria, Pakistan, Papua New Guinea, Somalia, Sudan, South Sudan and Yemen [8]. Of the countries that have eliminated MNT, only six used routine maternal immunization exclusively. Most countries that have successfully achieved MNT elimination status have utilized various additional vaccination strategies such as Supplementary Immunization Activities (SIA), which involve targeted immunization campaigns for women of reproductive age living in high risk districts, to facilitate three properly spaced doses of TTCV prior to pregnancy. Such targeted campaigns require additional effort, such as the identification of high risk populations and locations where access to vaccine is limited. They may occur outside of antenatal care visits and therefore require partnerships with local, regional and national stakeholders. In addition, active, ongoing surveillance of disease, including identification of new cases, is necessary to ensure the success of the program.

In regions where MNT elimination has been achieved [8], sustained elimination is dependent on:

- Ensuring high coverage of a routine childhood immunization schedule with DTP or DTaP (3 + 2 or 3 + 3—including adolescent Tdap dose—+ 1 adult dose);

- Establishing or strengthening tetanus surveillance systems to identify neonatal tetanus cases;
- Investigating all cases;
- Initiating detection campaigns in areas where there may be unreported cases;
- Focusing vaccination activities on women of childbearing age who live in high risk areas;
- Ensuring women to keep documentation of vaccination by way of a vaccination card or similar document;
- Ensuring that antenatal care providers participate in vaccination activities;
- Continuing to work on improving hygienic childbirth delivery and post-partum practices and procedures; and
- Implementing a high risk approach with SIA as needed.

While maternal tetanus immunization as a proof of concept constitutes a long established, and mostly successful possible model and platform for the administration of vaccines during pregnancy, it also provides an opportunity to identify the operational challenges that may support or hamper achieving high coverage of vaccination of pregnant women. Ongoing program evaluations in countries and further global implementation research efforts such as the WHO MIACSA project help policy makers and implementers to better understand the opportunities and challenges observed during implementation efforts in various settings to improve their maternal tetanus vaccination efforts as a public health strategy [12].

The importance of burden of diseases data in informing the introduction of maternal immunization

Inactivated influenza vaccines are universally recommended by WHO for use in pregnant women [16]. Unlike tetanus, which has been long established in low resource settings, influenza vaccination comes with unique challenges that affect not only the implementation of a maternal immunization program but also the decision-making process to introduce the vaccine.

Worldwide, influenza epidemics are estimated to result annually in about 3–5 million cases of severe illness and up to 650,000 deaths. Such figures come with substantial social and economic cost to countries [17]. The effect of seasonal influenza epidemics in low-resource settings are not fully known, but research estimates that 99% of deaths in children under 5 years of age with influenza-related lower respiratory tract infections occur in low-resource settings [18]. In addition to protecting against seasonal influenza outbreaks, influenza vaccination programs can support countries' planning efforts for a potential pandemic by increasing their capacity to produce (where applicable) or procure vaccines, to register and distribute them, to conduct targeted vaccine delivery, and to monitor vaccination coverage, safety and effectiveness [19,20].

In 2012, the WHO published a position paper identifying pregnant women as a priority group for countries planning to initiate or expand their seasonal influenza program, recommending vaccination of pregnant women at any stage during pregnancy [16]. Despite this, maternal influenza immunization has not been incorporated into routine immunization programs in many LMICs, with more than 95% of the uptake of all influenza vaccines in pregnant women worldwide occurring in the regions of the Americas, Europe, and the Western Pacific [9].

In LMICs, decision-making is hampered by lack of local data on influenza disease burden and subsequent lack of economic burden and cost effectiveness evaluations [20–23]. To help close this information gap, the WHO has developed several resource documents to support countries to evaluate the disease and economic burden of influenza, as well as a costing tool to estimate expected costs associated with maternal influenza vaccine introduction and sustained use of the vaccine [24–26]. First results from the use of this tool in Malawi, estimated that incremental delivery cost of a maternal influenza immunization program could be as low as estimated costs of childhood vaccination programs, as long as childhood immunization and antenatal care systems are capable of serving as platform for an additional vaccine [27]. Beyond lack of data on burden of disease, vaccine efficacy and cost effectiveness, WHO regions reported additional issues affecting decision making for vaccine introduction. Examples of these additional issues included the need for more safety data and uncertainty towards matching between vaccine containing strains and circulating influenza viruses.

In summary, compared to vaccination of pregnant women against tetanus, influenza vaccination is implemented far less across LMICs, requiring support focused on informing countries' decisions to introduce the vaccine, ideally based on the local benefit and reassuring safety profile of the vaccine. Identifying the optimal service delivery must include defining the optimal timing of vaccination, format of service delivery, and careful assessment of existing delivery capacity services.

The importance of identifying optimal service delivery and defining optimal timing of vaccine administration during pregnancy

Factors that have been identified to affect successful influenza vaccine delivery include a flexible approach to vaccine service delivery taking into consideration the local context, targeted education and community engagement, and attendance and access to antenatal care [28]. Routine antenatal visits are commonly seen as key to promote the benefit of newly introduced maternal vaccines, and their increased number and quality of content are seen as success factors [29,30]. Therefore, establishing a maternal influenza vaccination program can build on the successful experiences of introducing TTCV. However, unlike maternal tetanus programs, influenza vaccine programs targeting several risk groups will likely be implemented through campaigns in most countries, leading to increased demand for human resources within a finite period, having important implications for staff training and communication strategies to pregnant women, and putting stress on the cold chain prior to and during the influenza season [10]. The timing of influenza vaccination is a further challenge in tropical regions where influenza virus can circulate all year round, including outside the time when influenza vaccine may be available [31]. This poses challenges to vaccine procurement and distribution as evidenced by 6 out of 10 of the 54 Gavi, the Vaccine Alliance-eligible countries in 2015 [31–34]. To help countries identify their distinct influenza seasonality patterns, grouping countries into seasonality zones has been developed to inform the optimization and development of influenza vaccination strategies [31].

Pertussis vaccination, although not globally recommended for pregnant women due to lack of burden data in low resource settings, is included in national recommendations across several high-income countries. Unlike tetanus vaccine and influenza vaccine, recommendations for the optimal timing of administration during pregnancy vary but generally have a narrower time window in which it is optimally administered, for example between 27 and 36 weeks gestation in the United States (US) and from 20 weeks in Australia [35,36]. This is not consistent however across all settings with recommendations in the United Kingdom (UK) extending from 16 weeks gestation [37] and in Canada from 13 weeks.

A potential positive consequence of an earlier and broader window for vaccine administration during pregnancy is an increase in uptake and coverage. Another potential advantage of a strategy implementing vaccination earlier in pregnancy is that this may afford some protection to babies born prematurely, while, from a programmatic standpoint, women who present late in pregnancy may miss the opportunity for optimal protection. When the window for optimal vaccine delivery is narrow, it becomes more difficult to ensure that pregnant women are vaccinated within the specified timeframe. This may be particularly challenging in many low- and middle-income settings, where antenatal care coverage is low. In addition, inherent to adherence of administration at a specific time point in gestation is the need for accurate determination of gestational age, which itself is an important challenge in many settings. There are numerous methods utilized to estimate gestational age including last menstrual period, symphysio-pubis fundal height and Ballard score at delivery, but the gold standard that these are often compared to is ultrasound. The importance of accurate measurement of gestational age is not just relevant to ensuring optimal timing of maternal immunization when there are gestational specific recommendations, but also to measure pregnancy outcomes such as abortion, stillbirth and preterm birth. A recent meeting of key stakeholders identified strengthening of antenatal care services, including improving ultrasound facilities and skills, together with early antenatal attendance as essential elements for the successful implementation of health interventions in pregnancy [38].

These issues also are relevant to future vaccines, for example against RSV, which are currently in late phase clinical trials (see Chapter 10) and will likely be administered in a strict gestational window determined by stringent criteria to determine gestational age. WHO's objective is to promote the development of vaccines with optimal effectiveness and suitability for use in LMICs, thereby maximizing global vaccine impact. Preferred product characteristics (PPCs) are developed by WHO's Initiative for Vaccine Research (IVR) to provide guidance on preferences for new vaccines in priority disease areas. For RSV vaccines given to pregnant women, WHO PPC suggests a one dose regimen [39]. In low resource areas, if provided routinely through antenatal care, RSV vaccination may be challenging where limited capacity or access prevents pregnant women from benefiting from antenatal care. As seen with other vaccines such as pertussis, other limiting factors may be the need to conform to narrow time windows for optimal vaccine administration, challenges of accurate determination of gestational age, and the need for ongoing education and training of healthcare providers and communication to women, particularly as the number of recommended different maternal vaccines grows. There are also implications for women at risk of preterm birth. Depending on the service delivery system, strengthening of antenatal care services, including implementation of the increased number of antenatal care visits as per the updated WHO recommendations on antenatal care may help to increase the capacity for successful implementation of a future RSV vaccine [30].

Information needs for the introduction of new vaccines for pregnant women including RSV

When considering implementation of future vaccines such as RSV, particularly in LMICs, recommendations developed at a global level and reviewed by relevant WHO advisory bodies such as Strategic Advisory Group of Experts (SAGE) on Immunization and the Immunization Practices Advisory Committee (IPAC) would be necessary. With RSV being probably the first vaccine specifically developed for use in pregnant women, countries will need to determine the optimal service delivery system, to best integrate the vaccine into existing national immunization efforts. Furthermore, surveillance and safety monitoring, and program evaluation mechanisms need to be in place. To support such efforts, the Advancing Maternal Immunization (AMI) collaboration between WHO and PATH was conducted in 2018 to identify evidence gaps relevant to RSV vaccine service delivery and to provide advice in the form of a roadmap [14].

The success of any future maternal vaccine program incorporating an RSV vaccine will also be influenced by stakeholders' knowledge and attitude to vaccination and RSV itself. There is no published literature exploring pregnant women's knowledge and understanding of RSV disease in infants, nor on health care providers' knowledge or attitude towards RSV vaccination of pregnant women. As with any new vaccine, these will be important areas of future research to inform the development of effective national advocacy and communication strategies.

For any new vaccine, another important consideration in implementation is the development of robust local safety surveillance systems. This is particularly true for vaccines in pregnancy to ensure there is confidence in the safety of new vaccines for both the pregnant woman and her baby. This data is relevant to communicate messages for both women and health providers.

For new vaccines implementation, information on economic burden of disease, cost of the vaccine, and its cost effectiveness are essential to inform decision-making and implementation planning in countries. GAVI, The Vaccine Alliance, is the most likely source of external financing to support RSV vaccination in LMICs and considered RSV vaccination for its 2018 Vaccine Investment Strategy (Report to the Board, November 28–29, 2018). Global efforts are underway to support health economics analyses based on WHO tools and guidance to assess the impact, affordability, and financial sustainability of RSV maternal vaccination in low resource settings.

During 2015 and 2016, the WHO Product Development for Vaccines Advisory Committee (PDVAC), which informs SAGE on vaccine research and development priorities, identified as a priority the development of GBS vaccines suitable for immunization in pregnancy and use in LMICs [40]. The public health and economic value of GBS vaccination, however, needs to be better understood to spur investment into vaccine development, identify opportunities and challenges for future vaccine introduction and provide an analysis of data needs for future vaccine decision making. Modeling of cost effectiveness in low-income settings needs to be informed by data on burden of GBS disease in pregnancy, serotypes by region, estimates of disease preventable by intrapartum antibiotics and estimates of disease preventable by maternal vaccination.

In addition, other factors such as home birth rates, accurate determination of gestation, attendance for antenatal care, background prevalence of human immunodeficiency virus infection (which may lower response to vaccine) and diagnostic capacity for determination of endpoints are all challenges to implementation of a maternal GBS vaccine. In 2016, the largest number of newborn deaths occurred in South Asia, followed by sub-Saharan Africa, two regions with low antenatal care coverage. A widely accepted benchmark indicator for adequate access and use of primary health care services is four or more antenatal care visits. In 2016, only 62% of women worldwide benefited from at least four antenatal care visits (ANC4+), with gaps particularly in regions with the highest burden of GBS disease (sub-Saharan Africa 52% and South Asia 46% ANC4+, respectively) [41]. Clearly, an implementation challenge for any future GBS vaccine, is service delivery and a requirement to strengthen antenatal care access. Key aspects impacting on service delivery will include the optimal timing of delivery of the vaccine for maximal efficacy, safety when co-administered with other vaccines, human resources, and demand creation including communication to women addressing knowledge base and potential hesitancy.

To address these information gaps and to inform further GBS vaccine development, the London School of Hygiene and Tropical Medicine and the WHO jointly initiated a 3-year project at the end of 2017 to develop a comprehensive value proposition for GBS vaccination for pregnant women. This value proposition will be based on a thorough assessment of the preventable burden of disease, the costs and gains expected through vaccination of pregnant women, and the potential impact of operationalizing GBS vaccination programs particularly in low, middle and high-income contexts.

Cross cutting issues for implementation of any maternal vaccine

The initial decision to introduce a maternal vaccine, as with any vaccine, is based on the identified public health need matched against available resources, along with political and public will to support the delivery and acceptance of the vaccine. Decision makers need to select affordable and cost-effective interventions that are prioritized against other health interventions to prevent disease, if available.

The countries' Ministry of Health usually requests the National Immunization Technical Advisory Group (NITAG) [42] to conduct a review of local and global evidence that includes disease and vaccine characteristics, as well as economic and operational considerations (Fig. 2) [10,42]. Where available, the NITAGs review existing maternal vaccination activities to provide a better understanding of delivery mechanisms, the capacity for routine vaccine delivery, coverage rates that could be achieved, and operational challenges that may need to be addressed.

As provision of maternal vaccines during routine antenatal care services appears a targeted approach to reach pregnant women, the Ministry of Health should ensure an inclusive and informed decision-making process considering inputs from other ministries, academic, scientific and professional groups (e.g., obstetricians, midwives, or professional associations), civil society organizations, and the private sector.

Where studies cannot be carried out to detect the local burden of disease, decision makers may revert to using comparable data from other countries or regions as proxies. For economic estimations, direct costs and indirect costs (e.g., losses of productivity or income), as well as the impact on poorer or socially disadvantaged populations can be particularly relevant for countries where vaccine introduction is financed by national resources. Comparing such economic disease burden can help estimate the vaccine's direct impact (e.g., reducing of maternal and infant mortality, stillbirths, preterm births and costs of antibiotic use), and ultimately serve to understand the broader economic impact of maternal vaccines on long-term development, demographic dividend, household savings, and increased Gross Domestic Product.

After the evidence review, the Extended Program on Immunization (EPI) country manager and the program manager for reproductive, maternal, newborn, child and adolescent health (RMNCH) can draw on WHO guidance and toolboxes for current vaccines to provide recommendations on the planning of vaccine introduction to the Minister of Health who will ultimately decide whether to introduce the vaccine [10,43,44].

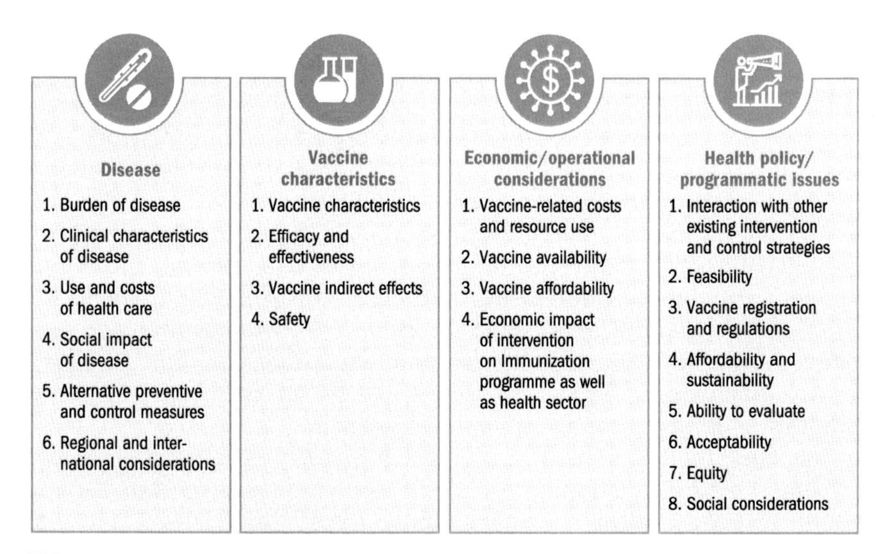

FIG. 2 Elements that should be assessed, discussed and addressed by NITAGS during the introduction of maternal vaccination. *Adapted from World Health Organization. How to implement influenza vaccination of pregnant women: an introduction manual for national immunization programme managers and policy makers; 2017. http://apps.who.int/iris/bitstream/10665/250084/1/WHO-IVB-16.06-eng.pdf.*

Developing a service delivery strategy for a new maternal vaccine is relevant to ensure alignment with the national immunization schedule, and to better understand the challenges and the benefits of integrating vaccination of pregnant women into existing delivery services and those able to best reach pregnant women [43].

Depending on the available resources, viable options include vaccination campaigns, routine vaccination, or vaccine delivery through outreach [45]. Pregnant women can be targeted through routine antenatal care, preconception or family planning visits, visits to health-care facilities with their children (e.g., routine childhood immunization), outpatient care (general practice, gynecology, family planning, high-risk clinics), and any other health settings where pregnant women might seek care for themselves or their children. Antenatal care visits, attended by a large majority of pregnant women, could provide the opportunity for health care workers to generate demand for maternal immunization and to address potential hesitation or concerns of pregnant mothers [46].

EPI programs due to their unique expertise in providing childhood vaccines, usually take charge of procuring and distributing maternal vaccines in the majority of LMICs. Other service delivery components such as planning and management of maternal tetanus vaccination and training of health staff may be led by immunization programs or benefit from joint management models and coordination mechanisms between both EPI and RMNCH. Collaboration on the optimal service delivery approach for a new vaccine can help managers to coordinate vaccine management responsibilities and to assess the impact of the vaccine introduction on the capacity of current antenatal care delivery mechanisms. Among others, this capacity includes the frequency and quality of care of antenatal care visits to ensure that vaccination can occur at the optimal gestation of pregnancy and be co-administered with other interventions to reduce logistical effort and provide a quality package of care to pregnant women [10,47].

Accurate documentation of vaccination may require the use of various methods of recording and reporting, for example, in addition to providing women with immunization cards, thorough record keeping in the health care facilities (immunization books or electronic record keeping) and regular reporting to local, regional and national authorities should be performed. Maintaining high vaccination coverage can only be achieved by monitoring vaccine administration in all settings.

The safety of vaccination in pregnancy is a key consideration in the process of implementation, for pregnant women themselves, but also for healthcare providers, vaccine manufacturers, policy and decision makers, regulatory agencies, and sponsors of vaccination programs. Therefore, critical components in the implementation of a maternal immunization program include the establishment of methods to accurately track vaccination coverage, safety surveillance for the identification and proper assessment of adverse events following immunization, and surveillance of cases of the disease prevented by the vaccine, to demonstrate the impact of the intervention.

Establishing mechanisms for safety surveillance or utilizing existing reporting systems that are adapted to incorporate maternal vaccinations is a complex issue. It is clear that using standard definitions of the relevant adverse events that may occur after vaccination is necessary to systematically collect accurate data, ensure precision and comparability, as well as harmonization across studies, reports, and surveillance systems. Efforts towards identifying and evaluating key events related to vaccination in pregnancy, which may affect the mother, the fetus, or the infant, have been led by the Brighton Collaboration and the GAIA project, supported by the Bill & Melinda Gates Foundation and WHO [48–50]. Evaluating, validating and implementing standard case definitions is an important step towards improving safety surveillance of vaccines during pregnancy, particularly given the added complexity by inherent risks of pregnancy. Lastly, establishing adequate case definitions and evaluation of disease cases, including the ability to establish an etiological diagnosis is important to determine the impact of maternal vaccines in reducing maternal and infant morbidity and mortality, depending on the pathogen; and to assess the success and potential deficiencies of the program.

There is also a need for a communication and engagement strategy to address concerns and vaccine hesitancy among pregnant women and health care workers.

Understanding and addressing factors affecting vaccine acceptance during pregnancy

Vaccine acceptance is influenced by a number of factors including maternal knowledge, attitudes and beliefs, healthcare provider recommendation and trust between the community and healthcare provider [28,51–53], socio-demographic factors [54,55], and the influence of news and social media [56,57]. In a literature review to identify factors influencing vaccination acceptance during pregnancy [54], some key factors that contribute to vaccine hesitancy in pregnancy were identified. They included safety concerns, a lack of perceived need for vaccination, concerns about effectiveness, a lack of recommendation by a healthcare provider, low knowledge of the disease, low knowledge of the vaccine or vaccine program, and access issues. Subsequent studies confirmed similar factors [58–66] and the importance of trust in health professionals [64,65,67,68]. Ethnicity and deprivation have also been found to be associated with vaccination in pregnancy, sometimes due to limited access, inadequate information or trust issues with the health providers or the health system [58,69–72].

The issue of vaccine hesitancy has become increasingly recognized as an important area that needs to be understood in its local context to inform the best response. The WHO SAGE working group on vaccine hesitancy has defined it as "delay in acceptance or refusal of vaccines despite availability of vaccination services" [73]. Vaccine hesitancy is context-specific, varying by time, place, vaccines, context and sub-population and includes factors such as complacency, convenience and confidence [73]. Complacency may occur if there is a lack of perceived need or value for the vaccine. A systematic review on determinants of influenza vaccination among pregnant women found that pregnant women were unaware that they were at high risk for influenza and its complications during pregnancy [55]. In a survey of 1025 pregnant women in Germany, 40.3% perceived that the influenza vaccine in pregnancy was unnecessary [74], similar to the reasons identified in a US study with non-vaccinated pregnant women [67]. And, in an online survey of pregnant women and women with children under 2 years of age in the UK, the perceived seriousness of the disease was a key consideration for pregnant women when deciding whether or not to vaccinate [66]. Convenience issues can be due to inconvenient location or time of vaccination, especially when the need for vaccination is under appreciated. Confidence in vaccination is another key domain influencing vaccine decisions and involves a number of factors mentioned above and is closely tied to level of confidence in the vaccine as well as the health provider and the health system.

Factors influencing vaccine acceptance

The role of information and its sources

In a survey of pregnant women in London ($N=200$), where uptake of pertussis vaccine during pregnancy was only 26% (52/200), information needs were reported as being a significant barrier to accept vaccination in pregnancy [61]. Of the women that had not been vaccinated during pregnancy, 51.3% indicated that they were not aware of the vaccination program and 32.6% cited that they did not vaccinate due to a lack of information [61]. The majority (91%) of women in the study believed that their healthcare provider should have provided additional, detailed information about vaccination during pregnancy [61]. In another study in the UK ($N=638$), 69.7% of the women surveyed reported that they had received enough information to make their decision about pertussis vaccination while 21.3% had some information, but would have liked more, and 8.9% reported not have any information or not enough [66]. A study in Canada that explored information-gathering and decision-making processes of immigrant mothers for scheduled childhood vaccines, vaccination during pregnancy, seasonal influenza and pandemic vaccination, Kowal and colleagues found that the current communication strategies did not always reach immigrant women [63].

Several studies identify a range of information sources that pregnant women look into when deciding whether or not to vaccinate [61–64]. These include personal contacts such as friends and family members, healthcare providers such as General Practitioners (GPs) and midwives, print material and online media [53,61,62,64,67]. Winslade and colleagues make the additional point that "Despite usage of the internet to look up medical information, women wanted to discuss vaccination with their midwives or general practitioners" [64]. In the study on immigrants in Canada, co-authors found that the study participants "were passive in both vaccine information gathering and decision-making, highlighting the importance of HCPs [healthcare practitioners] in promoting the uptake of immunization for immigrant families" [63].

Another study in Hackney, London, UK, investigated attitudes towards maternal vaccination among pregnant and recently pregnant women [65]. The most popular websites used by study participants were UK National Health Service's "NHS Choices"; online blogs and forums—especially Netmums and Mumsnet; Baby Centre; The Mayo Clinic; and Medscape [65]. These websites were particularly used by pregnant women to investigate possible vaccine side-effects rather than the diseases they aim to prevent [65]. Some women mocked the use of forums such as Netmums, criticizing incorrect information and the opinions of other users [65]. In a European study, Bouder and colleagues found that although the majority of study participants in Europe would go to the GP, pharmacy and Internet for information, participants trusted the GP and pharmacy more than the Internet [75]. In a US study by Healy and colleagues on knowledge and attitudes of pregnant women and their providers towards recommendations for immunization during pregnancy, 89.1% of the 825 women surveyed cited a provider as their most trusted source of information [62].

The role of addressing safety concerns

When it comes to vaccination, pregnant women have multiple safety concerns—concerns for the self, concerns for the fetus, and concerns for the baby [67]. A number of studies have identified these different safety concerns as factors impacting vaccination decisions [62,64,66,67]. The survey of pregnant women in Germany identified concerns of vaccine-related side effects as negatively associated with influenza vaccine uptake in pregnancy [74]. Perceptions also differ between influenza and pertussis. Women saw influenza as a disease affecting the mother however they viewed pertussis as a disease affecting their baby and were more likely to vaccinate against pertussis [53].

The role of the healthcare provider

There is a growing body of literature highlighting the importance of the healthcare provider in engaging with their patients, listening to their concerns and explaining the reasons for recommending vaccination in general [76] and particularly related to vaccinating in pregnancy [54,55,59–61,63, 67,74,77].

A study in Australia found that women were 20 times more likely to get vaccinated against influenza during pregnancy if their healthcare provider had recommended it [77]. A related study in Canada in 2013, cites one participant reporting a similar sentiment saying, "If they tell me to get it, I'll get it" [63]. In London, a survey of pregnant women found that only 24% of participants had a meaningful discussion with their GP about vaccinating during pregnancy, and a lack of professional encouragement was identified as a barrier to vaccine uptake [61].

The healthcare professional—patient relationship is a key influencer of pregnant women's decisions to accept pertussis and influenza vaccines [65]. The study in Hackney identified the importance of a trusting relationship between the healthcare provider and the pregnant woman [65]. This was also identified in a study in London on pertussis vaccination in pregnancy [64] and in a study in the US on influenza vaccination in pregnancy: "I trust my doctor. If you don't trust your doctor, you may as well not go to them. So, you know, he told me I should get it and I listened to him" [67].

Given their influence on vaccine confidence and acceptance, education and training of healthcare providers on the benefits and risks of the vaccine for both mother and neonate, and the need for revaccination with subsequent pregnancies, is critical.

Strategies to increase confidence and uptake of vaccines in pregnancy

As identified in the previous section, there are a multitude of reasons for low uptake of vaccination in pregnancy. It is important to understand the factors that influence the decision to vaccinate in order to inform the development of strategies and interventions to increase coverage of vaccination in pregnancy. It is also worth evaluating vaccination coverage by region to identify if there are specific areas of lower coverage. Once developed, strategies also need to be evaluated in order to identify which ones are effective in increasing uptake, as research has shown that some strategies to address myths can be ineffective and it is quite difficult to change beliefs [78,79]. A number of studies have been published exploring the effectiveness of various strategies to increase vaccination in pregnancy, including a couple of systematic reviews [80,81]. Below we describe strategies that have been found to be effective at an individual level, either as an individual intervention being evaluated, or as part of a multi-component intervention (two or more interventions). We then describe effective multi-component interventions. There may be other interventions, not discussed below, that may improve vaccination coverage. These have not been included here as either the strategies have not been evaluated, they have been evaluated but were not found to be effective or were found to be effective but within studies assessed to have weak-quality evidence.

In summary, no single intervention is a silver bullet in increasing vaccination uptake in pregnancy in all settings and contexts. Effective interventions will depend on the context and the reasons for low vaccination uptake.

Strategies found to be effective at an individual level to increase maternal vaccination uptake

Improving access and/or availability of vaccines

Two systematic reviews found improving access and/or availability of vaccines to be effective in improving uptake of vaccination in pregnancy [80,81]. A study in the US and a study in Australia illustrated that healthcare institutions that had implemented a standing order to allow midwives to administer pertussis and influenza vaccination, without seeking permission from a physician or referring to a physician or GP to administer the vaccine, was an effective intervention at improving vaccination rates [52,82].

In the UK, a change in recommended timing for pertussis vaccination to start from 16 weeks gestation facilitated opportunities for the vaccine to be discussed with the women earlier in the pregnancy [72].

Information and education

Information and education for pregnant women can be an effective intervention at improving vaccination uptake, especially when encompassing information on efficacy, safety, benefits and timing of vaccination, and the information and education materials are accessible within antenatal clinics and facilities, or provided by healthcare staff, and provide guidance on where to access vaccination [80]. A study in the US illustrated that handing out a patient information pamphlet with a statement about the importance of vaccination to protect the baby from influenza increased vaccination in pregnancy [83]. A Canadian study also found that an education pamphlet significantly increased seasonal influenza vaccine uptake [84]. A brief one-to-one education with pregnant women was found to increase vaccination rates in a study in Hong Kong [85]. The education session for pregnant women included information on safety, vaccine recommendations and benefits of vaccination [85]. Stockwell and colleagues identified that reminders and education via mobile phone text messages increased seasonal influenza vaccination uptake among urban, low-income pregnant women in the US, after adjusting for gestational age and the number of clinic visits [86]. In comparison, a study by Moniz and colleagues in the United States did not find text message prompts to be effective at increasing influenza vaccination rates in pregnant women [87]. These study authors suggest that the content or tone of the messages may influence the interventions effectiveness [87]. The effectiveness of this type of intervention would also depend on the initial reasons for non-vaccination and whether the intervention addresses those reasons.

Vaccine recommendation from a health care provider

The importance of a recommendation from a healthcare provider has been well documented in positively influencing a patient's decision to vaccinate [76]. This is also the case for vaccinations in pregnancy. The systematic review by Wong and colleagues identified healthcare providers informing pregnant women about the benefits of vaccination, and healthcare providers recommending the vaccine, as effective interventions to increase vaccination uptake [81]. The study by Meharry and colleagues in the United States illustrating a significant increase in vaccination uptake with an education pamphlet alone, also found an increase when this was combined with a verbalized benefit statement [83]. Education and information for healthcare providers, themselves, was also found to be important to ensure that they were equipped with the right information and current guidance to discuss vaccination with the pregnant women [80].

Also, alerts and/or reminders for health care providers have been shown to be an effective intervention to increase vaccination uptake in pregnant women [80]. Vaccination rates increased in an Obstetrics and Gynecology clinic with a best-practice alert on the electronic prenatal record, letting the health care provider know at each prenatal visit if the patient had not yet either received vaccination against influenza or voiced an informed refusal [88]. Another study in the United States also illustrated an increase in vaccination rates following a reminder being placed on each patient's chart [89]. A more recent study in the United States also identified the effectiveness of a paper based prompt on influenza vaccination rate in a resident continuity clinic for the underserved [90].

Multi-component interventions and demand creation

A number of studies have found that multicomponent strategies to address low uptake of vaccination are particularly effective [80,81]. One United States study reported that a combination of implementing a vaccine promotion intervention, that included education and reminders to both providers and pregnant women, and the provision of vaccine at antenatal clinics significantly increased influenza vaccine uptake [91]. The study by Ogburn and colleagues identified that education for staff as well as allowing midwives to provide vaccination increased vaccination rates [82].

The uptake issues described point to the relevance of demand creation strategies for the introduction of vaccines for use in pregnant women. A recent systematic review explored the evidence on the effectiveness of interventions to improve vaccination coverage among pregnant women [81]. Of the 11 studies included, the range of interventions included: provider reminder/recall, standing orders, provider education, pregnant women reminder/recall, pregnant women education and finally interventions to enhance vaccination access. The only study, within this recent systematic review, that demonstrated a statistically significant increase in uptake (a moderate quality randomized controlled trial) showed that providing an education pamphlet, with or without a verbalized benefit statement improved influenza vaccination rates among pregnant women [83]. Three cohort studies showed a positive effect of provider-focused interventions such as reminders and/or recalls [88,89,92]. Informing the mother may also help an ethical issue raised towards maternal vaccines primarily protecting the newborn against severe disease, where side effects may result in an asymmetry of burdens and benefits towards the mother versus the fetus. Ensuring that pregnant women are aware of the benefits and burdens of the vaccine and autonomously consent to the immunization by either obtaining informed consent or interest-based approaches may prevent instrumental approaches treating a woman as a "fetal container" [93–95].

Conclusion

The landscape of maternal immunization is rapidly evolving given the potential benefits for the mother, the fetus, and the newborn and further supported by the setting of new targets within the UN Strategic Development Goals. Although long recognized as a strategy for neonatal tetanus disease elimination, the benefit and ability of maternal immunization to reduce neonatal mortality from other infectious diseases such as influenza and pertussis across high and low resource settings is increasingly becoming recognized. Furthermore, new vaccines are under development promising to address a substantial burden of disease caused by GBS and RSV—two diseases costing lives predominantly in low resource settings.

With an increased number of maternal vaccines becoming available, there is urgent need to ensure their equitable operationalization across all countries to address a significant burden of vaccine preventable diseases. Among these, resource constrained contexts require careful consideration of local capacity and delivery options to ensure the maximization of the potential benefits of maternal immunization and to avoid overburdening of existing delivery systems.

Arguably the most relevant group to engage are pregnant women themselves. Taking a customer-oriented approach requires being inclusive and understanding the importance, benefit and most appropriate platform to access a range of vaccines and other important health services in pregnancy. Addressing pregnant women's concerns early will help generate demand, avoid potential misperceptions of the vaccine's safety, and ultimately prevent low uptake due to suboptimal implementation of this important prevention strategy. Healthcare workers as key influencers of program implementation and a source of trust for pregnant women need to be seen as customers and need to be empowered to drive communication and recommendation efforts.

Central to implementation of additional maternal vaccines—some of which require specific timing for administration during pregnancy or in relation to the circulation of pathogens—is the issue of service delivery and access to robust antenatal care services to also ensure equitable access to such vaccines in low resource settings. Implementation research is critical to understand effective mechanisms to generate demand in target groups, and to enable adequate resourcing to critical elements needed to operationalize service delivery in low resource settings. As with any new vaccine introduction, ensuring functionality and specificity of vaccine coverage and Adverse Event Following Immunization (AEFI) surveillance systems to capture pregnant women will be essential to understand the impact of the introduced vaccine. International efforts such as the AMI project, the MIACSA project and the GBS value proposition are undertaken to fill these existing information gaps and inform considerations on the introduction of future vaccines such as GBS and RSV vaccines [11,12].

In conclusion, global priorities for maternal immunization programs recommend strengthening investment in primary health care, addressing the critical health workforce gap in low and middle income countries, maintaining confidence in vaccines including addressing hesitancy issues (particularly with new vaccines) and integrating community health workers into multi-disciplinary health workforce teams as recommended in the WHO guidelines on community health workers.

Conflict of interest

Member of the Committee of Infectious Diseases of the American Academy of Pediatrics; Member of the influenza and the pertussis working groups of the Advisory Committee on Immunization Practices of the US Centers for Disease Control and Prevention (CDC); Member Data Safety Monitoring Committees for Moderna, Pfizer, NIH; Research grants from US National Institutes of Health (NIH), CDC, Bill and Melinda Gates Foundation (through World Health Organization and the Brighton Collaboration), Novavax, GlaxoSmithKline, Janssen, Biocryst; author and editor up to date have no interests to declare.

References

[1] Kollmann TR, Kampmann B, Mazmanian SK, Marchant A, Levy O. Protecting the newborn and young infant from infectious diseases: lessons from immune ontogeny. Immunity 2017;46:350–63.

[2] United Nations. Sustainable development goal 3, https://sustainabledevelopment.un.org/sdg3; 2016 [Last accessed 28 December 2017].

[3] World Health Organization. Child mortality and causes of death. In Global health observatory data, https://www.who.int/gho/child_health/mortality/en/; [Accessed 26 December 2017].

[4] UN Inter-agency Group for Child Mortality Estimation. Levels & trends in child mortality: report 2018, Estimates developed by the UN Inter-agency Group for Child Mortality Estimation, 2018.

[5] World Health Organization (WHO), United Nations Children's Fund (UNICEF), World Bank. State of the world's vaccines and immunization. 3rd ed. Geneva: WHO; 2009.

[6] Englund JA, Mbawuike IN, Hammill H, Holleman MC, Baxter BD, Glezen WP. Maternal immunization with influenza or tetanus toxoid vaccine for passive antibody protection in young infants. J Infect Dis 1993;168:647–56.

[7] Keller-Stanislawski B, Englund JA, Kang G, Mangtani P, Neuzil K, Nohynek H, et al. Safety of immunization during pregnancy: a review of the evidence of selected inactivated and live attenuated vaccines. Vaccine 2014;32:7057–64.

[8] World Health Organization. Maternal and neonatal tetanus elimination (MNTE)—the initiative and challenges, http://www.who.int/immunization/diseases/MNTE_initiative/en/. [Last seen 30 July 2018].

[9] Ortiz JR, Perut M, Dumolard L, Wijesinghe PR, Jorgensen P, Ropero AM, et al. A global review of national influenza immunization policies: analysis of the 2014 WHO/UNICEF joint reporting form on immunization. Vaccine 2016;34:5400–5.

[10] World Health Organization. How to implement influenza vaccination of pregnant women: an introduction manual for national immunization programme managers and policy makers, http://apps.who.int/iris/bitstream/10665/250084/1/WHO-IVB-16.06-eng.pdf; 2017.

[11] World Health Organization. Maternal immunization research and implementation portfolio, http://www.who.int/immunization/research/maternal_immunization/en/. [Accessed 4 January 2018].

[12] World Health Organization. Maternal immunization and antenatal care situation analysis (MIACSA), http://www.who.int/maternal_child_adolescent/epidemiology/mi-acsa-maternal-immunization/en/; 2017. [Last seen 19 June 2018].

[13] PATH. The advancing maternal immunization collaboration, https://www.path.org/resources/the-advancing-maternal-immunization-collaboration/; 2017. [Last seen 17 July 2018].

[14] PATH. Advancing RSV maternal immunization: a gap analysis report, https://www.path.org/resources/advancing-rsv-maternal-immunization-gap-analysis-report/; July 2018.

[15] Schofield FD, Tucker VM, Westbrook GR. Neonatal tetanus in New Guinea. Effect of active immunization in pregnancy. Br Med J 1961;2:785–9.

[16] World Health Organization. Vaccines against influenza WHO position paper—November 2012. Wkly Epidemiol Rec 2012;87:461–76.
[17] Iuliano AD, Roguski KM, Chang HH, Muscatello DJ, Palekar R, Tempia S, et al. Estimates of global seasonal influenza-associated respiratory mortality: a modelling study. Lancet 2018;391:1285–300.
[18] Nair H, Brooks WA, Katz M, Roca A, Berkley JA, Madhi SA, et al. Global burden of respiratory infections due to seasonal influenza in young children: a systematic review and meta-analysis. Lancet 2011;378:1917–30.
[19] Ziegler T, Mamahit A, Cox NJ. 65 Years of influenza surveillance by a WHO-coordinated global network. Influenza Other Respi Viruses 2018;12(5):558–65.
[20] de Francisco Shapovalova N, Donadel M, Jit M, Hutubessy R. A systematic review of the social and economic burden of influenza in low- and middle-income countries. Vaccine 2015;33:6537–44.
[21] Jit M, Newall AT, Beutels P. Key issues for estimating the impact and cost-effectiveness of seasonal influenza vaccination strategies. Hum Vaccin Immunother 2013;9:834–40.
[22] Ott JJ, Klein Breteler J, Tam JS, Hutubessy RC, Jit M, de Boer MR. Influenza vaccines in low and middle income countries: a systematic review of economic evaluations. Hum Vaccin Immunother 2013;9:1500–11.
[23] Peasah SK, Azziz-Baumgartner E, Breese J, Meltzer MI, Widdowson MA. Influenza cost and cost-effectiveness studies globally—a review. Vaccine 2013;31:5339–48.
[24] Chaiyakunapruk N, Kotirum S, Newall AT, Lambach P, Hutubessy RCW. Rationale and opportunities in estimating the economic burden of seasonal influenza across countries using a standardized WHO tool and manual. Influenza Other Respi Viruses 2018;12:13–21.
[25] Newall AT, Chaiyakunapruk N, Lambach P, Hutubessy RCW. WHO guide on the economic evaluation of influenza vaccination. Influenza Other Respi Viruses 2018;12:211–9.
[26] World Health Organization. Manual for estimating disease burden associated with seasonal influenza. World Health Organization; 2015.
[27] Pecenka C, Munthali S, Chunga P, Levin A, Morgan W, Lambach P, et al. Maternal influenza immunization in Malawi: piloting a maternal influenza immunization program costing tool by examining a prospective program. PLoS One 2017;12:e0190006.
[28] Fleming JA, Baltrons R, Rowley E, Quintanilla I, Crespin E, Ropero AM, et al. Implementation of maternal influenza immunization in El Salvador: experiences and lessons learned from a mixed-methods study. Vaccine 2018;36:4054–61.
[29] Ding H, Black CL, Ball S, Donahue S, Fink RV, Williams WW, et al. Influenza vaccination coverage among pregnant women—United States, 2014-15 influenza season. MMWR Morb Mortal Wkly Rep 2015;64:1000–5.
[30] World Health Organization. WHO recommendations on antenatal care for a positive pregnancy experience. 2016.
[31] Hirve S, Newman LP, Paget J, Azziz-Baumgartner E, Fitzner J, Bhat N, et al. Influenza seasonality in the tropics and subtropics—when to vaccinate? PLoS One 2016;11:e0153003.
[32] Hirve S, Lambach P, Paget J, Vandemaele K, Fitzner J, Zhang W. Seasonal influenza vaccine policy, use and effectiveness in the tropics and subtropics—a systematic literature review. Influenza Other Respi Viruses 2016;10(4):254–67.
[33] Lambach P, Alvarez AM, Hirve S, Ortiz JR, Hombach J, Verweij M, et al. Considerations of strategies to provide influenza vaccine year round. Vaccine 2015;33:6493–8.
[34] Ortiz JR, Neuzil KM. Influenza immunization of pregnant women in resource-constrained countries: an update for funding and implementation decisions. Curr Opin Infect Dis 2017;30:455–62.
[35] Centers for Disease Control and Prevention. Updated recommendations for use of tetanus toxoid, reduced diphtheria toxoid, and acellular pertussis vaccine (Tdap) in pregnant women—Advisory Committee on Immunization Practices (ACIP), 2012. MMWR Morb Mortal Wkly Rep 2013;62(7):131–5.

[36] Australian Technical Advisory Group on Immunisation. The Australian immunisation handbook. 2019. https://immunisationhandbook.health.gov.au/vaccine-preventable-diseases/pertussis-whooping-cough (Accessed September 16, 2019).

[37] Public Health England. Pertussis vaccination programme for pregnant women update: vaccine coverage in England, July to September 2017. London, UK: Public Health England Publications; 2017.

[38] Zuber P, Moran A, Chou D, et al. Mapping the landscape of global programmes to evaluate health interventions in pregnancy: the need for harmonised approaches, standards and tools. BMJ Glob Health 2018;e001053.

[39] World Health Organization. WHO preferred product characteristics for respiratory syncytial virus (RSV) vaccines. WHO/IVB/1711; 2017.

[40] Kobayashi M, Schrag SJ, Alderson MR, Madhi SA, Baker CJ, Sobanjo-Ter Meulen A, et al. WHO consultation on group B Streptococcus vaccine development: report from a meeting held on 27–28 April 2016. Vaccine 2016.

[41] UNICEF global databases 2014 boM, DHS and other national source, https://data.unicef.org/topic/maternal-health/antenatal-care/. [Last seen 19 July 2018].

[42] World Health Organization. National advisory committees on immunization, http://www-whoint/immunization/sage/national_advisory_committees. [Last viewed 26 March 2018].

[43] World Health Organization. Principles and considerations for adding a new vaccine to a national immunization programme: from decision to implementation and monitoring. 2014.

[44] World Health Organization. Maternal and neonatal immunization field guide for Latin America and the Caribbean. Washington, DC: PAHO; 2017.

[45] Wallace A, Dietz V, Cairns KL. Integration of immunization services with other health interventions in the developing world: what works and why? Systematic literature review. Trop Med Int Health 2009;14:11–9.

[46] Gerein N, Mayhew S, Lubben M. A framework for a new approach to antenatal care. Int J Gynaecol Obstet 2003;80:175–82.

[47] Abir T, Ogbo FA, Stevens GJ, Page AN, Milton AH, Agho KE. The impact of antenatal care, iron-folic acid supplementation and tetanus toxoid vaccination during pregnancy on child mortality in Bangladesh. PLoS One 2017;12:e0187090.

[48] Bonhoeffer J, Kochhar S, Hirschfeld S, Heath PT, Jones CE, Bauwens J, et al. Global alignment of immunization safety assessment in pregnancy—the GAIA project. Vaccine 2016;34:5993–7.

[49] Jones CE, Munoz FM, Kochhar S, Vergnano S, Cutland CL, Steinhoff M, et al. Guidance for the collection of case report form variables to assess safety in clinical trials of vaccines in pregnancy. Vaccine 2016;34:6007–14.

[50] Jones CE, Munoz FM, Spiegel HM, Heininger U, Zuber PL, Edwards KM, et al. Guideline for collection, analysis and presentation of safety data in clinical trials of vaccines in pregnant women. Vaccine 2016;34:5998–6006.

[51] Ding H, Black CL, Ball S, Donahue S, Fink RV, Williams WW, et al. Influenza vaccination coverage among pregnant women—United States, 2014-15 influenza season. MMWR Morb Mortal Wkly Rep 2015;64:1000–5.

[52] Krishnaswamy S, Wallace EM, Buttery J, Giles ML. Strategies to implement maternal vaccination: a comparison between standing orders for midwife delivery, a hospital based maternal immunisation service and primary care. Vaccine 2018;36:1796–800.

[53] Wiley KE, Cooper SC, Wood N, Leask J. Understanding pregnant women's attitudes and behavior toward influenza and pertussis vaccination. Qual Health Res 2015;25:360–70.

[54] Wilson RJ, Paterson P, Jarrett C, Larson HJ. Understanding factors influencing vaccination acceptance during pregnancy globally: a literature review. Vaccine 2015;33:6420–9.

[55] Yuen CY, Tarrant M. Determinants of uptake of influenza vaccination among pregnant women—a systematic review. Vaccine 2014;32:4602–13.

[56] Sakaguchi S, Weitzner B, Carey N, Bozzo P, Mirdamadi K, Samuel N, et al. Pregnant women's perception of risk with use of the H1N1 vaccine. J Obstet Gynaecol Can 2011;33:460–7.

[57] Fabry P, Gagneur A, Pasquier JC. Determinants of A (H1N1) vaccination: cross-sectional study in a population of pregnant women in Quebec. Vaccine 2011;29:1824–9.

[58] Myers KL. Predictors of maternal vaccination in the United States: an integrative review of the literature. Vaccine 2016;34:3942–9.

[59] Loubet P, Guerrisi C, Turbelin C, Blondel B, Launay O, Bardou M, et al. Influenza during pregnancy: incidence, vaccination coverage and attitudes toward vaccination in the French web-based cohort G-GrippeNet. Vaccine 2016;34:2390–6.

[60] Arriola CS, Vasconez N, Thompson M, Mirza S, Moen AC, Bresee J, et al. Factors associated with a successful expansion of influenza vaccination among pregnant women in Nicaragua. Vaccine 2016;34:1086–90.

[61] Donaldson B, Jain P, Holder BS, Lindsey B, Regan L, Kampmann B. What determines uptake of pertussis vaccine in pregnancy? A cross sectional survey in an ethnically diverse population of pregnant women in London. Vaccine 2015;33:5822–8.

[62] Healy CM, Rench MA, Montesinos DP, Ng N, Swaim LS. Knowledge and attitudes of pregnant women and their providers towards recommendations for immunization during pregnancy. Vaccine 2015;33:5445–51.

[63] Kowal SP, Jardine CG, Bubela TM. "If they tell me to get it, I'll get it. If they don't...": immunization decision-making processes of immigrant mothers. Can J Public Health 2015;106:e230–5.

[64] Winslade CG, Heffernan CM, Atchison CJ. Experiences and perspectives of mothers of the pertussis vaccination programme in London. Public Health 2017;146:10–4.

[65] Wilson RJCT, Lees S, Paterson P, Larson H. The patient—healthcare worker relationship: how does it affect patient views towards vaccination during pregnancy? In: Health and health care concerns among women and racial and ethnic minorities. Research in the sociology of health care, vol. 35. Emerald Publishing Limited; 2017. p. 59–77.

[66] Campbell H, Jan Van Hoek A, Bedford H, Craig L, Yeowell A-L, Green D, Yarwood J, Ramsay M, Amirthalingam G. Attitudes to immunisation in pregnancy among women in the UK targeted by such programmes. Br J Midwifery 2015;23(8):566–73.

[67] Meharry PM, Colson ER, Grizas AP, Stiller R, Vazquez M. Reasons why women accept or reject the trivalent inactivated influenza vaccine (TIV) during pregnancy. Matern Child Health J 2013;17:156–64.

[68] Arriola CS, Vasconez N, Bresee J, Ropero AM. Knowledge, attitudes and practices about influenza vaccination among pregnant women and healthcare providers serving pregnant women in Managua, Nicaragua. Vaccine 2018;36:3686–93.

[69] Khan AA, Varan AK, Esteves-Jaramillo A, Siddiqui M, Sultana S, Ali AS, et al. Influenza vaccine acceptance among pregnant women in urban slum areas, Karachi, Pakistan. Vaccine 2015;33:5103–9.

[70] Ding H, Black CL, Ball S, Fink RV, Williams WW, Fiebelkorn AP, et al. Influenza vaccination coverage among pregnant women—United States, 2016-17 influenza season. MMWR Morb Mortal Wkly Rep 2017;66:1016–22.

[71] Healy CM, Ng N, Taylor RS, Rench MA, Swaim LS. Tetanus and diphtheria toxoids and acellular pertussis vaccine uptake during pregnancy in a metropolitan tertiary care center. Vaccine 2015;33:4983–7.

[72] England PH. Vaccination against pertussis (Whooping cough) for pregnant women—2016. Information for healthcare professionals, https://assetspublishingservicegovuk/government/uploads/system/uploads/attachment_data/file/529956/FV_JUNE_2016_PHE_pertussis_in_pregnancy_information_for_HP_pdf; 2016. [Accessed 20 July 2018].

[73] SAGE. Report of the SAGE working group on vaccine hesitancy. 01 October 2014, http://www.who.int/immunization/sage/meetings/2014/october/1_Report_WORKING_GROUP_vaccine_hesitancy_final.pdf; 2014. [Accessed 10 October 2014].

[74] Bodeker B, Walter D, Reiter S, Wichmann O. Cross-sectional study on factors associated with influenza vaccine uptake and pertussis vaccination status among pregnant women in Germany. Vaccine 2014;32:4131–9.

[75] Bouder F, Way D, Lofstedt R, Evensen D. Transparency in Europe: a quantitative study. Risk Anal 2015;35:1210–29.
[76] Paterson P, Meurice F, Stanberry LR, Glismann S, Rosenthal SL, Larson HJ. Vaccine hesitancy and healthcare providers. Vaccine 2016;34:6700–6.
[77] Wiley KE, Massey PD, Cooper SC, Wood NJ, Ho J, Quinn HE, et al. Uptake of influenza vaccine by pregnant women: a cross-sectional survey. Med J Aust 2013;198:373–5.
[78] Berman CJ, O'Brien JD, Juarez L, Kahn R, Miller J, Zong M, Ariely D. Increasing vaccination. A behavioural science approach. [22 February 2018] Duke Centre of Advanced Hindsight; 2018.
[79] World Health Organization. The guide to tailoring immunization programmes (TIP) increasing coverage of infant and child vaccination in the WHO European region. World Health Organization Regional Office for Europe; 2013. http://wwweurowhoint/en/health-topics/communicable-diseases/influenza/vaccination. [Last viewed 31 July 2018].
[80] Bisset KA, Paterson P. Strategies for increasing uptake of vaccination in pregnancy in high-income countries: a systematic review. Vaccine 2018;36:2751–9.
[81] Wong VW, Lok KY, Tarrant M. Interventions to increase the uptake of seasonal influenza vaccination among pregnant women: a systematic review. Vaccine 2016;34:20–32.
[82] Ogburn T, Espey EL, Contreras V, Arroyo P. Impact of clinic interventions on the rate of influenza vaccination in pregnant women. J Reprod Med 2007;52:753–6.
[83] Meharry PM, Cusson RM, Stiller R, Vazquez M. Maternal influenza vaccination: evaluation of a patient-centered pamphlet designed to increase uptake in pregnancy. Matern Child Health J 2014;18:1205–14.
[84] Yudin MH, Salripour M, Sgro MD. Impact of patient education on knowledge of influenza and vaccine recommendations among pregnant women. J Obstet Gynaecol Can 2010;32:232–7.
[85] Wong VWY, Fong DYT, Lok KYW, Wong JYH, Sing C, Choi AY, et al. Brief education to promote maternal influenza vaccine uptake: a randomized controlled trial. Vaccine 2016;34:5243–50.
[86] Stockwell MS, Westhoff C, Kharbanda EO, Vargas CY, Camargo S, Vawdrey DK, et al. Influenza vaccine text message reminders for urban, low-income pregnant women: a randomized controlled trial. Am J Public Health 2014;104(Suppl 1):e7–12.
[87] Moniz MH, Hasley S, Meyn LA, Beigi RH. Improving influenza vaccination rates in pregnancy through text messaging: a randomized controlled trial. Obstet Gynecol 2013;121:734–40.
[88] Klatt TE, Hopp E. Effect of a best-practice alert on the rate of influenza vaccination of pregnant women. Obstet Gynecol 2012;119:301–5.
[89] Sherman MJ, Raker CA, Phipps MG. Improving influenza vaccination rates in pregnant women. J Reprod Med 2012;57:371–6.
[90] Pierson RC, Malone AM, Haas DM. Increasing influenza vaccination rates in a busy urban clinic. J Nat Sci 2015;1:e57.
[91] Panda B, Stiller R, Panda A. Influenza vaccination during pregnancy and factors for lacking compliance with current CDC guidelines. J Matern Fetal Neonatal Med 2011;24:402–6.
[92] Mouzoon ME, Munoz FM, Greisinger AJ, Brehm BJ, Wehmanen OA, Smith FA, et al. Improving influenza immunization in pregnant women and healthcare workers. Am J Manag Care 2010;16:209–16.
[93] Purdy LM. Are pregnant women fetal containers? Bioethics 1990;4:273–91.
[94] Verweij M, Lambach P, Ortiz JR, Reis A. Maternal immunisation: ethical issues. Lancet Infect Dis 2016;16:e310–4.
[95] Chamberlain AT, Lavery JV, White A, Omer SB. Ethics of maternal vaccination. Science (New York, NY) 2017;358:452–3.

PART II

Vaccines with current recommendations for use in pregnancy

CHAPTER

6

Tetanus

Inci Yildirim[a,b], *Saad B. Omer*[a,b,c]

[a]Division of Infectious Diseases, Department of Pediatrics, Emory University, Atlanta, GA, United States, [b]Department of Epidemiology, Rollins School of Public Health, Atlanta, GA, United States, [c]Hubert Department of Global Health, Rollins School of Public Health, Atlanta, GA, United States

Introduction

Maternal and neonatal tetanus is an important cause of maternal and neonatal mortality in many developing countries. The *Maternal and Neonatal Tetanus Elimination Program (MNTE)* launched by World Health Assembly aims to eliminate MNT through promotion of clean birth and neonatal care practices, as well as maternal immunization with tetanus toxoid. Immunization of pregnant women or women of childbearing age with tetanus toxoid is an inexpensive, effective and safe strategy to prevent maternal and neonatal tetanus.

Clinical burden of disease for women and infants, and epidemiology

Tetanus is caused by an exotoxin (tetanus toxin) produced by *Clostridium tetani,* and is unique among vaccine-preventable diseases in that it is not transmissible from person to person (not communicable) [1,2]. *C. tetani* is a Gram-positive, spore forming, anaerobic bacillus. If not exposed to sunlight, spores can persist in the soil for months to years. Since tetanus spores are widespread in the soil worldwide, complete eradication from environment is not possible. Therefore, herd immunity will not contribute

https://doi.org/10.1016/B978-0-12-814582-1.00007-3

to tetanus prevention through vaccination. Recovery from clinical tetanus does not result in immunity to tetanus toxin and those who recover can be infected again [3,4]. Therefore, continuing immunization against tetanus is needed to eliminate disease. Intestinal colonization has also been reported among different animals such as horses, dogs, small mammals and human beings. Spores are resistant to boiling and certain disinfectants.

Tetanus toxin, tetanospasmin, is one of the most potent neurotoxins that causes neurotransmitter blockade following inoculation of spores into the wounds [5,6]. It is estimated that the lethal dose to humans is less than 2.5 ng per kg. It is produced as a single polypeptide protoxin chain of 150,000 kDa molecular weight, and has affinity to the ganglioside containing receptors at the nerve termini. Peripheral neuromuscular junctions, autonomic nervous system, spinal cord and brain are affected forming a clinical syndrome similar to strychnine poisoning. Once inside the neurons, tetanospasmin cannot be neutralized with tetanus antitoxin. It does not cross brain-blood barrier, but is carried in via neuronal transport, and accumulates in the central nervous system [7]. Tetanus can present in different clinical patterns i.e. generalized, local, cephalic and neonatal. The diagnosis is usually based on clinical findings and epidemiological links such as history of a contaminated wound.

Maternal and neonatal tetanus (MNT) is preventable, but when MNT occurs, it is often fatal and is characterized by muscular rigidity and spasm [8,9]. Maternal tetanus is associated with abortion, miscarriages and unhygienic delivery conditions, whereas neonatal tetanus occurs secondary to poor post-partum cord care practices [8,10]. Both clinical syndromes cluster in poor, rural, resource-limited communities. Tetanus is vastly under-reported due to home births and suboptimum recording of births and deaths, particularly in early life. The World Health Organization (WHO) case definition for confirmed neonatal tetanus is "any neonate with the normal ability to suck and cry during the first 2 days of life, and who, between 3 and 28 days of age cannot suck normally, and becomes stiff or has spasms (i.e. jerking of the muscles)" [11]. Neonatal tetanus was estimated to be responsible for over half a million neonatal deaths globally in early 1980s [10]. In 1988, the WHO estimated that 787,000 newborns died of neonatal tetanus corresponding to approximately 6.7 deaths per 1000 live births. Without medical care, case fatality ratio for neonatal tetanus is close to 100%. Community based surveys in early 1980s reported neonatal mortality rates ranging from 50 to 110 per 1000 live births in low-middle income countries contemplating 23–72% of all neonatal deaths [12]. In contrast, in high-income countries neonatal tetanus incidence has been lower, and mortality rates were reported as 0.01 per 1000 live births [13]. Following widespread use of tetanus vaccination, maternal and neonatal tetanus have become exceedingly rare in high-income communities (0.01/100,000 in 2015 and 0/100,000 in 2016 in WHO Europe countries) [14–17].

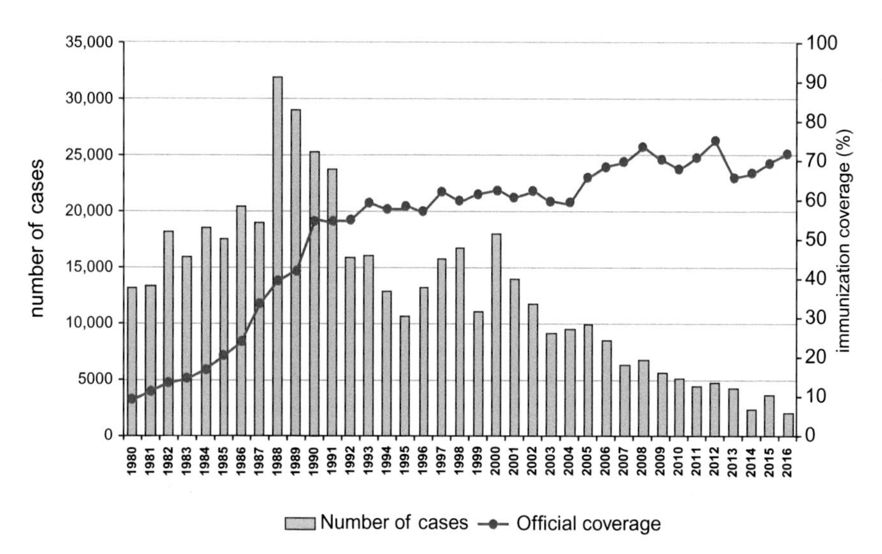

FIG. 1 Estimated global annual neonatal tetanus cases, and coverage with at least two doses of tetanus toxoid among pregnant women, 1985–2017. *Source: WHO/IVB database, 2017 Data as of 19 July 2017. http://www.who.int/immunization/monitoring_surveillance/burden/vpd/surveillance_type/active/neonatal_tetanus/en/.*

The World Health Assembly launched *The Neonatal Tetanus Elimination Program* in 1989 to eliminate maternal and neonatal tetanus by the year 2000 via promotion of clean birth and neonatal care practices [18]. This program was later expanded to involve maternal immunization, birth hygiene and surveillance; and renamed as the *Maternal and Neonatal Tetanus Elimination Program (MNTE)* [10,19]. Following these initiatives and increasing coverage with at least two doses of tetanus toxoid among pregnant women, the incidence of neonatal tetanus declined substantially worldwide (Fig. 1). The proportion of neonatal deaths caused by tetanus fell from 14% in 1993 to 2% in 2008, and number of deaths from tetanus among infants declined from 787,000 deaths to 34,019 in 2015, worldwide [10,20]. MNTE defines elimination as less than one case per 1000 live births in every district in every country. Thirty-four out of the 59 countries targeted had achieved elimination in 2014, and the most recent strategic plan aimed to achieve tetanus elimination in additional 11 countries (mainly in Asia and Africa) by 2015 (Fig. 2). By March 2018, there were still 14 countries that have not yet attained elimination. WHO estimates that in 2015, 34,019 newborns died from tetanus which represents a 96% reduction from the burden of tetanus-related mortality in the late 1980s.

In the United States, mortality from neonatal tetanus declined from 0.64 per 1000 live births in 1900 to 0.01 per 1000 live births in 1960s following the widespread use of tetanus toxoid vaccination in late 1940s [21]. In late 1960s, an average of 26 neonatal tetanus cases was reported each year,

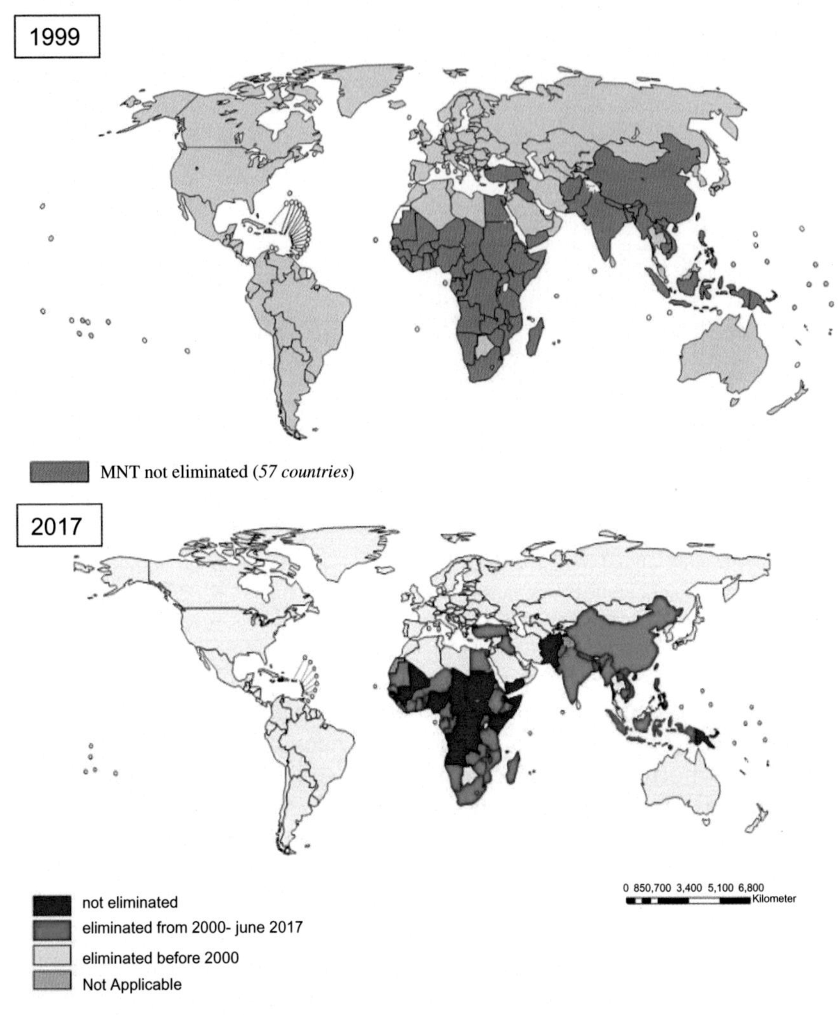

FIG. 2 Progress toward the global elimination of maternal and neonatal tetanus. The 14 countries where maternal and neonatal tetanus is still a public health problem include: Afghanistan, Angola, Central African Republic, Chad, Congo DR, Guinea, Mali, Nigeria, Pakistan, Papua New Guinea, Somalia, Sudan, South Sudan and Yemen. As of March 2018, 14 countries have not yet eliminated maternal and neonatal tetanus. *Source: http://www.who.int/immunization/diseases/MNTE_initiative/en/index4.html.*

whereas a total of 34 cases was reported over 4 decades between 1972 and 2012, and approximately 90% of these were infants born outside of the hospital [21–23].

Maternal tetanus is defined as tetanus during pregnancy or within 6 weeks of birth, miscarriage or abortion [11]. It was estimated that 15,000–30,000 cases occurred annually in 1993, and 27% of these were secondary to abortion [10]. Case fatality ratio was 52% (range 16–80) [24]. Maternal tetanus is still an important cause of mortality and morbidity in low-income communities [25], and was responsible for 5.4% of the maternal deaths in a tertiary medical center in Nigeria in 2008 [26].

Maternal and neonatal tetanus are associated with home births, unclean birthing surfaces, untrained birth attendants, and unhygienic tools used to cut the umbilical stump or abortion practices. Neonatal tetanus is more common in communities where traditional practices such as using animal dung for umbilical cord care or having domestic animals are prevalent. Approximately 60% of the births in low/middle-income countries occur with trained birth attendants [16]. China was able to eliminate maternal and neonatal tetanus in 2012 as a result of promotion of birth hygiene and increasing in-hospital delivery rates to 98% even without administering any tetanus toxoid after childhood [10]. Recently, India also has been declared free of maternal and neonatal tetanus in 2015, as a result of strong commitment to improve access to immunization, hygienic obstetric and postnatal care practices, and skilled birth attendance in the most vulnerable populations [27].

Current recommendations tetanus vaccination in pregnancy, and evidence of effectiveness/efficacy in women and infants

The MNTE initiative involves; (a) strengthening routine immunization of pregnant women with tetanus toxoid vaccine, (b) Supplementary Immunization Activities (SIAs) in selected high risk areas, targeting women of reproductive age with three properly-spaced doses of the vaccine, (c) promotion of clean deliveries and clean cord care practices, and (d) reliable surveillance [9–11]. Immunization of pregnant women or women of childbearing age is an inexpensive and efficacious strategy to prevent maternal and neonatal tetanus. If the mother is not immunized per WHO guidelines, neither she, nor the newborn, has protection against tetanus at delivery (Table 1). WHO has recommended that unimmunized pregnant women or those women without documentation of previous vaccination should receive two doses of tetanus toxoid given 4 weeks apart (Table 2). The first dose should be given as early as possible during pregnancy. For prevention of maternal tetanus that can be caused by the unclean birth practices, the last dose of tetanus toxoid must be given at least two weeks prior delivery. A total of 5 doses is considered sufficient for life-long immunity so further doses should be given during subsequent pregnancies or at least at intervals of at least 1 year. For all women of childbearing age in areas at high risk of neonatal tetanus, additional immunization with 3 doses of tetanus toxoid should be given. WHO also recommends vaccinating the mother as soon as possible if a case of neonatal tetanus is identified. In the United States, the Centers for Disease Control and Prevention

TABLE 1 WHO recommendations for tetanus immunization to prevent maternal and neonatal tetanus for women of childbearing age and pregnant women who were immunized during infancy, childhood or adolescence.

		Recommended immunizations	
Age at vaccination	Previous immunizations	At present contact/ pregnancy	Later (at intervals of at least 1 year)
Infancy	3 DTP	2 doses of TT/Td[a]	1 dose of TT/Td
Childhood	4 DTP	1 dose of TT/Td	1 dose of TT/Td
School age	3 DTP+1 DT/Td	1 dose of TT/Td	1 dose of TT/Td
School age	4 DTP+1 DT/Td	1 dose of TT/Td	None
Adolescence	4 DTP+1 DT at 4–6 years+1 TT/Td at 14–16 years	None	None

[a] *Minimum 4 weeks interval between doses.*
TT tetanus toxoid, Td tetanus toxoid-reduced diphtheria toxoid, DTP diphtheria-tetanus-pertussis vaccine.
Source: Galazka AM. The immunological basis for immunization series. Module 3: Tetanus. Geneva: World Health Organization; 1993 (WHO/EPI/GEN/93.13), p. 17.

TABLE 2 WHO recommendations for tetanus immunization to prevent maternal and neonatal tetanus for women of childbearing age and pregnant women without previous exposure to TT, Td or DTP.

Dose of TT or Td	When to give	Expected duration of protection
1	At first contact or as early as possible in pregnancy	None
2	At least 4 weeks after TT1	1–3 years
3	At least 6 months after TT2 or during subsequent pregnancy	At least 5 years
4	At least 6 months after TT3 or during subsequent pregnancy	At least 10 years
5	At least 6 months after TT4 or during subsequent pregnancy	For all childbearing age years and possibly longer

TT tetanus toxoid, Td tetanus toxoid-reduced diphtheria toxoid, DTP diphtheria-tetanus-pertussis vaccine.
Source: Core information for the development of immunization policy. 2002 update. Geneva: World Health Organization; 2002 (document WHO/V&B/02.28), p. 130.

(CDC) Advisory Committee on Immunization Practices (ACIP) has recommended using tetanus, reduced diphtheria toxoid, and acellular pertussis (Tdap) during the third or late second trimester (after 20 weeks' gestation) to women who have not previously received Tdap [28,29], in order to prevent pertussis among young infants.

The protection against neonatal tetanus gained by maternal immunization depends on the ability of the mother to produce antibody following vaccination and transplacental transfer of these antibodies to the fetus. Fetal immunoglobulin G (IgG) antibody levels start increasing from the fourth month of pregnancy until term, and neonates born to immunized women usually have a total tetanus antibody concentration equal to, or sometimes higher than the mother does. Tetanus vaccination produces protective antibody levels in more than 80% of the pregnant women after 2 doses lasting 1–3 years, although some studies indicate even longer protection (Table 2) [9,24,30]. The intervals between doses of tetanus toxoid as well as the interval between the last dose of tetanus and the delivery affect the tetanus antitoxin antibody level in the cord sera (Fig. 3). The ratio of antitoxin in cord sera to maternal sera increases with longer intervals between the doses and prolonged intervals between the second dose and delivery [4,12,32]. Starting tetanus toxoid immunization as early as possible in the pregnancy is essential to achieve an optimal immunological response and therefore clinical protection, especially in developing countries where women may present to health care late in pregnancy with no prior tetanus immunization. Reduction in passive antibody transfer has been reported in certain settings. Maternal HIV can cause up to 52% lower (95% CI, 30–67) placental transfer of tetanus antitoxin antibodies [33]. The reports on the impact of placental malaria infection on transplacental transfer have been conflicting. Several studies from the Republic of the Gambia [34] and the Republic of Malawi [35] found no effect, however in a recent study from Kenya, Cumberland et al. reported that tetanus antibody levels were lower by 48% (95% CI, 26–62) in newborns whose mothers had active-chronic or past placental malaria [33]. Hypergammaglobulinemia and prematurity are also associated with substantial reductions in transfer of anti-tetanus antibodies, which may be confounding factors in published studies [8,34,36].

Tetanus toxoid administered to pregnant women has been shown to be efficient and effective for protection against neonatal tetanus in many studies including field trials and hospital based studies (Table 3). In a non-randomized controlled study by Schofield et al. carried in New Guinea in 1959–1960, vaccine efficacy for tetanus toxoid administered during pregnancy was estimated as 94% for 3 doses, and 65% for 2 doses, whereas 1 dose had no effect against neonatal tetanus [37]. Toxoid injection during pregnancy was also shown to be safe and cost was not high. This study provided the initial evidence for the substantial protection against tetanus that is provided to the infants by active immunization of pregnant women. A double-blind, randomized controlled field trial from a rural area of the Colombia reported a neonatal tetanus mortality rate of 78 per 1000 live births, whereas no neonatal tetanus cases occurred in babies of mothers given two or three doses of tetanus toxoid (1961–1966)

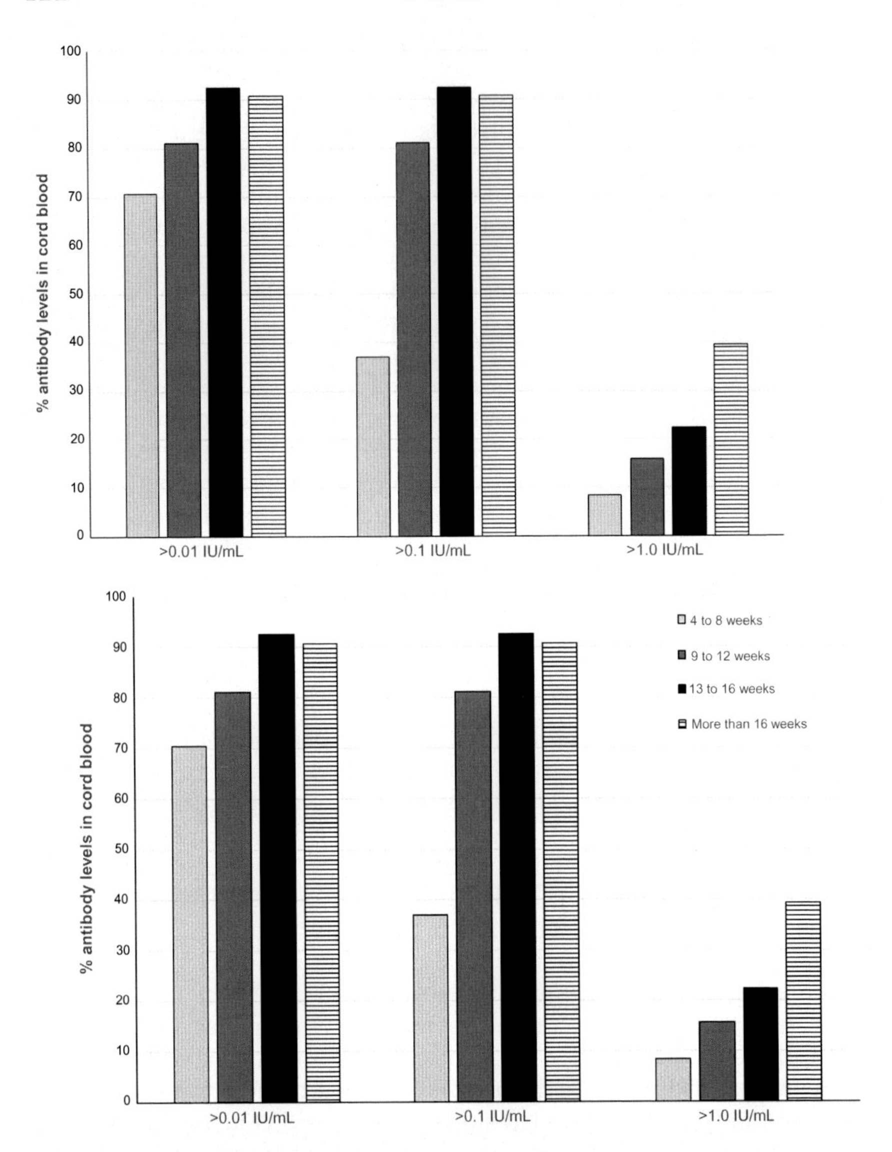

FIG. 3 Levels of tetanus antitoxin antibody in cord blood following two doses of maternal immunization with tetanus toxoid during pregnancy [31]. *Modified from Galazka AM. 2 doses how many weeks apart? The immunological basis for immunization series. Module 3: Tetanus. Geneva: World Health Organization; 1993 (WHO/EPI/GEN/93.13), p. 17.*

[38,39]. In a recent systematic review, Blencowe et al. estimated that immunization of pregnant women or women of childbearing age with at least two doses of tetanus toxoid reduces the mortality from neonatal tetanus by 94% (95% CI, 80–98) [43]. Prevention of neonatal tetanus depends on the transplacental transfer of specific IgG antibodies induced

TABLE 3 Impact of tetanus immunization in pregnant women on neonatal tetanus burden.

Year (Ref)	Study type/population	Outcome	Results		
1959 [37]	Nonrandomized, controlled trial New Guinea, 585 live births	NT	- 3 doses - 2 doses - 1 dose	94% 65% None	
1961–1966 [38, 39]	Double-blind field trial Colombia, 1618 women	Mortality from NT	No NT cases in infant born to women with >1 dose, whereas 78 deaths per 000 live births among controls		
1974 [30]	Observational study Bangladesh, 41,571 women	Mortality from NT	Suspected NT - 3 years post-imm. - 10 years post-imm. NT mortality - 10 years post-imm.	1 dose 91 (33–99) NS NS	2 doses 56 (17–76) 48 (3–73) 74 (23–91)
1985 [40]	Community-based survey Burma, 4257 live births	Mortality from NT	2 doses	66%	
1988 [41]	House-to-house survey Ethiopia, 2010 live births	Mortality from NT	2 doses	100%	
1988 [42]	Community-based survey India, 4344 live births	Mortality from NT	2 doses	79%	

NT neonatal tetanus, post-imm post-immunization.

by maternal immunization [4]. Duration of protection with specific tetanus vaccination schedules may vary, and there is not enough evidence for comparison. Historically, due to concerns of waning immunity over years tetanus boosters were recommended every 10 years in the US, and ACIP currently recommends a dose of tetanus containing vaccine (Tdap) during each pregnancy [29]. WHO does not recommend any further TT doses following the five doses given during childhood, and one booster given in adolescence unless indicated for wound management. Booster doses given in adolescents and adults are shown to induce high antibody levels that can persist for decades. The second booster during adolescence is thought to provide protection for women through their childbearing years. However, in some settings childhood and adolescent vaccination will be incomplete and vaccination is then indicated in pregnancy.

Evidence of safety of tetanus vaccination in pregnancy

Tetanus toxoid vaccine has an excellent safety profile in all age groups including pregnant women [44]. Mild local reactions are the most commonly reported adverse events after immunization. More severe reactions such as neurologic and hypersensitivity reactions are very rare [23,45]. Local reactions are reported in up to 95% of the recipients [23,44]. The frequency and severity of local adverse events has been reported to increase with increasing number of prior dosing [46]. In addition, subcutaneous administration is associated with more frequent local reactions compared to intramuscular administration. Aluminium adjuvants were reported to be linked to more frequent local erythema, induration and local pain in children, but not to serious reactions, and clinical significance of this association is unclear [47]. Tetanus toxoid has been used in combination with diphtheria toxoid and pertussis vaccines without causing significant increase in adverse effects [29,44]. Reactogenicity data indicate that minor adverse events such as swelling and pain occur more often after tetanus toxoid-diphtheria combination than tetanus toxoid alone [29,44,48]. Diphtheria-tetanus toxoid and pertussis (dTap, Tdap) immunization is also reported to be associated with increased frequency of local reactions, but is generally well tolerated, also in pregnancy. Systemic reactions to TT (i.e. fever, malaise, headache, irritability, lymphadenopathy) are less common than local reactions, and reported in 10% of the adults, and up to 25% of the infants who received tetanus toxoid in combined formulations [4,49–51]. Recipients with pre-existing high levels of antitoxin experience more severe fever accompanied with local reactions or malaise. Although exceedingly rare, neurologic reactions including brachial plexus palsy has been reported after tetanus toxoid immunization [47,52].

In 1994, the Vaccine Safety Committee of United States Institute of Medicine concluded that there is adequate evidence to suggest a causal relationship between the tetanus toxoid immunization and brachial plexus neuropathy, and estimated that 0.5–1 cases per 100,000 vaccine recipients was attributable to tetanus toxoid immunization [53]. However, passive surveillance data from United States for 1991 to 2003 suggested that the true frequency may be much less (0.69 cases per 10,000 doses) and initial data was limited to reach a conclusion. Guillain-Barré Syndrome is another severe adverse event that is rarely reported following tetanus toxoid injection. Pollard and Selby reported a 42-year-old patient from Australia who developed Guillain-Barré Syndrome in three separate occasions following each tetanus toxoid injection [54]. However, available data is nonconclusive with some of the following studies reporting similar or even lower frequency of Guillain-Barré Syndrome after tetanus toxoid immunization than expected background rates [55,56]. ACIP recommends that Guillain-Barré Syndrome occurring <6weeks after receipt of a tetanus toxoid–containing vaccine is a precaution for subsequent administration of tetanus toxoid–containing vaccines. Other neurological adverse events such as acute encephalopathy or seizures have been reported, but there is not adequate data to support a causal relationship. Allergic reactions following tetanus toxoid administration has been reported rarely. The Vaccine Safety Committee of United States Institute of Medicine has concluded in 2011 that there is adequate evidence regarding an association between tetanus toxoid vaccine and anaphylaxis based on six cases presenting temporality and clinical symptoms consistent with anaphylaxis [47].

Tetanus toxoid has been used extensively in pregnant women worldwide, and is considered safe and effective in pregnancy [57]. Initial studies from early 1960s, reported no adverse events among pregnant women receiving three injections of fluid formalinized tetanus toxoid during pregnancy, and substantial protection against neonatal tetanus [37]. Aggregated data from 68,770 infants in the Latin American Collaborative Study of Congenital Malformations (ECLAMC) failed to find any statistical association between any birth defect and tetanus toxoid vaccine [58].

Catindig et al. reviewed data from the National Maternal and Child Health Service on the number of pregnant women who had received tetanus toxoid and spontaneous abortions in the Philippines between 1980 and 1994, and in addition conducted a case control study with 418 pregnant women [59]. They found that even as tetanus toxoid coverage steadily increased over years, there was no similar trend in the number of abortions, and there was no significant association between tetanus toxoid administration and spontaneous abortion. In a recent review, there were 132 pregnant women reported to the Vaccine Adverse Event Reporting System (VAERS) who received tetanus toxoid in combination with reduced diphtheria toxoid, and acellular pertussis (Tdap) vaccine in January

1, 2005, through June 30, 2010 period and no adverse event was identified in 55 (42%) of the recipients [60]. Local reactions were the most frequent adverse event (4.5%). The most frequent pregnancy-specific adverse event was spontaneous abortion (16.7%) and one case with a major congenital anomaly (gastroschisis) was identified. No maternal or infant deaths were reported. The ACIP concluded that available data did not suggest any increased risk of clinically significant severe adverse events in pregnant women who received tetanus toxoid alone or in Tdap, and the potential benefit of preventing neonatal tetanus outweighs the theoretical concerns of possible severe adverse events [29,44,60]. ACIP also stated the need for safety studies of severe adverse events after multiple Tdap given in subsequent pregnancies. Safety of the combined Tdap vaccine has been studied worldwide with reassuring evidence that Tdap in pregnancy is safe (see Chapter 8).

Future prospects

Tetanus spores are ubiquitous in the environment and elimination of exposure to tetanus toxoid is impossible. Herd immunity through high coverage rates of infant and adolescent vaccination is therefore not possible to achieve and tetanus immunization programs continue to be needed even in developed countries. Although substantial improvement has been achieved since the WHO MNTE was established, by March 2018, there were still 14 targeted countries where maternal and neonatal tetanus is still a preventable public health problem. In addition, countries that have been declared to be free of disease are unlikely to have completely eradicated the disease and will be vulnerable to natural disasters or conflicts. Reinforcement of programs to promote hygienic birth practices, to improve coverage with primary series during childhood, and maternal immunization with tetanus toxoid must continue.

Conclusion

Maternal immunization with tetanus toxoid vaccine has potential benefits to mother and infant. Immunization of pregnant women or women of childbearing age is an inexpensive, very effective and safe strategy to prevent maternal and neonatal tetanus. Enhanced neonatal tetanus surveillance, and monitoring the impact of the immunization programs are essential for identification and implementation of targeted interventions to eliminate maternal and neonatal tetanus. Home births contribute to the underreporting of cases, and impact the accurate validation of elimination interventions. Improving local reporting systems with education of local community workers and traditional birth attendants should be promoted. Once elimination of maternal and neonatal tetanus has been achieved, maintaining high access to clean delivery and sustaining high routine immunization coverage with tetanus toxoid in pregnant women and women of childbearing age in addition to childhood vaccines are essential.

References

[1] Bleck TP, Brauner JS. Tetanus. In: Scheld WM, Whitley RJ, Marra CM, editors. Infections of the central nervous system. 3rd ed. Philadelphia, PA: Lippincott, Williams & Williams; 2004. p. 625–48.

[2] Hatheway CL, Johnson EA. Clostridium: the spore-bearing anaerobes. In: Collier L, Balows A, Sussman M, editors. Topley & Wilson's microbiology and microbial infections. 2. London: Arnold/Wiley; 1998. p. 731–82.

[3] World Health Organisation. Tetanus. Available from: http://www.who.int/immunization/diseases/tetanus/en/.

[4] Borrow R, Balmer P, Roper MH. The immunological basis for immunization series. Module 3: Tetanus update 2006. Geneva: World Health Organization; 2006.

[5] Lalli G, Bohnert S, Deinhardt K, Verastegui C, Schiavo G. The journey of tetanus and botulinum neurotoxins in neurons. Trends Microbiol 2003;11(9):431–7.

[6] Gill DM. Bacterial toxins: a table of lethal amounts. Microbiol Rev 1982;46(1):86–94.

[7] Caleo M, Schiavo G. Central effects of tetanus and botulinum neurotoxins. Toxicon 2009;54(5):593–9.

[8] Thwaites CL, Beeching NJ, Newton CR. Maternal and neonatal tetanus. Lancet 2015;385(9965):362–70.

[9] World Health Organisation. Standards for maternal and neonatal care; 2007.

[10] World Health Organisation. Maternal and neonatal tetanus elimination (MNTE). Available from: http://www.who.int/immunization/diseases/MNTE_initiative/en.

[11] World Health Organisation. WHO-recommended surveillance standard of neonatal tetanus. Available from: http://www.who.int/immunization/monitoring_surveillance/burden/vpd/surveillance_type/active/NT_Standards/en.

[12] Stanfield JP, Galazka A. Neonatal tetanus in the world today. Bull World Health Organ 1984;62(4):647–69.

[13] Heath Jr CW, Zusman J, Sherman IL. Tetanus in the United States, 19501960. Am J Public Health Nations Health 1964;54:769–79.

[14] (HFA-DB) WHOROfEEHfAD. Available from: https://gateway.euro.who.int/en/datasets/european-health-for-all-database/.

[15] Liang JL, Tiwari T, Moro P, Messonnier NE, Reingold A, Sawyer M, et al. Prevention of pertussis, tetanus, and diphtheria with vaccines in the United States: recommendations of the Advisory Committee on Immunization Practices (ACIP). MMWR Recomm Rep 2018;67(2):1–44.

[16] World Health Organisation. World health statistics; 2013.

[17] Pascual FB, McGinley EL, Zanardi LR, Cortese MM, Murphy TV. Tetanus surveillance—United States, 1998–2000. MMWR Surveill Summ 2003;52(3):1–8.

[18] World Health Assembly. Resolution 42.32: expanded programme on immunization, 1989. In: Handbook of resolutions and decisions of the World Health Assembly and the Executive Board (1985–1992). Geneva: World Health Organization; 1993.

[19] United Nations Children's Fund/World Health Organization/United Nations Population Fund. Maternal and neonatal tetanus elimination by 2005: Strategies for achieving and maintaining elimination, Nov 2000. Geneva: World Health Organization; 2002.

[20] Boschi-Pinto C, Young M, Black RE. The Child Health Epidemiology Reference Group reviews of the effectiveness of interventions to reduce maternal, neonatal and child mortality. Int J Epidemiol 2010;39(Suppl. 1):i3–6.

[21] Hinman AR, Foster SO, Wassilak SG. Neonatal tetanus: potential for elimination in the world. Pediatr Infect Dis J 1987;6(9):813–6.

[22] Centers for Disease Control Prevention. Neonatal tetanus—Montana, 1998. MMWR Morb Mortal Wkly Rep 1998;47(43):928–30.

[23] Roper M, Wassilak SG, Scobie HM, Ridpath AD, Orenstein WA. Tetanus toxoid. In: Plotkin SA, Orenstein WA, Offit PA, Edwards KM, editors. Plotkin's vaccines. 7th ed. Philedelphia, PA: Elsevier; 2018. p. 1052–79.

[24] Fauveau V, Mamdani M, Steinglass R, Koblinsky M. Maternal tetanus: magnitude, epidemiology and potential control measures. Int J Gynaecol Obstet 1993;40(1):3–12.

[25] Namasivayam A, Osuorah DC, Syed R, Antai D. The role of gender inequities in women's access to reproductive health care: a population-level study of Namibia, Kenya, Nepal, and India. Int J Womens Health 2012;4:351–64.

[26] Okusanya BO, Aigere EO, Abe A, Ibrahim HM, Salawu RA. Maternal deaths: initial report of an on-going monitoring of maternal deaths at the Federal Medical Centre Katsina, Northwest Nigeria. J Matern Fetal Neonatal Med 2013;26(9):885–8.

[27] World Health Organisation. WHO congratulates India on maternal and neonatal tetanus elimination. Available from: http://www.searo.who.int/mediacentre/features/2015/maternal-and-neonatal-tetanus-elimination/en/.

[28] Centers for Disease Control and Prevention. Updated recommendations for use of tetanus toxoid, reduced diphtheria toxoid and acellular pertussis vaccine (Tdap) in pregnant women and persons who have or anticipate having close contact with an infant aged 12 months—Advisory Committee on Immunization Practices (ACIP), 2011. Morb Mortal Wkly Rep 2011;60:1424–6.

[29] Centers for Disease Control and Prevention. Updated recommendations for use of tetanus toxoid, reduced diphtheria toxoid and acellular pertussis vaccine (Tdap) in pregnant wome—Advisory Committee on Immunization Practices (ACIP), 2011. Morb Mortal Wkly Rep 2013;60:131–5.

[30] Koenig MA, Roy NC, McElrath T, Shahidullah M, Wojtyniak B. Duration of protective immunity conferred by maternal tetanus toxoid immunization: further evidence from Matlab, Bangladesh. Am J Public Health 1998;88(6):903–7.

[31] Galazka A. The immunological basis for immunization series module 3: Tetanus. Geneva: World Health Organization; 1993.

[32] Stanfield JP, Gall D, Bracken PM. Single-dose antenatal tetanus immunisation. Lancet 1973;1(7797):215–9.

[33] Cumberland P, Shulman CE, Maple PA, Bulmer JN, Dorman EK, Kawuondo K, et al. Maternal HIV infection and placental malaria reduce transplacental antibody transfer and tetanus antibody levels in newborns in Kenya. J Infect Dis 2007;196(4):550–7.

[34] Okoko JB, Wesumperuma HL, Hart CA. The influence of prematurity and low birthweight on transplacental antibody transfer in a rural West African population. Trop Med Int Health 2001;6(7):529–34.

[35] de Moraes-Pinto MI, Verhoeff F, Chimsuku L, Milligan PJ, Wesumperuma L, Broadhead RL, et al. Placental antibody transfer: influence of maternal HIV infection and placental malaria. Arch Dis Child Fetal Neonatal Ed 1998;79(3):F202–5.

[36] Gendrel D, Richard-Lenoble D, Massamba MB, Picaud A, Francoual C, Blot P. Placental transfer of tetanus antibodies and protection of the newborn. J Trop Pediatr 1990;36(6):279–82.

[37] Schofield FD, Tucker VM, Westbrook GR. Neonatal tetanus in New Guinea. Effect of active immunization in pregnancy. Br Med J 1961;2(5255):785–9.

[38] Newell KW, Leblanc DR, Edsall G, Levine L, Christensen H, Montouri MH, et al. The serological assessment of a tetanus toxoid field trial. Bull World Health Organ 1971;45(6):773–85.
[39] Newell KW, Duenas Lehmann A, LeBlanc DR, Garces Osorio N. The use of toxoid for the prevention of tetanus neonatorum. Final report of a double-blind controlled field trial. Bull World Health Organ 1966;35(6):863–71.
[40] Stroh G, Kyu UA, Thaung U, Lwin UK. Measurement of mortality from neonatal tetanus in Burma. Bull World Health Organ 1987;65(3):309–16.
[41] Maru M, Getahun A, Hosana S. A house-to-house survey of neonatal tetanus in urban and rural areas in the Gondar region, Ethiopia. Trop Geogr Med 1988;40(3):233–6.
[42] Kumar V, Kumar R, Mathur VN, Raina N, Bhasin M, Chakravarty A. Neonatal tetanus mortality in a rural community of Haryana. Indian Pediatr 1988;25(2):167–9.
[43] Blencowe H, Lawn J, Vandelaer J, Roper M, Cousens S. Tetanus toxoid immunization to reduce mortality from neonatal tetanus. Int J Epidemiol 2010;39(Suppl. 1):i102–9.
[44] Centers for Disease Control and Prevention. Prevention of pertussis, tetanus, and diphtheria with vaccines in the United States: recommendations of the Advisory Committee on Immunization Practices (ACIP). Morb Mortal Wkly Rep 2018;67(2):1–44.
[45] Myers MG, Beckman CW, Vosdingh RA, Hankins WA. Primary immunization with tetanus and diphtheria toxoids. Reaction rates and immunogenicity in older children and adults. JAMA 1982;248(19):2478–80.
[46] Peebles TC, Levine L, Eldred MC, Edsall G. Tetanus-toxoid emergency boosters: a reappraisal. N Engl J Med 1969;280(11):575–81.
[47] Stratton KR, Ford A, Rusch E, Clayton EW. Diphtheria toxoid–, tetanus toxoid–, and acellular pertussis-containing vaccines. Washington, DC: Committee to Review Adverse Effects of Vaccines; Institute of MEdicine; 2011.
[48] Macko MB, Powell CE. Comparison of the morbidity of tetanus toxoid boosters with tetanus-diphtheria toxoid boosters. Ann Emerg Med 1985;14(1):33–5.
[49] Levine L, Edsall G. Tetanus toxoid: what determines reaction proneness? J Infect Dis 1981;144(4):376.
[50] Lloyd JC, Haber P, Mootrey GT, Braun MM, Rhodes PH, Chen RT, et al. Adverse event reporting rates following tetanus-diphtheria and tetanus toxoid vaccinations: data from the Vaccine Adverse Event Reporting System (VAERS), 1991-1997. Vaccine 2003;21(25–26):3746–50.
[51] Mallet E, Belohradsky BH, Lagos R, Gothefors L, Camier P, Carriere JP, et al. A liquid hexavalent combined vaccine against diphtheria, tetanus, pertussis, poliomyelitis, *Haemophilus influenzae* type B and hepatitis B: review of immunogenicity and safety. Vaccine 2004;22(11–12):1343–57.
[52] Hamati-Haddad A, Fenichel GM. Brachial neuritis following routine childhood immunization for diphtheria, tetanus, and pertussis (DTP): report of two cases and review of the literature. Pediatrics 1997;99(4):602–3.
[53] Stratton KR, Howe CJ, Johnston RB. Diphtheria and tetanus toxoids. Washington, DC: Institute of Medicine Vaccine Safety Committee; 1994.
[54] Pollard JD, Selby G. Relapsing neuropathy due to tetanus toxoid. Report of a case. J Neurol Sci 1978;37(1–2):113–25.
[55] Rantala H, Cherry JD, Shields WD, Uhari M. Epidemiology of Guillain-Barre syndrome in children: relationship of oral polio vaccine administration to occurrence. J Pediatr 1994;124(2):220–3.
[56] Tuttle J, Chen RT, Rantala H, Cherry JD, Rhodes PH, Hadler S. The risk of Guillain-Barre syndrome after tetanus-toxoid-containing vaccines in adults and children in the United States. Am J Public Health 1997;87(12):2045–8.

[57] Omer SB, Jamieson DJ. Maternal immunization. In: Plotkin SA, Orenstein WA, Offit PA, Edwards KM, editors. Plotkin's vaccines. 7th ed. Philedelphia, PA: Elsevier; 2018. p. 567–79.
[58] Silveira CM, Caceres VM, Dutra MG, Lopes-Camelo J, Castilla EE. Safety of tetanus toxoid in pregnant women: a hospital-based case-control study of congenital anomalies. Bull World Health Organ 1995;73(5):605–8.
[59] Catindig N, Abad-Viola G, Magboo F, Roces MC, Dayrit M. Tetanus toxoid and spontaneous abortions: is there epidemiological evidence of an association? Lancet 1996;348(9034):1098–9.
[60] Zheteyeva YA, Moro PL, Tepper NK, Rasmussen SA, Barash FE, Revzina NV, et al. Adverse event reports after tetanus toxoid, reduced diphtheria toxoid, and acellular pertussis vaccines in pregnant women. Am J Obstet Gynecol 2012;207(1):59. e1–7.

CHAPTER

7

Influenza

Deshayne B. Fell[a,b,c], Milagritos D. Tapia[d,e], Marta C. Nunes[f,g]

[a]School of Epidemiology and Public Health, University of Ottawa, Ottawa, Canada, [b]Children's Hospital of Eastern Ontario Research Institute, Ottawa, Canada, [c]ICES, Ottawa, Canada, [d]Center for Vaccine Development, Bamako, Mali, [e]University of Maryland, School of Medicine, Center for Vaccine Development and Global Health, Baltimore, MD, United States, [f]Medical Research Council: Respiratory and Meningeal Pathogens Research Unit, Faculty of Health Sciences, University of the Witwatersrand, Johannesburg, South Africa, [g]Department of Science and Technology/National Research Foundation: Vaccine Preventable Diseases Unit, University of the Witwatersrand, Johannesburg, South Africa

Introduction

Influenza infection affects all age groups and causes mild to severe illness. The World Health Organization (WHO) estimates that during normal seasonal epidemics 5–15% of the population is typically infected by influenza viruses, with 3 to 5 million cases of severe illness and up to 650,000 influenza-associated deaths per year globally [1,2]. The highest burden of severe illness is concentrated in those aged <1 year or ≥65 years, adults infected with human immunodeficiency virus (HIV) and pregnant women [3–5]. Influenza infection attack rates and disease severity vary considerably from season to season and across world regions. In tropical areas, influenza circulates throughout the year albeit with peaks; in temperate regions, influenza is highly seasonal and typically circulates during winter months.

Although influenza vaccines have less than optimal effectiveness against influenza infection, vaccination nevertheless remains the best strategy to prevent influenza illness. Current seasonal influenza vaccines

https://doi.org/10.1016/B978-0-12-814582-1.00008-5

need to be reformulated annually, in order to elicit a protective antibody response that recognizes viral genetic variants that are predicted to arise through antigenic drift. There is currently no influenza vaccine approved by regulators in any country for use in infants <6 months old. Infants can, however, be protected from influenza illness in their first months of life via maternal influenza vaccination during pregnancy [6–9].

Influenza virus

Influenza viruses belong to the orthomyxoviridae family and have a segmented, negative-sense, single-stranded RNA genome with helical symmetry and different size ribonucleoproteins. The influenza virus genus is divided into three types—A, B and C—as defined by the antigenicity of the nucleoproteins and matrix proteins in the viral core [10]. Influenza-A viruses are further divided into subtypes according to the antigenic properties of the surface glycoproteins haemagglutinin and the neuraminidase. Eighteen antigenically different haemagglutinins (H1–18) and eleven different neuraminidases (N1–11) have been identified, and their combination designates the virus subtype. Currently, H1 and H3 subtypes of influenza-A are endemic in humans. Circulating influenza-B viruses are classified into two groups, i.e., Yamagata-like or Victoria-like lineages [11]. Influenza-A and -B viruses are responsible for seasonal epidemics and current vaccines are designed to target these viruses. Influenza type C viruses are detected much less frequently and cause mild infections, not presenting significant public health concerns. Influenza types differ in the range of animal hosts that they can infect; humans are the only known reservoir of influenza-B and -C, whereas influenza-A is also found in other animals [12]. Given the multiple subtypes, higher mutation rates and diverse hosts, influenza-A poses the greatest pandemic threat. People are susceptible to influenza throughout their lives, due to the virus's ability to continually mutate by antigenic drift and antigenic shift (only influenza-A viruses) mechanisms [13].

While inactivated influenza vaccines (IIV) and live attenuated influenza vaccines are the most widely available vaccines, recombinant influenza vaccines have recently been developed and, in the United States of America (US), a vaccine produced in insect cells was approved by the Food and Drug Administration on October 2016 [14]. The cell-based products may have improved match to circulating influenza strains as they avoid egg-adaption issues. Traditionally, influenza vaccines (both IIV and live attenuated) have been produced to protect against three different seasonal influenza viruses; however, recently, vaccines protecting against four different viruses, including both influenza B lineage viruses (quadrivalent vaccines), have become available in most industrialized countries.

Clinical burden of disease and epidemiology

Clinical burden of influenza illness during pregnancy

A systematic review of studies published up until February 2015 by Katz et al. reporting laboratory-confirmed influenza outcomes in pregnant women, found a limited number of studies, a wide range of incidence estimates, and limited representation from low- and middle-income countries [15]; (Fig. 1). Among the nine eligible studies identified [6,16–23], five reported data exclusively from the 2009 H1N1 pandemic period [16–18,22,23], three covered seasonal influenza epidemic time periods [6,19,21], only one was conducted in a middle-income country [6], and no studies were identified from low-income countries. Among three older sero-epidemiological studies reporting maternal influenza virus infection (including asymptomatic infection) in the United Kingdom (UK) and US [19–21], rates ranged from 483 cases per 10,000 pregnant women (95% CI: 399–614) [19] to 1,097 per 10,000 pregnant women (95% CI: 957–1258) [21], in the two UK studies. Rates of symptomatic influenza virus infection were reported by three studies [6,17,23] and ranged from 0.10 per 10,000 pregnant women (95% CI: 0.07–0.14) in the US [23] to 486 per 10,000 pregnant women (95% CI: 375–630) in South Africa [6]. The latter estimate

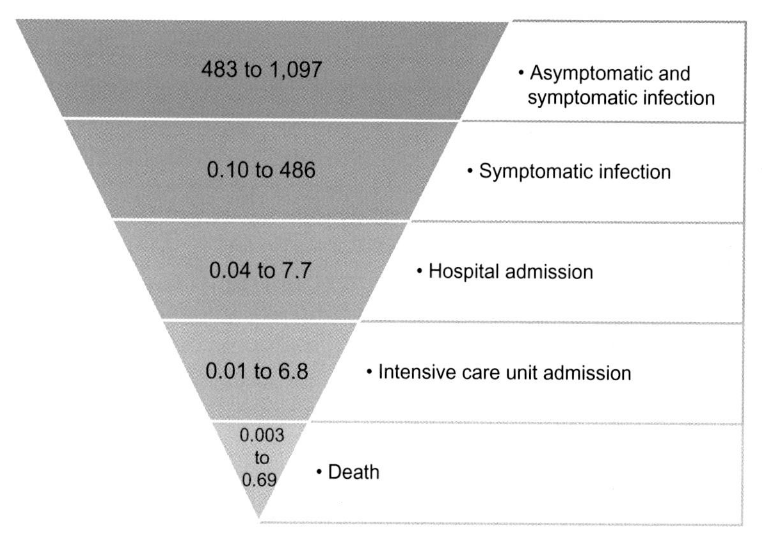

FIG. 1 Published incidence rates for laboratory-confirmed influenza outcomes among pregnant women. Numbers represent cases per 10,000 pregnant women. *Data as reported in a review by Katz MA, Gessner BD, Johnson J, Skidmore B, Knight M, Bhat N, et al. Incidence of influenza virus infection among pregnant women: a systematic review. BMC Pregnancy Childbirth 2017;17(1):155 and Fell DB, Azziz-Baumgartner E, Baker MG, Batra M, Beaute J, Beutels P, et al. Influenza epidemiology and immunization during pregnancy: final report of a World Health Organization working group. Vaccine 2017;35(43):5738–50.*

was from the combined placebo-groups (pregnant women living with and without HIV) from a randomized controlled trial (RCT) of trivalent IIV conducted in South Africa [pregnant women living without HIV: 361 per 10,000 (95% CI: 259–490); pregnant women living with HIV: 1,702 per 10,000 (95% CI: 1008–2705)] [6]. Rates of influenza-associated hospital admission (reported by four case series studies [17,18,22,23]) ranged from 0.04 per 10,000 pregnant women (95% CI: 0.02–0.07) [23] in the US to 7.7 per 10,000 pregnant women (95% CI: 6.7–8.7) [18] in the UK, and estimates of influenza-associated intensive care unit (ICU) admissions (reported by four case series studies [16,17,22,23]) ranged from 0.01 per 10,000 pregnant women (95% CI: 0.00–0.03) [23] to 6.8 per 10,000 pregnant women (95% CI: 5.2–8.8) [16]. Although influenza-associated mortality was included as an outcome in the review, the number of cases reported by any given study was low (between one and eight deaths) yielding variable rates ranging from 0.003 cases per 10,000 pregnant women (95% CI: 0–0.021) [23] to 0.69 per 10,000 pregnant women (95% CI: 0.26–1.51) [16].

Since the Katz et al. systematic review was published [15], additional studies have reported incidence rate estimates for influenza during pregnancy. Regan et al. performed a retrospective cohort study to evaluate the epidemiology of seasonal influenza during pregnancy in Western Australia, during the 2012–2014 influenza season period [24]. This population-based study, using a data linkage of notified laboratory-confirmed influenza cases with the routine state-wide perinatal data collection, reported an incidence of 0.22 laboratory-confirmed influenza cases per 10,000 pregnancies [24]. A similar retrospective cohort study among linked perinatal and laboratory databases in New Zealand, between 2012 and 2015, reported an adjusted incidence rate of laboratory-confirmed influenza hospitalization among pregnant women of 56 per 10,000 women-weeks (95% CI: 40–73) [25]. Two additional RCTs of IIV during pregnancy were also published — the first, a prospective phase IV RCT conducted in Mali, reported an incidence of laboratory-confirmed influenza type A and type B among subjects randomized to the control group of 328 and 182 per 10,000 person-years of follow-up, respectively [8]. The second RCT, conducted in Nepal, reported an incidence rate of laboratory-confirmed influenza infection among pregnant women in the placebo group of 571 per 10,000 person-years [7].

While the Katz et al. systematic review was limited to studies reporting laboratory-confirmed influenza outcomes, it is worth noting the availability of other studies that have reported non-laboratory confirmed influenza outcomes among pregnant women. In general, these studies often report higher rates, likely due to the use of non-specific diagnostic codes often used in health administrative databases [i.e., International Classification of Disease (ICD) codes]. The clinical illnesses identified by such codes may or may not have been confirmed by laboratory testing and, as such, may misclassify non-specific influenza-like illnesses (ILI) as influenza. For example, a Canadian study estimated that one hospitalization per 1,000 healthy pregnant women was attributed to influenza (as identified from the diagnosis codes in a population-based hospitalization database) [26]. Rates of influenza-associated mortality have also been reported in ecological studies using statistical modeling approaches. For example, in an ecologic modeling study conducted using Vital Statistics data from 1999 to 2009 in South Africa, the mean annual seasonal influenza–associated mortality rate was estimated to be 1.26 per 10,000 person-years (95% CI: 0.72–1.8); and 1.93 per 10,000 person-years (95% CI: 1.1–2.76) during the 2009 H1N1 influenza pandemic [5].

Overall, robust estimates of laboratory-confirmed influenza outcomes among pregnant women are limited by the low number of studies, along with the predominance of data from small case series, from high-income settings, and from the 2009 H1N1 pandemic time period [27].

Clinical burden of influenza illness among infants under 6 months of age

Infants under 6 months of age are at high risk of severe influenza and associated complications; they have high rates of influenza-associated hospitalization [28–32] and mortality [33]. However, as they are not eligible for influenza vaccination, maternal immunization is considered the primary strategy to protect this age group. Immunization of pregnant women has been shown to reduce influenza virus infection among young infants through transplacental transfer of maternal influenza antibodies [6–9]. To fully appreciate the impact that maternal immunization might have on infant influenza, age-specific data over the first year of life are needed [27].

In 2011, estimates of the global incidence of influenza outcomes among children under the age of 5 years demonstrated that influenza in young children results in significant utilization of health services, particularly among infants younger than one year [34]. Age-specific estimates for infants under 6 months were not included in the report [34], yet are necessary to inform evidence-based decision-making regarding vaccination programs, evaluate the impact of maternal immunization programs, provide of appropriate health services, and prioritize future research. In 2015, the WHO published a manual to guide countries on how to conduct influenza surveillance [35]. In addition to other age-specific groupings, the guide suggested reporting influenza burden data in the 0 to <2 years and 2 to <5 years age groups and, when feasible, to report among those <2 years of age in the following groups: 0 to <6 months, 6 months to <1 year of age, and 1 to <2 years.

In 2014, a WHO working group systematically reviewed the evidence and estimated incidence rates of laboratory-confirmed influenza outcomes, including ambulatory and hospital settings among infants less than 6 months of age, in 27 studies that met the eligibility criteria [36]; Fig. 2. Subsequent to the WHO review, more recently, influenza surveillance data from diverse settings worldwide were highlighted [37]. Though 18 studies representing 16 different countries were presented, data for infants less than 6 months of age were only reported from Kenya [38]. Additional recent studies presenting incidence rates in this age group include data from Nepal and South Africa [7,39].

The majority of studies included in the WHO systematic review [36] reported laboratory-confirmed influenza outcomes among infants under 6 months of age originating from the US [4,30,31,33,40–46], with fewer from lower-middle-income [32,38,47–49], upper-middle-income [6,39,50–55], other non-US high-income [56–59] and low-income countries [7,8] (see Table 1 in Fell et al. [36]). The studies assessing influenza in a variety of settings predominantly used reverse transcription-polymerase chain reaction (RT-PCR) as the laboratory testing method.

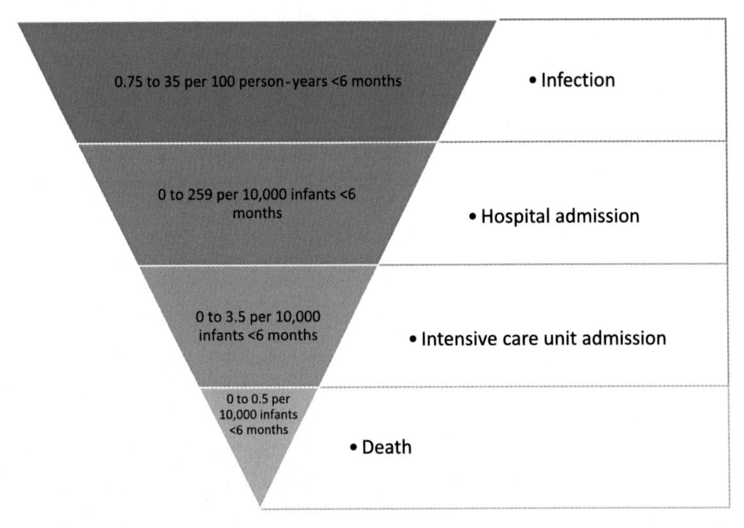

FIG. 2 Published incidence rates for laboratory-confirmed influenza outcomes among infants <6 months of age. *Data as reported in a review by Fell DB, Johnson J, Mor Z, Katz MA, Skidmore B, Neuzil KM, et al. Incidence of laboratory-confirmed influenza disease among infants under 6 months of age: a systematic review. BMJ Open 2017;7(9):e016526 and Fell DB, Azziz-Baumgartner E, Baker MG, Batra M, Beaute J, Beutels P, et al. Influenza epidemiology and immunization during pregnancy: final report of a World Health Organization working group. Vaccine 2017;35(43):5738–50.*

Laboratory-confirmed influenza illness in ambulatory care settings

Community-based active household surveillance conducted between 2009 and 2011, in the Cajamarca region of Peru, found an adjusted incidence of laboratory-confirmed influenza illness among infants less than 6 months of age of 35 per 100 person-years of follow-up (95% CI: 26–48) [50]. Using similar active surveillance methods, the RCT from South Africa reported an incidence of laboratory-confirmed influenza illness among infants born to women living without HIV in the placebo group of 3.6% (95% CI: 2.6–5.0) [6], and the RCT from Mali reported an incidence of 8.3 per 100 person-years among infants in the control arm [8]. Over three influenza seasons in the Suzhou District of China, the incidence of laboratory-confirmed influenza illness among infants under 6 months ranged from 2.3 per 100 in 2013–2014 to 2.9 per 100 in 2012–2013 [55]. In Kenya, surveillance data collected at the Siaya County Referral hospital in the Nyanza region from 2012–2014 were used to extrapolate an overall average annual national rate of non-hospitalized influenza-associated severe acute respiratory infections of 358.7 per 100 000 persons [38]. Finally, data from the New Vaccine Surveillance Network in the US estimated a rate of laboratory-confirmed influenza illness, based on outpatient clinic visits among infants under 6 months, of 2.8 per 100 infants (95% CI: 0.7–11.1) in 2002–2003 and 5.9 per 100 infants (95% CI: 2.8–12.8) in 2003–2004 [31].

Laboratory-confirmed influenza hospitalization

Of 25 available studies included in the WHO review that provided estimates of laboratory-confirmed influenza hospitalization for infants less than 6 months of age, 10 originated from the US. Rates of laboratory-confirmed influenza hospitalization during seasonal epidemics varied from a low of 9.3 per 10,000 infants (95% CI: 7.9–10.9) in 2006–2007 [4] to a high of 91.2 per 10,000 infants (95% CI: 67–145) in 2003–2004 [43] (see Fig. 2 and Supplementary Table S3 in Fell et al. [36] for details of individual studies).

The 15 non-US studies reported similar laboratory-confirmed influenza hospitalization rates for seasonal influenza. Most incidence rates ranged from 6.2 per 10,000 infants (95% CI: 3.1–9.3) in China in 2007 [52] to 73.0 per 10,000 infants (95% CI: 40.6–121.7) in Spain in 2003–2004 [32]. However, a higher rate was reported from one post-pandemic study of seasonal influenza from China [250 per 10,000 infants (95% CI: 213–292) under 6 months in 2010–2011] [54]. The highest estimate from non-US based studies from the 2009 pandemic H1N1 influenza time period was 259 per 10,000 person-years (95% CI: 97.0–689) in Kenya [49], and the only US-based estimate for the 2009 pandemic H1N1 time period was 20.2 per 10,000 infants (95% CI: 18.1–22.5) [41].

Laboratory-confirmed influenza and intensive care unit admissions

Laboratory-confirmed influenza ICU admission rates for infants under 6 months are available from seven studies [31,40,44,52,53,58,59]. All rates were, however, computed by the WHO review authors after publication, either due to non-reporting in the original study or due to graphical presentation of rates in a figure only [36]. Estimated rates of laboratory-confirmed influenza ICU admission for seasonal influenza ranged from 0.5 per 10,000 infants (95% CI: 0.8–16.5) between 2000–2001 and 2003–2004 in the Salt Lake City area of the US [31] to 3.5 per 10,000 (95% CI: 1.7–6.4) between 2001 and 2004 in the surveillance counties covered by the New Vaccine Surveillance Network [40]. Nonetheless, the absolute number of laboratory-confirmed influenza ICU admissions of infants under 6 months was very low in all study populations. During the 2009 H1N1 pandemic period, the rate of laboratory-confirmed influenza ICU admission was 2.9 per 10,000 infants (95% CI: 1.6–5.0) in Argentina [53] and 2.5 per 10,000 infants (95% CI: 0.79–6.0) in Israel [59].

Laboratory-confirmed influenza death

Nine studies included in the WHO systematic review included laboratory-confirmed influenza death among infants under 6 months of age as an outcome [4,6,8,32,33,40,48,53,57]. In six of these, no laboratory-confirmed influenza deaths were observed (see Table 5 in Fell et al. [36] for details) [6, 32, 40, 48, 57]. In the 2003–2004 season, enhanced national-level surveillance of pediatric laboratory-confirmed influenza deaths in the US found a rate of 0.88 per 100,000 infants (95% CI: 0.52–1.39)—the highest among all pediatric age groups up to 18 years [33]. In a smaller surveillance study using data from the Emerging Infections Program operating in 10 American states, three influenza deaths of infants under 6 months were recorded during 2003–2004 to 2007–2008 combined, resulting in a rate of 0.41 per 100,000 person-years (95% CI: 0.11–1.12) [4]. Among all nine studies, the highest rate of laboratory-confirmed influenza deaths in infants was reported in Buenos Aires, Argentina, for the 2009 pandemic H1N1 time period and was 5 per 100,000 infants (95% CI: 0.82–16.1) [53].

Overall, limited data for influenza-confirmed outcomes for infants under 6 months are available, particularly from non-US settings. Worldwide representative incidence data are necessary to completely evaluate influenza disease burden and the potential impact of maternal influenza on morbidity and mortality in young infants.

Influenza during pregnancy and birth outcomes

The risk of preterm birth, small-for-gestational-age birth, and fetal death among women with clinical influenza disease and/or laboratory-confirmed influenza virus infection during pregnancy, compared with women with no influenza during pregnancy, was assessed in a WHO-initiated systematic review of 21 comparative studies published up to December 2014 [60]. Thirteen of the included studies were from pre-2009 seasonal epidemic time periods [19,21,61–71], five exclusively encompassed the 2009 H1N1 pandemic period [17,72–75], two reported on both [76,77] and only one reported on influenza seasons after the 2009 H1N1 pandemic [6].

Fifteen studies assessed the risk of preterm birth following influenza illness during pregnancy [17,61–68,72–77]. Out of six studies considered to have high methodological quality [17,63,72,73,75,77], two found that mothers with severe H1N1 2009 pandemic influenza (H1N1pdm09) infection (i.e., illness requiring hospitalization) were more likely to give birth to preterm infants, with adjusted odds ratios (aOR) of 2.39 (95% CI: 1.64–3.49) [17] to 4.00 (95% CI: 2.71–5.90) [72]. No significant association with preterm birth was reported for less severe H1N1pdm09 illness or for seasonal influenza. The pooled adjusted estimate for the risk of small-for-gestational-age birth across five studies [21,63,73,74,77] indicated no significant association with maternal influenza infection (pooled aOR: 1.24, 95% CI: 0.96–1.59) [60]. Although nine studies reported on the risk of fetal death [21,61,67,69–72,74,75], heterogeneity in study methods (including definitions of fetal death) and small sample sizes precluded meta-analysis of the results [60]. Of two studies considered high quality, that were conducted during the 2009 H1N1 pandemic [72,75], one reported a 4.2-times higher odds of fetal death for hospitalized women with severe maternal influenza (95% CI: 1.4–12.4) [72], while the other reported an increased risk of fetal death for pregnant women with mild-to moderate influenza illness severity [hazard ratio (HR): 1.91, 95% CI: 1.07–3.41] [75].

Subsequent to the systematic review [60], three observational studies have evaluated the impact of maternal influenza on birth outcomes [24,78,79]. Regan et al. reported similar risk of adverse birth outcomes between women with laboratory-confirmed influenza infection (only 14% of whom were admitted to hospital for influenza) and the general obstetrical population. The only exception was a 4.0% (95% CI: 0.3–7.6%) reduction in the mean percent of optimal birth weight for infants born to women with laboratory-confirmed influenza B infection, compared to infants born to the general population [24]. A retrospective cohort study from Ontario, Canada evaluated the association between clinical H1N1pdm09 illness (i.e., influenza-coded health care encounters, of which 5% were influenza-associated hospital admissions) and preterm birth [78]. In the overall obstetrical population, no increased risk of preterm birth was observed for pregnant women with documented H1N1pdm09 illness, compared to women with no influenza [adjusted hazard ratio (aHR): 1.0, 95% CI: 0.88–1.02]. However, in a sub-group analysis of women with pre-existing medical conditions such as asthma, diabetes, and cardiopulmonary conditions, H1N1pdm09 illness was associated with an increased risk of preterm birth (aHR:1.5, 95% CI: 1.1–2.2) [78]. Lastly, Newsome et al. conducted a matched retrospective cohort study comparing outcomes, such as low birth weight, preterm birth, small-for-gestational-age birth and low Apgar scores, among infants born to women with laboratory-confirmed H1N1pdm09 illness, with two unexposed groups: 1) infants born to women without H1N1pdm09 during the same year as the exposed group, and 2) infants born to women during the year prior to the 2009 H1N1 pandemic [79]. Most of the women (420/490; 86%) with laboratory-confirmed H1N1pdm09 infection were hospitalized. In this study, infants born to women with laboratory-confirmed H1N1pdm09 were more likely to be preterm [adjusted risk ratio (aRR): 1.4, 95% CI: 1.1–1.9 for prior year comparisons; aRR: 1.7, 95% CI: 1.3–2.2 for same year comparisons] and to have low Apgar scores (aRR: 2.3, 95% CI: 1.3–3.9 for prior year pregnancies; aRR: 4.0, 95% CI: 2.1–7.6 for same year pregnancies). No differences were seen among the groups for the other infant outcomes. However, in a sub-group analysis, infants born to women with severe H1N1pdm09 (i.e., 62 women who were admitted to intensive care), were more likely to be preterm (aRR: 3.9, 95% CI: 2.7–5.6 for same year pregnancies), with low birth weight (aRR: 4.6, 95% CI: 2.9–7.5 for same year pregnancies) and have lower Apgar scores (aRR: 8.7, 95% CI: 3.6–21.2 for same year pregnancies) [79].

Overall, the current evidence indicates that pregnant women with severe H1N1pdm09 illness, particularly those women who required hospitalization, had an increased risk of preterm delivery and fetal death. Robust evidence on milder H1N1pdm09 illness and on seasonal influenza illness during pregnancy is much more limited, and findings from the existing studies are conflicting.

Current recommendations for influenza vaccination during pregnancy

Recommendations for vaccination of pregnant women against influenza have existed since the 1960s in the US [80], and since 2004, the Advisory Committee on Immunization Practices (ACIP) and the American College of Obstetricians and Gynecologists (ACOG) have recommended that women who will be pregnant during the influenza season should receive the IIV in any trimester of pregnancy [81,82]. The WHO has also recommended influenza vaccination for all pregnant women since 2005 [83]. In contrast, most European countries introduced seasonal influenza vaccination for pregnant women only after the 2009 H1N1 influenza pandemic. Moreover, in 2012, the WHO recommended that pregnant women should be prioritized above other groups for influenza vaccination in countries considering the initiation or expansion of their programs for seasonal influenza vaccination [84].

Although the initial recommendations for influenza vaccination during pregnancy were based on limited safety and immunogenicity studies, a number of observational studies and RCTs have since corroborated the safety of this strategy, including on fetal outcomes. Also, observational studies and RCTs have demonstrated the effectiveness of maternal influenza vaccination in protecting the women and their infants from influenza illness [6–9,75,85–90].

Evidence of effectiveness of influenza vaccination in pregnancy

Efficacy of influenza vaccination during pregnancy against influenza in women

Four RCTs from Bangladesh, South Africa, Mali and Nepal, have investigated the effect of seasonal influenza vaccination during pregnancy in preventing illness, both in the mothers during pregnancy up to 6 months post-delivery and in the infants [6–9], (Table 1). In all trials, active weekly surveillance was performed until 6 months post-partum to assess clinical respiratory symptoms in the mothers and the infants following birth. In the earlier trial in Bangladesh, 340 pregnant women in the third

TABLE 1 Efficacy of maternal influenza vaccination in preventing influenza illness in the women until 6 months post-partum.

Period, country	Control group	Population	Outcomes	Vaccine efficacy
2004–2005 Bangladesh [9]	23-valent pneumococcal vaccine	IIV—172 Control—168	Febrile respiratory illness	36% (95%CI: 4–57%)
2011–2012 South Africa [6]	Saline placebo	IIV—1062 AR: 1.8% Control—1054 AR: 3.6%	PCR-confirmed influenza	50% (95%CI: 15–71%)
2011–2013 Mali [8]	Meningococcal vaccine	IIV—2108 AR: 0.5% Control—2085 AR: 1.9%	PCR-confirmed influenza	70% (95%: 42–86%)
2011–2013 Nepal [7]	Saline placebo	IIV—1847 AR: 1.7% Control—1846 AR: 2.4%	PCR-confirmed influenza	31% (95%CI: −10%, 56%)

IIV: inactivated influenza vaccine; *AR*: attack rate; *PCR*: polymerase chain reaction.

trimester were randomized to receive either pneumococcal polysaccharide vaccine or IIV, from August 2004 through May 2005. In this study, the 2004 IIV Southern Hemisphere vaccine formulation was used. While vaccine efficacy against laboratory-confirmed influenza was only assessed in the infants, mothers who received IIV had a reduced rate of respiratory illness with fever of 36% (95% CI: 4–57) [9]. The trial from South Africa included 2116 HIV-uninfected pregnant women in their second or third trimester randomized to IIV or placebo in 2011 and 2012. It was conducted in Soweto, an urban Black-African township outside of Johannesburg, and used the influenza vaccines recommended for the Southern Hemisphere, which had the same formulation for both 2011 and 2012 influenza seasons [6]. During the follow-up period, 19 episodes of PCR-confirmed influenza were detected in the IIV-group compared to 38 cases in the women who received placebo, resulting in a vaccine efficacy of 50.4% (95% CI: 14.5–71.2) in the mothers [6]. In Mali, West-Africa, 4193 third trimester pregnant women were randomly assigned to receive IIV or quadrivalent meningococcal vaccine [8]. Continuous recruitment occurred from September 2011 to April 2013, and both Northern and Southern Hemisphere vaccine formulations were used. An overall vaccine efficacy of 70.3% (95%CI: 42.2–85.8) against PCR-confirmed influenza in the women was reported [8]. The last RCT was undertaken in Nepal, where influenza viruses circulate perennially. In this trial, two consecutive annual cohorts of pregnant women were enrolled from April 2011 to April 2012 (cohort 1) and from April 2012 through September 2013 (cohort 2) [7]. In total, 3693 pregnant women (17–34 weeks' gestation) were randomized to IIV or placebo. The incidence of maternal PCR-confirmed influenza was lower in the vaccine-group than in the placebo-group in cohort 1, for a vaccine efficacy of 45% (95% CI: 1–70), but not in cohort 2 or the two cohorts combined (31%, 95% CI: -10, 56). In Nepal there were, however, fewer cases of ILI in vaccinated women compared to the placebo-group (vaccine efficacy: 19%, 95% CI: 1–34) in the two cohorts combined. The authors acknowledged that the lack of effect on PCR-confirmed influenza illness might be due to the fact that they required fever as part of the diagnosis of maternal ILI [7].

Two other observational studies evaluated the vaccine effectiveness of either seasonal [90] or monovalent H1N1pdm09 [75] vaccines in preventing influenza disease during pregnancy. A test-negative case-control study, over two influenza seasons from 2010 to 2012 of IIV in the US, estimated the adjusted vaccine effectiveness in pregnancy to be between 44% and 53% against acute respiratory illness associated with PCR-confirmed influenza [90]. Similarly, in Norway, a nationwide registry-based retrospective cohort study reported that vaccination with pandemic influenza vaccines reduced the risk of a clinical diagnosis of influenza (with or without laboratory-confirmation) among pregnant women by 70% [75].

The four RCTs performed to date were not designed to assess the effect of maternal vaccination on influenza-associated hospitalizations. The effectiveness of influenza vaccination in preventing severe illness associated with hospitalization during pregnancy has, however, been estimated by two observational studies [91,92]. The first was a retrospective cohort study conducted in Australia during the 2012 and 2013 influenza seasons, which investigated the association of IIV during pregnancy and hospital attendance for acute respiratory illness [91]. Overall, 34,701 pregnant women were included and maternal vaccination resulted in an 81% (95% CI: 31–95) reduction in emergency department visits and 65% (95% CI: 3–87%) reduction in inpatient hospital admissions during influenza season. Laboratory-confirmed influenza hospitalizations were also lower in the vaccinated group compared to the unvaccinated group, but the difference was not significant (aHR: 0.16, 95% CI: 0.01–1.76) [91]. Furthermore, in an international multi-site study using administrative data from healthcare systems with integrated laboratory, medical, and vaccination records in four countries (Australia, Canada, Israel, and the US), Thompson et al. identified pregnant women whose pregnancies overlapped with the local influenza seasons from 2010 to 2016 [92]. Although 19,450 hospitalizations for acute respiratory or febrile illness among pregnant women were noted, based on ICD codes on the hospital record, only 6% had a clinician-ordered PCR testing for influenza viruses. Adjusted overall vaccine effectiveness, assessed using a test-negative design, was 40% (95% CI: 12–59) against PCR-confirmed influenza-associated hospitalization during pregnancy. A sub-analysis by trimester at admission found a significant effect of 55% among women who were hospitalized in their first and second trimester. Interestingly, this study, using a broad acute respiratory or febrile illness case definition that included other diagnoses besides the typical acute respiratory illnesses, suggested that there may be a larger vaccine-preventable PCR-confirmed influenza burden among hospitalized pregnant women [92].

Efficacy of influenza vaccination during pregnancy against influenza in infants

There is currently no influenza vaccine approved by regulators and recommended in any country for use in infants <6 months old, even if they have the same high-risk conditions underlying recommendations for influenza vaccination in older children. Infants can, however, be protected from influenza illness in their first months of life by maternal antibodies transferred via the placenta during pregnancy and possibly through breastmilk post-partum [6–9]. The four RCTs described above and numerous observational studies have demonstrated infant protection from maternal influenza vaccination.

In the trial conducted in Bangladesh, vaccine efficacy against laboratory-confirmed influenza in the infants was 63% (95% CI: 5–85); in addition, there was a significant reduction in respiratory illnesses with fever, with a vaccine efficacy of 29% (95% CI: 7–46) [9]. In the South African trial, IIV administered during pregnancy resulted in a 49% (95% CI: 12–70) reduction in PCR-confirmed influenza in the infants up to 24 weeks of age [6]. In Mali, during the entire study period, an overall vaccine efficacy of 33% (95% CI: 4–54) was demonstrated in infants <6 months old [8]. In the placebo randomized controlled trial in Nepal, the infant vaccine efficacy was 30% (95% CI: 5–48) [7]. In a meta-analysis of the four trials, the calculated pooled efficacy of maternal influenza vaccination in preventing laboratory-confirmed influenza in infants <6 months old was 36% (95% CI: 22–48) [93], Fig. 3. Moreover, three observational studies, two from the US and one from England, also evaluated the effect of maternal IIV in preventing medically-attended laboratory-confirmed influenza in infants <6 months old [86,89,94]. These three studies reported a significant protective effect that was generally

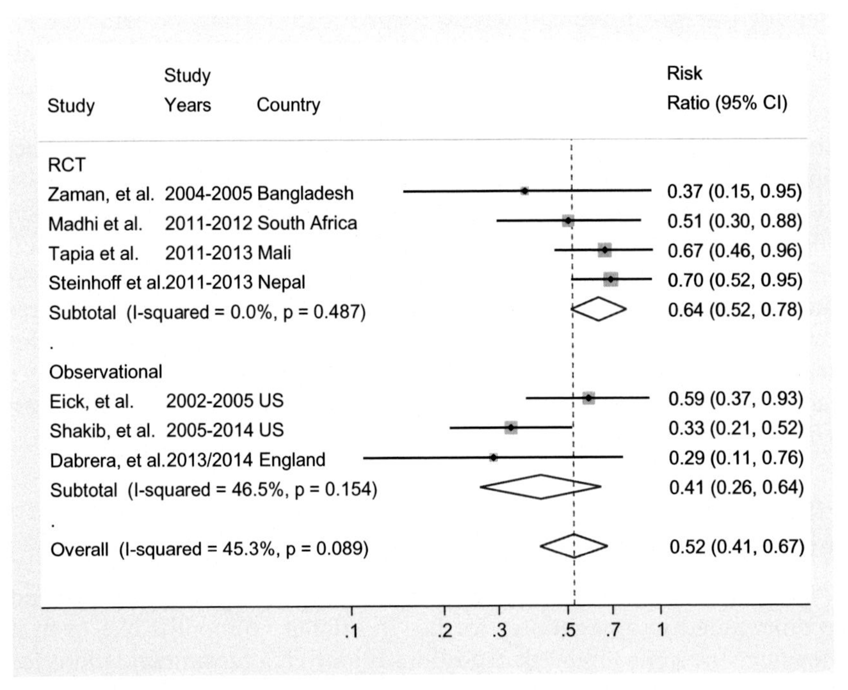

FIG. 3 Forest plot of influenza vaccination during pregnancy in preventing laboratory-confirmed influenza infection in infants younger than 6 months stratified by study type. RCT: randomized control trials. Dotted line represents the overall measure of effect. *Source: Nunes MC, Madhi SA. Influenza vaccination during pregnancy for prevention of influenza confirmed illness in the infants: a systematic review and meta-analysis. Human Vaccines Immunother 2017.*

higher than that reported in the clinical trials (71–41% reduction in the risk of laboratory-confirmed influenza for infants born to vaccinated women). The pooled effect of all clinical trials and observational studies was a 48% (95% CI: 33–59) against influenza-confirmed infection among infants whose mothers received vaccine during pregnancy [93].

To measure the impact of influenza vaccination during pregnancy, on infant hospitalizations that could be precipitated by influenza, infection is of crucial public health importance. Although the RCTs did not have individual power to detect an effect of maternal vaccination on infant hospitalizations for laboratory-confirmed influenza, in a post-hoc analysis of the South African trial, vaccination was associated with 57% (95% CI: 7–81) reduction in all-cause acute lower respiratory tract infection among infants <3 months old [95]. Notably, this observation was independent of identifying influenza virus among the hospitalized cases, suggesting that preventing influenza infection may prevent subsequent lower respiratory tract infection in young infants. These data were subsequently corroborated in the Nepal study [i.e., 31% (95% CI: 6–50) lower rate of severe pneumonia], although not in the Malian study. A pooled analysis across the three RCTs yielded an overall vaccine efficacy of 20% (95% CI: 1–34) against severe pneumonia [96]. Additionally, four observational studies covering 14 different influenza seasons, reported on the impact of maternal vaccination on laboratory-confirmed influenza hospitalizations in infants [85,86,88,89]. The pooled vaccine effectiveness of influenza vaccination during pregnancy in preventing infant laboratory-confirmed influenza hospitalizations was 72% (95% CI: 39–87) [93].

Immunogenicity of influenza vaccines during pregnancy

The most common method used to determine responses following influenza vaccination is the hemagglutination-inhibition (HAI) assay. This assay measures the concentration of antibody required to prevent influenza virus from agglutinating red blood cells. A titer of 1:40 has been recognized as protective based on a 50% reduction in disease in healthy adults [97]. Thus, normally, the term seroprotection refers to those individuals with a titer of 1:40 or higher, while seroconversion refers to an increase in HAI titer of at least 4-fold from pre- to post-vaccination.

Pregnant women mount robust immune responses to influenza vaccination. In the four RCTs, vaccinated women had a wide range of significant increases in antibody titres to all vaccine strains, with HAI titer increases in Bangladesh and Nepal ranging from 2 to 18-times from pre-IIV to delivery [7,98], and in South Africa, HAI fold-increases of 6 to 10-times were observed from pre-IIV to one month post-IIV [6].

It has been hypothesized that influenza vaccination may result in less robust immune responses during pregnancy due to the alterations in immune function. The immune responses to influenza vaccination in pregnant women were directly compared to those in non-pregnant women in a few small studies [99–105]. Two studies from the 1960s and 70s found that pregnant and non-pregnant women achieved similar antibody titers post-vaccination with either polyvalent [102] or monovalent [104] vaccines. More recently, Christian et al. also reported similar seroconversion (70–74% for A/H1N1, 59–63% for A/H3N2, and 63–74% for influenza-B) and seroprotection rates (85–89% for A/H1N1, 81–93% for A/H3N2, and 83–100% for influenza-B) in pregnant and non-pregnant women, who received IIV during the 2011–2012 influenza season [101]. Adequate immune responses to IIV were also detected in pregnant and post-partum women, who were vaccinated as part of their routine clinical care from 2006 to 2010, with a non-significant decreased response being observed just in the first trimester [103]. On the other hand, while Kay et al. found similar post-vaccination HAI titers in both pregnant and non-pregnant women in 2012–2013, pregnant women achieved higher fold-increases for A/H1N1 and influenza-B; in this study, however, 100% of the non-pregnant women had been vaccinated in the previous year in comparison to only 43% of the pregnant women [105]. While these results indicate that pregnancy does not appear to compromise influenza vaccines immunogenicity, slightly lower responses to vaccination in pregnant compared to non-pregnant women were detected in two studies [99,100,106]. Schlaudecker et al. found that, although the percentage of participants who seroconverted and achieved seroprotective levels were similar in pregnant and non-pregnant women, pregnant women had lower post-vaccination titers, with similar high previous vaccination history and comparable baseline titers between the two groups [99,106]. Additionally, when the authors assessed immune responses according to the gestational timing of vaccination, they detected a gradual decline in titers with progression of pregnancy, with an average 4% decline with every additional week of gestational age [106]. Bischoff et al. also compared the immune responses in pregnant and non-pregnant women conferred by a MF59-adjuvanted vaccine [100]. Despite high seroconversion rates detected in both groups (95–98%), the number of non-pregnant women who seroconverted was 2.5-fold higher than among pregnant women, suggesting that the humoral immune responses to this vaccine in pregnant women might be diminished [100].

As described above, influenza vaccines provide only partial protection against infection. Attempts have been made to use adjuvants or higher doses of antigen to improve influenza vaccine immunogenicity. During the 2009 H1N1 pandemic, multiple studies evaluated the immunogenicity of monovalent H1N1pdm09 non-adjuvanted [107–110] or adjuvanted vaccines [100,111] during pregnancy. Two studies from Japan, that used two doses of monovalent H1N1pdm09 non-adjuvanted vaccines in 149 and 128 pregnant women each across the three pregnancy trimesters, reported seroconversion rates of 82% and 91%. In both of these studies, the second vaccination conferred little additional benefit [109,110]. Two studies in Europe, evaluating the antibody responses in women immunized with an H1N1pdm09 MF59-adjuvanted vaccine during pregnancy from gestational week 20 [100] or during the third trimester [111], reported 96% and 100%, respectively, seroprotection post-vaccination. However, while seroprotection levels were maintained by 100% of the women 5 months post-delivery in one study, only 82% displayed this protection three months post-vaccination in the other study [100,111]. In an RCT, Jackson et al. evaluated whether a higher hemagglutinin content of an H1N1pdm09 non-adjuvanted vaccine would benefit pregnant women [107]. In this study, 120 pregnant women, in their second or third trimester, were randomized to receive two doses of H1N1pdm09 vaccines containing either 25 μg or 49 μg of hemagglutinin. Following the first vaccination, HAI titers ≥1:40 were detected in 93% of the women who received the 25 μg dose; the second vaccination also did not significantly increase the antibody levels, and there was no benefit in using the higher hemagglutinin concentration [107]. More recently, the immunogenicity of two seasonal IIVs, one non-adjuvanted and one adjuvanted (the immunoadjuvant being Polyoxidonium), were compared in 37 and 42 pregnant Russian women, respectively [112]. Post-vaccination HAI titer ≥1:40 were detected in 65–95% of women after immunization with the non-adjuvanted vaccine and in 73–90% of participants after the administration of the Polyoxidonium-containing vaccine. Comparing the response to the two vaccines, only for A/H3N2 post-vaccination seroprotective levels were lower in the non-adjuvanted group [112].

Transplacental antibody transfer

Influenza vaccination during pregnancy, with either trivalent or monovalent vaccines, results in efficient transplacental antibody transfer from the mother to the fetus, with detectable rise in newborn antibody levels to the vaccine strains starting approximately two weeks after maternal vaccination [113,114]. Good correlation between maternal antibodies and cord-blood or infant blood antibodies has been shown in multiple studies [107,108,115,116]. Due to the active transport process, fetal antibody concentrations can equal that of the mother and, in some cases, exceed maternal levels at term gestation. Studies have, however, reported a wide range of infant-to-maternal antibody ratios with studies that evaluated the immunogenicity of monovalent H1N1pdm09 vaccines [107,108], normally detecting higher ratios (1.4–2.9) compared to the RCTs that used IIV, where ratios of HAI antibodies ranged from 0.7 to 0.8 in the South African trial and between 0.8 and 1.1 in the trial from Bangladesh [6,98]. One study surprisingly reported a transplacental antibody transfer of only 0.55 after H1N1pdm09 MF59-adjuvanted vaccination [111].

The amount of maternal antibodies transferred to the fetus is influenced by the interval between immunization and delivery [107,113]. In South Africa, higher transplacental transfer was associated with longer intervals between vaccination and delivery, with cumulative transfer of antibodies to the fetus (0.004 titer increase for each day of pregnancy after vaccination) [116]. Conversely, a study using a H1N1pdm09 vaccine observed a trend toward lower cord-blood titers with longer intervals between the time of vaccination and delivery [107]. Eick et al. found, that for the eight influenza strains tested, there was no difference in the cord-blood titers between the 123 women vaccinated in the second trimester and the 390 in the third trimester [94].

Despite influenza vaccination at, any time during pregnancy, conferring some protection to the infants, it has been suggested that vaccination needs to occur at least 15 days before birth, for a significant increase in antibody titers in the newborn to be realized [113]. The programmatic distribution of influenza vaccines to pregnant women can be challenging, by trying to reconcile the time of vaccine availability with the best time to vaccinate during pregnancy, to optimize protection both of the women and their infants. Most results suggest that pregnant women should be immunized as soon as vaccine is available, first to protect themselves as early as possible in the influenza season, and second to take into consideration the possible cumulative transfer of antibodies to the fetus.

Duration of protection

The duration of passive protection in the infant, conferred by vaccination during pregnancy, depends on the concentration of antibody achieved by vaccination, which is contingent on the efficiency of the transport across the placenta and how quickly the passively-acquired antibodies wane. Half-lives of 42–50 days have been described in South Africa and Bangladesh for the different influenza-specific hemagglutinin antibodies in infants whose mothers received influenza vaccine during pregnancy [98,116]. The effect of this decay in antibodies was evident from the highest vaccine efficacy observed during the first 8 weeks of life in the South African trial (86% in infants <8 weeks old vs. 49% in infants <6 months old) and the first 4 months of life in Mali (68% in infants <4 months old vs. 33% in infants <6 months old) [8,117].

Safety of influenza vaccination during pregnancy

The seasonal vaccines evaluated in the RCTs were well tolerated among pregnant women, with the majority of injection site and systemic reactogenicity reported as mild to moderate and self-limited [6–9]. The reactogenicity profile of influenza vaccination during pregnancy has been reported to be similar to that observed in non-pregnant adults [118]. Unsolicited adverse events and serious adverse events were also reported at similar frequency among the intervention groups in the RCTs [6–9], and assessment of these endpoints will be extended in pooled analyses of the RCTs from Nepal, Mali and South Africa [119].

Prior to the 2009 H1N1 pandemic, safety evidence of influenza vaccination in pregnancy relied mainly on post-marketing pharmacovigilance. There are numerous limitations to these passive reporting surveillance systems, which lack internal comparison groups, denominator information, and are unable to causally attribute adverse events to vaccination [120]. Following the accelerated licensure of the H1N1pdm09 vaccines and vaccination of large numbers of pregnant women, several large studies exploring the safety of influenza vaccination during pregnancy were conducted [120–123]. In the US alone, 2.4 million pregnant women were vaccinated against H1N1pdm09, and 294 adverse events were reported to the Vaccine Adverse Event Reporting System. Medical review of these events, however, did not identify any concerning patterns of maternal or fetal outcomes associated with the administration of pandemic vaccine during pregnancy [120].

Safety of influenza vaccination and birth outcomes

Notwithstanding the limitations of data generated by passive surveillances systems, these studies have generally concluded that rates of adverse pregnancy outcomes, following influenza vaccination of pregnant women, were lower than background rates in the obstetrical population, and no concerning patterns of adverse events have been reported from these systems [124]. Post-licensure vaccine safety studies have also demonstrated the safety of IIV administration during pregnancy to the mother, fetus and newborn [125,126]. Several studies have specifically assessed the association between maternal influenza vaccination and adverse birth outcomes, including preterm birth, small-for-gestational-age birth, fetal death and congenital anomalies and malformations, with no increase in adverse outcomes being observed following vaccination [76,87,127–133].

Since early pregnancy represents a particularly vulnerable gestational time window for fetal organogenesis and development, it is extremely important to monitor birth outcomes after maternal vaccination in the first trimester. McMillan et al. systematically reviewed 12 studies of maternal influenza vaccination, including approximately 4000 women with first trimester exposure. Across these studies, ORs for first trimester IIV and birth defects ranged from 0.67 to 2.18, and none of these estimates was statistically significant [87]. In 2015, Polyzos et al.

also conducted a meta-analysis of largely overlapping studies of 4733 women exposed to IIV in the first trimester and reported an OR of 0.98 (95% CI: 0.95–1.20) for major birth defects in the offspring [134]. More recently, studies from US and Swedish cohorts, including nearly 15,000 first trimester IIV exposures, along with one US-based case-control study, have demonstrated safety of first trimester influenza vaccination with respect to birth defects [135–137]. A large observational study using electronic health data from seven Vaccine Safety Datalink (VSD) sites, examined risks for more than 50 pre-specified major structural birth defects among singleton, live births. Among 425,944 women, including 12% with IIV exposure in their first trimester, no increased risk for selected major structural birth defects following vaccination was observed [138].

One case-control study using VSD data from the US conducted during the 2010–11 and 2011–12 seasons (when the H1N1pdm09 strain was included in the vaccine compositions) noted, however, an association between receipt of IIV containing H1N1pdm09 and risk for spontaneous abortion (miscarriage) in the 28-days after IIV, when an H1N1pdm-09-containing vaccine had also been received the previous season [139]. It is important to note that the post-hoc sub-group analysis only included 18 women (14 cases and 4 controls); moreover, there was no association with spontaneous abortion, if vaccination occurred more than 28-days before the pregnancy loss, or even within the 28-day window, if vaccination was not received 2 years in a row. This was an unexpected finding, since previous studies [132,135,140–142] and systematic reviews [87,143] have not found any association between influenza vaccination and spontaneous abortion. Despite the limited conclusions that can be drawn from this study, given the case-control design as well as the small number of women included in the sub-analysis, further research is warranted. In February 2019, Donahue and colleagues released their findings from a follow-up VSD study on this topic [172]. The follow-up study was designed to address limitations of the earlier study, including a larger sample size and pre-specified sub-group analyses by influenza season and according to prior-season influenza vaccination status. In this case-control study of 1,236 matched pairs, during three influenza seasons (2012–2013 to 2014–2015) in the US, adjusted odds ratios for spontaneous abortion were not statistically significant and were close to, or less than, 1.0 in all analyses, including season-specific and when stratified according to prior-season influenza vaccination status. It is worth noting that not only are these findings congruent with the existing literature on this topic [87,143], they are also similar to a pre-H1N1 pandemic study of IIV and spontaneous abortion conducted by VSD [132] and support the safety of early pregnancy influenza vaccination.

Safety of influenza vaccination and pediatric health outcomes

An increasing number of studies have assessed influenza vaccination in pregnancy and its association with pediatric health outcomes beyond the first 6 months of life. Two studies assessed outcomes among infants up to one year of age, comparing those born to H1N1pdm09 vaccinated mothers and unvaccinated mothers, and found no differences in developmental scores or infection-related physician visits in a small cohort from the Netherlands [144], nor in rates of health services utilization in a large Canadian cohort study [145]. In a large retrospective cohort study using Danish registry data on 61,359 infants (10% of whom were born to H1N1pdm09 influenza-vaccinated mothers), there was no increased risk of childhood morbidities (e.g., infectious diseases, neurologic, autoimmune, or behavioural conditions) up to the age of 5 years; however, the authors observed a significant inverse association between H1N1pdm09 influenza vaccination and pediatric gastrointestinal infections (aRR 0.84, 95% CI: 0.74–0.94) [146]. Similar findings were reported by a large Canadian study of infant outcomes up to the age of 5 years, including a small reduction in the rate of gastrointestinal infections [adjusted incidence risk ratio (aIRR): 0.94, 95% CI: 0.91–0.98] [147]. The latter study also found a small but statistically significant increased association with asthma (aHR: 1.05, 95% CI: 1.02–1.09), but no increase in under-5 mortality [147], similar to a Swedish study also from the 2009 H1N1 pandemic time period [148]. Finally, no overall increase in the risk of autism spectrum disorder was observed, following seasonal influenza vaccination during pregnancy, in a California-based cohort study of infants born between 2000 and 2010 [149].

Safety of influenza vaccination and maternal health outcomes

Several studies have evaluated the association between influenza vaccination and maternal health and obstetrical outcomes. Naleway and colleagues [150] performed a narrative review of eight studies evaluating maternal obstetrical outcomes, six of which focused on the 2009 pH1N1 vaccine [123,127,141,151–155]. Among the included studies, five were retrospective matched cohorts [127,151,152,154,155], and three used a prospective cohort design [123,141,153]. Four different maternal outcomes were assessed: gestational diabetes, gestational hypertension, preeclampsia and chorioamnionitis. Four studies assessed gestational diabetes, with two finding no significant difference in rates between vaccinated and unvaccinated pregnant women [127,152], and the other two studies reporting significantly lower odds (aOR: 0.48, 95% CI: 0.29–0.80) [141], or significantly reduced risks, in a 42-day post-vaccination window (aIRR: 0.89, 95% CI: 0.82–0.96) and until the end of pregnancy (aHR: 0.88,

95% CI: 0.83, 0.92) in the vaccinated groups [155]. Gestational hypertension was not found to be associated with vaccination with a monovalent H1N1pdm09 MF59-adjuvanted vaccine [154] or with IIV in a 42-day window (aIRR: 0.99, 95% CI: 0.89–1.11) and until the end of pregnancy (aHR: 1.04, 95% CI: 0.98–1.11) [155]. Using a pre-specified conservative significance threshold of $P < .005$, one study in the Naleway review assessed chorioamnionitis and found a non-significant small increase in the risk of chorioamnionitis (aHR: 1.08, 95% CI: 1.02–1.15) [155]. Lastly, seven studies included in the review assessed preeclampsia/eclampsia and found no statistically significant increase in odds or incidence [127,141,151–155]. To address some of the identified limitations of the primary studies included in the review, the authors additionally incorporated primary analyses using data from the VSD and the Pregnancy and Influenza Project (PIP) cohorts, and found no significant associations with any of the four outcomes [150].

Kharbanda and colleagues conducted a retrospective observational cohort study at 7 VSD sites [155]. The sample included 74,292 IIV-exposed pregnancies matched to 144,597 unvaccinated pregnancies. The risk of adverse maternal outcomes was assessed using two risk windows – one within 42 days of the vaccination, and a second from vaccination to the end of pregnancy. In the 42-day window, there was no increased risk of hyperemesis (aIRR: 0.88, 95% CI: 0.79–0.99), gestational hypertension (aIRR: 0.67, 95% CI: 0.34–1.34), proteinuria (aIRR: 0.73, 95% CI: 0.54–0.99) or urinary tract infection (aIRR: 0.92, 95% CI: 0.84–1.02). In the second risk window, there was also no significant increased risk of proteinuria (aHR: 0.88, 95% CI: 0.78–1.00), urinary tract infection (aHR: 1.0, 95% CI: 0.94–1.05), venous complications (aHR: 0.64, 95% CI: 0.39–1.05), pulmonary embolus (aHR: 1.23, 95% CI: 0.58–2.64), or peripartum cardiomyopathy (aHR: 0.66, 95% CI: 0.35–1.24) [155].

Lastly, Nordin et al. conducted a retrospective matched observational study, also using the VSD cohort [156]. They evaluated five maternal outcomes over a 42-day post-vaccination window using two cohorts: (1) monovalent H1N1pdm09 influenza vaccine versus unvaccinated, and (2) monovalent H1N1pdm09 vaccine versus seasonal IIV. The researchers found no significant association for: (i) hyperemesis (monovalent H1N1pdm09 vs. unexposed, aIRR: 0.87, 95% CI: 0.72−1.05; monovalent H1N1pdm09 vs. IIV, aIRR: 1.09, 95% CI: 0.85−1.39), (ii) benign essential hypertension of pregnancy (monovalent H1N1pdm09 vs. unexposed, aIRR: 0.69, 95% CI: 0.13−3.58; monovalent H1N1pdm09 vs. IIV, aIRR: 0.96, 95% CI: 0.14−6.8), (iii) gestational diabetes (monovalent H1N1pdm09 vs. unexposed, aIRR: 1.03, 95% CI: 0.82−1.30; monovalent H1N1pdm09 vs. IIV, aIRR: 1.31, 95% CI: 0.98−1.76), (iv) thrombocytopenia (monovalent H1N1pdm09 vs. unexposed, aIRR: 1.22, 95% CI: 0.65−2.28; monovalent H1N1pdm09 vs. IIV, aIRR: 1.42, 95% CI: 0.61−3.31), or (v) proteinuria (monovalent H1N1pdm09 vs. unexposed, aIRR: 1.23, 95% CI: 0.61–2.47; monovalent H1N1pdm09 vs. IIV, aIRR:1.26, 95% CI: 0.47–3.39) [156].

Influenza vaccination during pregnancy and fetal outcomes

Another compelling benefit of maternal influenza vaccination would be the potential to protect the fetus from the consequences of maternal infection. However, clear evidence that vaccination during pregnancy improves fetal outcomes is still lacking, and inconsistent observations have been reported.

Evidence from randomized controlled trials

In the primary analyses of the Bangladesh RCT, no differences were observed in mean gestational age or birth weight between the intervention and control groups [9]. However, in a subsequent secondary analysis to further evaluate perinatal outcomes among the sub-group of infants born during the influenza season, maternal influenza vaccination was associated with a 56% (95% CI: 1–81) reduction in the adjusted odds of delivering a small-for-gestational-age infant, and a 193 gram (95% CI: 9–378) increase in mean birth weight [157]. The RCT in Nepal included an *a priori* co-primary objective of assessing the impact of maternal vaccination on low birth weight and, therefore, has been the only trial adequately powered to detect a difference in mean birth weight between the intervention groups [7]. In that trial, maternal vaccination was associated with a 15% reduction (95% CI: 3–25) in the proportion of infants born with low birth weight (i.e., <2500 grams), and a 43 gram (95% CI: 9–77) increase in mean birth weight [7]. The vaccination effect on birth weight was not observed in either the South African or Malian trials; however, neither was specifically powered to evaluate secondary endpoints of birth outcomes [6,8]. These inconsistent data from the RCTs suggest a marginal effect of seasonal IIV on birth weight, which may be heterogenous by both timing of vaccination and by underlying characteristics of the population [158]. None of the trials observed any reduction in the risk of preterm birth.

Evidence from observational epidemiological studies

Observational studies have played a central role in the ongoing assessment of fetal outcomes following maternal immunization, yet they also pose numerous methodological challenges [159,160]. Unlike RCTs, which benefit from random distribution of baseline characteristics between intervention and control groups, important differences in underlying risk profiles, between pregnant women who become vaccinated and those who do not, can bias the results in observational studies, typically in the direction of exaggerating apparent benefits [159]. Moreover, observational studies on this topic are also prone to complex time-related biases that lead to overestimation of any potential beneficial effects of influenza vaccination on adverse birth outcomes [161].

Comparison of some results between the RCTs and observational studies is possible and raises concern about both the adequacy of measuring and adjusting for covariates in some epidemiological studies, as well as of accounting for temporal factors. For instance, although none of the four maternal influenza immunization trials has reported any reduction in preterm birth, many observational studies have reported significant reductions, up to as high as 70% [129,162]. A recent simulation study estimated expected effect sizes for the relationship between influenza vaccination during pregnancy and preterm birth [161]. Using a range of plausible scenarios for rates of influenza illness during pregnancy, vaccine effectiveness and vaccine uptake, any protective benefits of influenza vaccination on preterm birth would be expected to be small and difficult to detect. Another recent methodological study subjected the assessment of influenza vaccination and preterm birth to a series of different analytical methodologies accounting for confounding and time-related bias (immortal time-bias) [163]. The investigators found the impact of the time-related factors (e.g., accounting for time-varying exposure and outcome) to be responsible for a greater degree of bias than confounding [163].

Similarly, although no differences in stillbirth rates were observed in the Nepalese or South African RCTs, undertaken in settings where background stillbirth rates were >20 per 1000 births, and which enrolled >5,500 women in total, a meta-analysis of the association of maternal influenza vaccination and stillbirths, which included seven epidemiological studies, reported a 27% (95% CI: 4–45) lower risk of stillbirth (31% reduction, 95% CI: 10–47 for H1N1pdm09 vaccines) [143]. Fewer data are available on stillbirth than other fetal outcomes owing to challenges with ascertaining this rare outcome; however, confounding and time-biases are also likely to have contributed to the magnitude of the apparent protective effects estimated by several observational studies [164].

A meta-analysis in 2016 observed significant heterogeneity across studies which reported on the effect of influenza vaccination during pregnancy and fetal outcomes, with maternal H1N1pdm2009 vaccination being associated with 8% (95% CI: 1–15) and 12% (95% CI: 2–21) lower risk for preterm birth and low birth weight, respectively. Although similar trends were observed for studies reporting on seasonal IIV in pregnant women, the effect was only significant for low birth weight (26% reduction, 95% CI: 12–39 based on only two studies) and differences were not significant for preterm birth (OR: 0.94, 95% CI: 0.87–1.01) [165].

Influenza vaccination in pregnant women living with HIV

Women who are pregnant and HIV-infected combine two high-risk conditions, resulting in higher influenza-related complications and attenuation of immune responses to vaccines [5,116]. Moreover, infants born to mothers living with HIV have an increased risk of hospitalization and death from respiratory virus-associated lower respiratory tract infection, including influenza virus, even if themselves not HIV-infected (HIV-exposed-uninfected) [51,166]. In parallel to the South African trial described above, during 2011, a smaller immunogenicity trial was performed in 194 pregnant women living with HIV [6]. Before vaccination and one month post-vaccination, pregnant women living with HIV, compared to those without HIV, had lower levels of HAI antibodies and a decreased likelihood of seroconversion (41% vs. 92%, respectively, to at least one strain) [116]. Similar observations were reported from a smaller study in the US, where increased regulatory T cells numbers were associated with attenuated immune responses to influenza vaccination in women living with HIV, resulting in reduced immunogenicity to at least one influenza vaccine strain, despite similar pre-vaccination HAI titers, compared to women without HIV [167]. Correlation analyses of post-vaccination HAI titers and HIV viral load in pregnant women living with HIV, most of whom were on antiretroviral therapy, generally have not shown any association [116,167,168]. Improved immune responses were, however, detected against at least one IIV strain in women with CD4+ T lymphocyte counts of >350 cells/μL [116].

The immunogenicity of an unadjuvanted monovalent H1N1pdm09 vaccine was also evaluated in pregnant women living with HIV [168,169]. In that US-based study, a second dose of a high-concentration (30 μg/dose) vaccine marginally improved the percentage of women with HAI titers ≥1:40 and who demonstrated seroconversion (73% and 66% after dose 1 and 80% and 72% after dose 2, respectively). Although HAI titers were increased in women with higher CD4+ T lymphocyte counts, women living with HIV still had poorer immune responses than historical controls without HIV after a single 15 μg/dose vaccine [168].

Although the transplacental antibody transfer was similar in the HIV-infected and HIV-uninfected cohorts in South Africa for two of the three vaccine strains, due to the lower antibody levels post-vaccination among the women living with HIV, their newborns had lower antibody titers at birth and were less likely to have HAI titers ≥1:40 compared to HIV-unexposed infants (ranging from 43–79% vs. 82–95%, respectively, for the different vaccine strains) [116]. No direct comparison is available for infant HAI titers after maternal H1N1pdm09 monovalent influenza vaccination between HIV-exposed and HIV-unexposed infants [168].

In the South African RCTs, the vaccine efficacy against PCR-confirmed influenza was similar in women living with HIV (58%; 95% CI: 0.2–82) and without HIV (50%; 95% CI: 15–71), despite the reduced immune response to vaccination in the former [6,116]. While the study was not powered to detect vaccine efficacy in the HIV-exposed infants, a modest vaccine efficacy point estimate was reported (26.7%; 95% CI: −132, 77) [6]. Based on the immunogenicity data from the two South African trials, despite vaccine efficacy in the mothers, alternate strategies to protect the HIV-exposed infants need to be considered. Whilst vaccine-induced cell mediated immunity might play a role in adults, protection in the infants will only be mediated by the presence of antibodies; hence, higher antibody concentrations need to be elicited in the women.

Conclusion

Since influenza vaccination during pregnancy has been demonstrated to confer protection to both mothers and young infants against influenza infection, this strategy should be used as the first line of protection. Nonetheless, there are still some considerations and open questions for future research in this field. These include the need for more immunogenic influenza vaccines for pregnant women to increase the concentration of the antibodies transferred transplacentally, which would last for a longer period in the infants. Infant passive protection beyond the first 2–4 months of life is desirable, as only after 6 months of age are infants eligible to receive influenza vaccine themselves. More immunogenic vaccines might be achieved with higher antigen concentrations or the use of adjuvants. During the 2009 H1N1 pandemic, newer formulations of inactivated antigens administered with adjuvants were found to enhance the immune responses, compared with non-adjuvanted vaccines [123,141,152,154,170]. Based on the immunogenicity data from the two South African trials, despite vaccine efficacy observed in mothers living with HIV, alternate strategies to protect the HIV-exposed infants need to be considered [6]. Whilst vaccine-induced cell mediated immunity might play a role in adults, protection in the infants will only be mediated by the presence of antibodies; hence, higher antibody concentrations need to be elicited in the women to prevent disease in their HIV-exposed infants.

The development of universal influenza vaccines containing common antigens that could provide protection against different virus strains, reducing the need for annual vaccine update and hopefully provide protection across multiple pregnancies, is also being explored. These strategies would be greatly beneficial from a programmatic perspective, especially for low and middle-income countries.

Furthermore, evidence regarding an increased susceptibility or severe disease outcome in young infants born to influenza virus infected mothers is still to be shown. Animal models, such as a murine pregnancy model which mimics key clinical findings of the 2009 pandemic H1N1 influenza [171], could help to demonstrate any causal relationship between maternal infection during pregnancy and offspring´s disease outcome.

The paucity of high-quality data on the effect of vaccination during pregnancy, against hospitalizations associated with laboratory-confirmed influenza in infants, is still a limitation for truly estimating the effect size of this intervention on more severe outcomes. Large, prospective, multi-season studies, with long-term follow up powered to address this endpoint, should be performed. Further data is also needed to better define the burden of influenza in young children, especially those presenting with non-respiratory symptoms, or those that might have cleared the virus at the time of laboratory testing. Additionally, in future large studies of influenza vaccination during pregnancy, long-term health outcomes in children should be evaluated, for which there are currently almost no data.

References

[1] Lafond KE, Nair H, Rasooly MH, Valente F, Booy R, Rahman M, et al. Global role and burden of influenza in pediatric respiratory hospitalizations, 1982-2012: a systematic analysis. PLoS Med 2016;13(3):e1001977.

[2] WHO. n.d. http://www.who.int/mediacentre/factsheets/fs211/en/.

[3] Zhou H, Thompson WW, Viboud CG, Ringholz CM, Cheng PY, Steiner C, et al. Hospitalizations associated with influenza and respiratory syncytial virus in the United States, 1993-2008. Clin Infect Dis 2012;54(10):1427–36.

[4] Dawood FS, Fiore A, Kamimoto L, Bramley A, Reingold A, Gershman K, et al. Burden of seasonal influenza hospitalization in children, United States, 2003 to 2008. J Pediatr 2010;157(5):808–14.

[5] Tempia S, Walaza S, Cohen AL, von Mollendorf C, Moyes J, McAnerney JM, et al. Mortality associated with seasonal and pandemic influenza among pregnant and non-pregnant women of childbearing age in a High-HIV-prevalence setting-South Africa, 1999-2009. Clin Infect Dis 2015;61(7):1063–70.

[6] Madhi SA, Cutland CL, Kuwanda L, Weinberg A, Hugo A, Jones S, et al. Influenza vaccination of pregnant women and protection of their infants. N Engl J Med 2014;371(10):918–31.

[7] Steinhoff MC, Katz J, Englund JA, Khatry SK, Shrestha L, Kuypers J, et al. Year-round influenza immunisation during pregnancy in Nepal: a phase 4, randomised, placebo-controlled trial. Lancet Infect Dis 2017;17(9):981–9.

[8] Tapia MD, Sow SO, Tamboura B, Teguete I, Pasetti MF, Kodio M, et al. Maternal immunisation with trivalent inactivated influenza vaccine for prevention of influenza in infants in Mali: a prospective, active-controlled, observer-blind randomised phase 4 trial. Lancet Infect Dis 2016;16(9):1026–35.

[9] Zaman K, Roy E, Arifeen SE, Rahman M, Raqib R, Wilson E, et al. Effectiveness of maternal influenza immunization in mothers and infants. N Engl J Med 2008;359(15):1555–64.

[10] Zhang J, Lamb RA. Characterization of the membrane association of the influenza virus matrix protein in living cells. Virology 1996;225(2):255–66.

[11] Tan Y, Guan W, Lam TT, Pan S, Wu S, Zhan Y, et al. Differing epidemiological dynamics of influenza B virus lineages in Guangzhou, southern China, 2009-2010. J Virol 2013;87(22):12447–56.
[12] Zambon MC. Epidemiology and pathogenesis of influenza. J Antimicrob Chemother 1999;44(Suppl B):3–9.
[13] Smith DJ, Lapedes AS, de Jong JC, Bestebroer TM, Rimmelzwaan GF, Osterhaus AD, et al. Mapping the antigenic and genetic evolution of influenza virus. Science 2004;305(5682):371–6.
[14] Dunkle LM, Izikson R, Patriarca P, Goldenthal KL, Muse D, Callahan J, et al. Efficacy of recombinant influenza vaccine in adults 50 years of age or older. N Engl J Med 2017;376(25):2427–36.
[15] Katz MA, Gessner BD, Johnson J, Skidmore B, Knight M, Bhat N, et al. Incidence of influenza virus infection among pregnant women: a systematic review. BMC Pregnancy Childbirth 2017;17(1):155.
[16] Knight M, Pierce M, Seppelt I, Kurinczuk JJ, Spark P, Brocklehurst P, et al. Critical illness with AH1N1v influenza in pregnancy: a comparison of two population-based cohorts. BJOG 2011;118(2):232–9.
[17] Doyle TJ, Goodin K, Hamilton JJ. Maternal and neonatal outcomes among pregnant women with 2009 pandemic influenza A(H1N1) illness in Florida, 2009-2010: a population-based cohort study. PLoS ONE 2013;8(10):e79040.
[18] Yates L, Pierce M, Stephens S, Mill AC, Spark P, Kurinczuk JJ, et al. Influenza A/H1N1v in pregnancy: an investigation of the characteristics and management of affected women and the relationship to pregnancy outcomes for mother and infant. Health Technol Assess 2010;14(34):109–82.
[19] Griffiths PD, Ronalds CJ, Heath RB. A prospective study of influenza infections during pregnancy. J Epidemiol Community Health 1980;34(2):124–8.
[20] Hardy JM, Azarowicz EN, Mannini A, Medearis Jr DN, Cooke RE. The effect of Asian influenza on the outcome of pregnancy, Baltimore, 1957-1958. Am J Public Health Nations Health 1961;51:1182–8.
[21] Irving WL, James DK, Stephenson T, Laing P, Jameson C, Oxford JS, et al. Influenza virus infection in the second and third trimesters of pregnancy: a clinical and seroepidemiological study. BJOG 2000;107(10):1282–9.
[22] Creanga AA, Johnson TF, Graitcer SB, Hartman LK, Al-Samarrai T, Schwarz AG, et al. Severity of 2009 pandemic influenza A (H1N1) virus infection in pregnant women. Obstet Gynecol 2010;115(4):717–26.
[23] Jamieson DJ, Honein MA, Rasmussen SA, Williams JL, Swerdlow DL, Biggerstaff MS, et al. H1N1 2009 influenza virus infection during pregnancy in the USA. Lancet 2009;374(9688):451–8.
[24] Regan AK, Moore HC, Sullivan SG. N DEK, Effler PV. Epidemiology of seasonal influenza infection in pregnant women and its impact on birth outcomes. Epidemiol Infect 2017;145(14):2930–9.
[25] Prasad N, Huang QS, Wood T, Aminisani N, McArthur C, Baker MG, et al. Influenza associated outcomes among pregnant, post-partum, and non-pregnant women of reproductive age. J Infect Dis 2019;219(12):1893–903.
[26] Schanzer DL, Langley JM, Tam TW. Influenza-attributed hospitalization rates among pregnant women in Canada 1994-2000. J Obstet Gynaecol Can 2007;29(8):622–9.
[27] Fell DB, Azziz-Baumgartner E, Baker MG, Batra M, Beaute J, Beutels P, et al. Influenza epidemiology and immunization during pregnancy: final report of a World Health Organization working group. Vaccine 2017;35(43):5738–50.
[28] Izurieta HS, Thompson WW, Kramarz P, Shay DK, Davis RL, DeStefano F, et al. Influenza and the rates of hospitalization for respiratory disease among infants and young children. N Engl J Med 2000;342(4):232–9.

[29] Neuzil KM, Mellen BG, Wright PF, Mitchel Jr EF, Griffin MR. The effect of influenza on hospitalizations, outpatient visits, and courses of antibiotics in children. N Engl J Med 2000;342(4):225–31.
[30] Poehling KA, Edwards KM, Griffin MR, Szilagyi PG, Staat MA, Iwane MK, et al. The burden of influenza in young children, 2004-2009. Pediatrics 2013;131(2):207–16.
[31] Poehling KA, Edwards KM, Weinberg GA, Szilagyi P, Staat MA, Iwane MK, et al. The underrecognized burden of influenza in young children. N Engl J Med 2006;355(1):31–40.
[32] Montes M, Vicente D, Perez-Yarza EG, Cilla G, Perez-Trallero E. Influenza-related hospitalisations among children aged less than 5 years old in the Basque Country, Spain: a 3-year study (July 2001-June 2004). Vaccine 2005;23(34):4302–6.
[33] Bhat N, Wright JG, Broder KR, Murray EL, Greenberg ME, Glover MJ, et al. Influenza-associated deaths among children in the United States, 2003-2004. N Engl J Med 2005;353(24):2559–67.
[34] Nair H, Brooks WA, Katz M, Roca A, Berkley JA, Madhi SA, et al. Global burden of respiratory infections due to seasonal influenza in young children: a systematic review and meta-analysis. Lancet 2011;378(9807):1917–30.
[35] World Health Organization. A Manual for Estimating Disease Burden Associated With Seasonal Influenza. Geneva: World Health Organization; 2015. http://apps.who.int/iris/bitstream/10665/178801/1/9789241549301_eng.pdf?ua=1&ua=1.
[36] Fell DB, Johnson J, Mor Z, Katz MA, Skidmore B, Neuzil KM, et al. Incidence of laboratory-confirmed influenza disease among infants under 6 months of age: a systematic review. BMJ Open 2017;7(9):e016526.
[37] Lee VJ, Ho ZJM, Goh EH, Campbell H, Cohen C, Cozza V, et al. Advances in measuring influenza burden of disease. Influenza Other Respir Viruses 2018;12(1):3–9.
[38] Dawa JA, Chaves SS, Nyawanda B, Njuguna HN, Makokha C, Otieno NA, et al. National burden of hospitalized and non-hospitalized influenza-associated severe acute respiratory illness in Kenya, 2012-2014. Influenza Other Respir Viruses 2018;12(1):30–7.
[39] McMorrow ML, Tempia S, Walaza S, Treurnicht FK, Moyes J, Cohen AL, et al. The role of human immunodeficiency virus in influenza- and respiratory syncytial virus-associated hospitalizations in South African Children, 2011-2016. Clin Infect Dis 2019;68(5):773–80.
[40] Ampofo K, Gesteland PH, Bender J, Mills M, Daly J, Samore M, et al. Epidemiology, complications, and cost of hospitalization in children with laboratory-confirmed influenza infection. Pediatrics 2006;118(6):2409–17.
[41] Cox CM, D'Mello T, Perez A, Reingold A, Gershman K, Yousey-Hindes K, et al. Increase in rates of hospitalization due to laboratory-confirmed influenza among children and adults during the 2009-10 influenza pandemic. J Infect Dis 2012;206(9):1350–8.
[42] Grijalva CG, Craig AS, Dupont WD, Bridges CB, Schrag SJ, Iwane MK, et al. Estimating influenza hospitalizations among children. Emerg Infect Dis 2006;12(1):103–9.
[43] Grijalva CG, Weinberg GA, Bennett NM, Staat MA, Craig AS, Dupont WD, et al. Estimating the undetected burden of influenza hospitalizations in children. Epidemiol Infect 2007;135(6):951–8.
[44] Iwane MK, Edwards KM, Szilagyi PG, Walker FJ, Griffin MR, Weinberg GA, et al. Population-based surveillance for hospitalizations associated with respiratory syncytial virus, influenza virus, and parainfluenza viruses among young children. Pediatrics 2004;113(6):1758–64.
[45] Proff R, Gershman K, Lezotte D, Nyquist AC. Case-based surveillance of influenza hospitalizations during 2004-2008, Colorado, USA. Emerg Infect Dis 2009;15(6):892–8.
[46] Griffin MR, Walker FJ, Iwane MK, Weinberg GA, Staat MA, Erdman DD, et al. Epidemiology of respiratory infections in young children: insights from the new vaccine surveillance network. Pediatr Infect Dis J 2004;23(11 Suppl):S188–92.

[47] Ali A, Akhund T, Warraich GJ, Aziz F, Rahman N, Umrani FA, et al. Respiratory viruses associated with severe pneumonia in children under 2 years old in a rural community in Pakistan. J Med Virol 2016;88(11):1882–90.
[48] Broor S, Dawood FS, Pandey BG, Saha S, Gupta V, Krishnan A, et al. Rates of respiratory virus-associated hospitalization in children aged <5 years in rural northern India. J Infect 2014;68(3):281–9.
[49] McMorrow ML, Emukule GO, Njuguna HN, Bigogo G, Montgomery JM, Nyawanda B, et al. The Unrecognized Burden of Influenza in Young Kenyan Children, 2008-2012. PLoS ONE 2015;10(9):e0138272.
[50] Budge PJ, Griffin MR, Edwards KM, Williams JV, Verastegui H, Hartinger SM, et al. A household-based study of acute viral respiratory illnesses in Andean children. Pediatr Infect Dis J 2014;33(5):443–7.
[51] Cohen C, Moyes J, Tempia S, Groome M, Walaza S, Pretorius M, et al. Epidemiology of acute lower respiratory tract infection in HIV-exposed uninfected infants. Pediatrics 2016;137(4).
[52] Ji W, Zhang T, Zhang X, Jiang L, Ding Y, Hao C, et al. The epidemiology of hospitalized influenza in children, a two year population-based study in the People's Republic of China. BMC Health Serv Res 2010;10:82.
[53] Libster R, Bugna J, Coviello S, Hijano DR, Dunaiewsky M, Reynoso N, et al. Pediatric hospitalizations associated with 2009 pandemic influenza A (H1N1) in Argentina. N Engl J Med 2010;362(1):45–55.
[54] Yu H, Huang J, Huai Y, Guan X, Klena J, Liu S, et al. The substantial hospitalization burden of influenza in central China: surveillance for severe, acute respiratory infection, and influenza viruses, 2010-2012. Influenza Other Respir Viruses 2014;8(1):53–65.
[55] Zhang T, Zhang J, Hua J, Wang D, Chen L, Ding Y, et al. Influenza-associated outpatient visits among children less than 5 years of age in eastern China, 2011-2014. BMC Infect Dis 2016;16:267.
[56] Bennet R, Hamrin J, Wirgart BZ, Ostlund MR, Ortqvist A, Eriksson M. Influenza epidemiology among hospitalized children in Stockholm, Sweden 1998-2014. Vaccine 2016;34(28):3298–302.
[57] Nelson EA, Ip M, Tam JS, Mounts AW, Chau SL, Law SK, et al. Burden of influenza infection in hospitalised children below 6 months of age and above in Hong Kong from 2005 to 2011. Vaccine 2014;32(49):6692–8.
[58] Silvennoinen H, Peltola V, Vainionpaa R, Ruuskanen O, Heikkinen T. Incidence of influenza-related hospitalizations in different age groups of children in Finland: a 16-year study. Pediatr Infect Dis J 2011;30(2):e24–8.
[59] Stein M, Tasher D, Glikman D, Shachor-Meyouhas Y, Barkai G, Yochai AB, et al. Hospitalization of children with influenza A(H1N1) virus in Israel during the 2009 outbreak in Israel: a multicenter survey. Arch Pediatr Adolesc Med 2010;164(11):1015–22.
[60] Fell DB, Savitz DA, Kramer MS, Gessner BD, Katz MA, Knight M, et al. Maternal influenza and birth outcomes: systematic review of comparative studies. BJOG 2017;124(1):48–59.
[61] Martin A, Cox S, Jamieson DJ, Whiteman MK, Kulkarni A, Tepper NK. Respiratory illness hospitalizations among pregnant women during influenza season, 1998-2008. Matern Child Health J 2013;17(7):1325–31.
[62] Morken NH, Gunnes N, Magnus P, Jacobsson B. Risk of spontaneous preterm delivery in a low-risk population: the impact of maternal febrile episodes, urinary tract infection, pneumonia and ear-nose-throat infections. Eur J Obstet Gynecol Reprod Biol 2011;159(2):310–4.
[63] McNeil SA, Dodds LA, Fell DB, Allen VM, Halperin BA, Steinhoff MC, et al. Effect of respiratory hospitalization during pregnancy on infant outcomes. Am J Obstet Gynecol 2011;204(6 Suppl 1):S54–7.

[64] Rogers VL, Sheffield JS, Roberts SW, McIntire DD, Luby JP, Trevino S, et al. Presentation of seasonal influenza A in pregnancy: 2003-2004 influenza season. Obstet Gynecol 2010;115(5):924–9.
[65] Cox S, Posner SF, McPheeters M, Jamieson DJ, Kourtis AP, Meikle S. Hospitalizations with respiratory illness among pregnant women during influenza season. Obstet Gynecol 2006;107(6):1315–22.
[66] Acs N, Banhidy F, Puho E, Czeizel AE. Pregnancy complications and delivery outcomes of pregnant women with influenza. J Matern Fetal Neonatal Med 2006;19(3):135–40.
[67] Hartert TV, Neuzil KM, Shintani AK, Mitchel Jr EF, Snowden MS, Wood LB, et al. Maternal morbidity and perinatal outcomes among pregnant women with respiratory hospitalizations during influenza season. Am J Obstet Gynecol 2003;189(6):1705–12.
[68] Tuyishime JD, De Wals P, Moutquin JM, Frost E. Influenza-like illness during pregnancy: results from a study in the eastern townships, Province of Quebec. J Obstet Gynaecol Can 2003;25(12):1020–5.
[69] Korones SB, Todaro J, Roane JA, Sever JL. Maternal virus infection after the first trimester of pregnancy and status of offspring to 4 years of age in a predominantly Negro population. J Pediatr 1970;77(2):245–51.
[70] Wilson MG, Stein AM. Teratogenic effects of asian influenza. An extended study. JAMA 1969;210(2):336–7.
[71] Stanwell-Smith R, Parker AM, Chakraverty P, Soltanpoor N, Simpson CN. Possible association of influenza A with fetal loss: investigation of a cluster of spontaneous abortions and stillbirths. Commun Dis Rep CDR Rev 1994;4(3):R28–32.
[72] Pierce M, Kurinczuk JJ, Spark P, Brocklehurst P, Knight M, UKOSS. Perinatal outcomes after maternal 2009/H1N1 infection: national cohort study. BMJ 2011;342:d3214.
[73] Naresh A, Fisher BM, Hoppe KK, Catov J, Xu J, Hart J, et al. A multicenter cohort study of pregnancy outcomes among women with laboratory-confirmed H1N1 influenza. J Perinatol 2013;33(12):939–43.
[74] Nieto-Pascual L, Arjona-Berral JE, Marin-Martin EM, Munoz-Gomariz E, Ilich I, Castelo-Branco C. Early prophylactic treatment in pregnant women during the 2009-2010 H1N1 pandemic: obstetric and neonatal outcomes. J Obstet Gynaecol 2013;33(2):128–34.
[75] Haberg SE, Trogstad L, Gunnes N, Wilcox AJ, Gjessing HK, Samuelsen SO, et al. Risk of fetal death after pandemic influenza virus infection or vaccination. N Engl J Med 2013;368(4):333–40.
[76] Ahrens KA, Louik C, Kerr S, Mitchell AA, Werler MM. Seasonal influenza vaccination during pregnancy and the risks of preterm delivery and small for gestational age birth. Paediatr Perinat Epidemiol 2014;28(6):498–509.
[77] Hansen C, Desai S, Bredfeldt C, Cheetham C, Gallagher M, Li DK, et al. A large, population-based study of 2009 pandemic Influenza A virus subtype H1N1 infection diagnosis during pregnancy and outcomes for mothers and neonates. J Infect Dis 2012;206(8):1260–8.
[78] Fell DB, Platt RW, Basso O, Wilson K, Kaufman JS, Buckeridge DL, et al. The relationship between 2009 Pandemic H1N1 influenza during pregnancy and preterm birth: a population-based cohort study. Epidemiology 2018;29(1):107–16.
[79] Newsome K, Alverson CJ, Williams J, McIntyre AF, Fine AD, Wasserman C, et al. Outcomes of infants born to women with influenza A(H1N1)pdm09. Birth Defects Res 2019;111(2):88–95.
[80] Burney LE. Influenza immunization: Statement. Public Health Rep 1960;75(10):944.
[81] ACOG Committee Opinion No. 732: Influenza Vaccination During Pregnancy. Obstet Gynecol 2018;131(4). e109-e14.
[82] Harper SA, Fukuda K, Uyeki TM, Cox NJ, Bridges CB, Centers for Disease C, et al. Prevention and control of influenza: recommendations of the Advisory Committee on Immunization Practices (ACIP). MMWR Recomm Rep 2004;53(RR-6):1–40.

[83] Influenza vaccines. Wkly Epidemiol Rec 2005;80(33):279–87.
[84] Vaccines against influenza WHO position paper - November 2012. Wkly Epidemiol Rec 2012;87(47):461–76.
[85] Benowitz I, Esposito DB, Gracey KD, Shapiro ED, Vazquez M. Influenza vaccine given to pregnant women reduces hospitalization due to influenza in their infants. Clin Infect Dis 2010;51(12):1355–61.
[86] Dabrera G, Zhao H, Andrews N, Begum F, Green H, Ellis J, et al. Effectiveness of seasonal influenza vaccination during pregnancy in preventing influenza infection in infants, England, 2013/14. Euro Surveill 2014;19(45):20959.
[87] McMillan M, Porritt K, Kralik D, Costi L, Marshall H. Influenza vaccination during pregnancy: a systematic review of fetal death, spontaneous abortion, and congenital malformation safety outcomes. Vaccine 2015;33(18):2108–17.
[88] Poehling KA, Szilagyi PG, Staat MA, Snively BM, Payne DC, Bridges CB, et al. Impact of maternal immunization on influenza hospitalizations in infants. Am J Obstet Gynecol 2011;204(6 Suppl 1):S141–8.
[89] Shakib JH, Korgenski K, Presson AP, Sheng X, Varner MW, Pavia AT, et al. Influenza in Infants Born to Women Vaccinated During Pregnancy. Pediatrics 2016;137(6).
[90] Thompson MG, Li DK, Shifflett P, Sokolow LZ, Ferber JR, Kurosky S, et al. Effectiveness of Seasonal Trivalent Influenza Vaccine for Preventing Influenza Virus Illness Among Pregnant Women: A Population-Based Case-Control Study During the 2010-2011 and 2011-2012 Influenza Seasons. Clin Infect Dis 2013;58(4):449–57.
[91] Regan AK, Klerk N, Moore HC, Omer SB, Shellam G, Effler PV. Effectiveness of seasonal trivalent influenza vaccination against hospital-attended acute respiratory infections in pregnant women: A retrospective cohort study. Vaccine 2016;34(32):3649–56.
[92] Thompson MG, Kwong JC, Regan AK, Katz MA, Drews SJ, Azziz-Baumgartner E, et al. Influenza Vaccine Effectiveness in Preventing Influenza-associated Hospitalizations During Pregnancy: A Multi-country Retrospective Test Negative Design Study, 2010-2016. Clin Infect Dis 2019;68(9):1444–53.
[93] Nunes MC, Madhi SA. Influenza vaccination during pregnancy for prevention of influenza confirmed illness in the infants: a systematic review and meta-analysis. Human Vaccines Immunother 2018;14(3):758–66.
[94] Eick AA, Uyeki TM, Klimov A, Hall H, Reid R, Santosham M, et al. Maternal influenza vaccination and effect on influenza virus infection in young infants. Arch Pediatr Adolesc Med 2011;165(2):104–11.
[95] Nunes MC, Cutland CL, Jones S, Downs S, Weinberg A, Ortiz JR, et al. Efficacy of maternal influenza vaccination against all-cause lower respiratory tract infection hospitalizations in young infants: Results from a randomized controlled trial. Clin Infect Dis 2017;65(7):1066–71.
[96] Omer SB, Clark DR, Aqil AR, Tapia MD, Nunes MC, Kozuki N, et al. Maternal influenza immunization and prevention of severe clinical pneumonia in young infants: analysis of randomized controlled trials conducted in Nepal, Mali, and South Africa. Pediatr Infect Dis J 2018;37(5):436–40.
[97] Hobson D, Curry RL, Beare AS, Ward-Gardner A. The role of serum haemagglutination-inhibiting antibody in protection against challenge infection with influenza A2 and B viruses. J Hyg 1972;70(4):767–77.
[98] Steinhoff MC, Omer SB, Roy E, Arifeen SE, Raqib R, Altaye M, et al. Influenza immunization in pregnancy—antibody responses in mothers and infants. N Engl J Med 2010;362(17):1644–6.
[99] Schlaudecker EP, McNeal MM, Dodd CN, Ranz JB, Steinhoff MC. Pregnancy modifies the antibody response to trivalent influenza immunization. J Infect Dis 2012;206(11):1670–3.
[100] Bischoff AL, Folsgaard NV, Carson CG, Stokholm J, Pedersen L, Holmberg M, et al. Altered response to A(H1N1)pnd09 vaccination in pregnant women: a single blinded randomized controlled trial. PLoS ONE 2013;8(4):e56700.

[101] Christian LM, Porter K, Karlsson E, Schultz-Cherry S, Iams JD. Serum proinflammatory cytokine responses to influenza virus vaccine among women during pregnancy versus non-pregnancy. Am J Reprod Immunol 2013;70(1):45–53.
[102] Hulka JF. Effectiveness of polyvalent influenza vaccine in pregnancy. Report of a controlled study during an outbreak of Asian Influenza. Obstet Gynecol 1964;23:830–7.
[103] Sperling RS, Engel SM, Wallenstein S, Kraus TA, Garrido J, Singh T, et al. Immunogenicity of trivalent inactivated influenza vaccination received during pregnancy or postpartum. Obstet Gynecol 2012;119(3):631–9.
[104] Murray DL, Imagawa DT, Okada DM, St Geme Jr. JW. Antibody response to monovalent A/New Jersey/8/76 influenza vaccine in pregnant women. J Clin Microbiol 1979;10(2):184–7.
[105] Kay AW, Bayless NL, Fukuyama J, Aziz N, Dekker CL, Mackey S, et al. Pregnancy does not attenuate the antibody or plasmablast response to inactivated influenza vaccine. J Infect Dis 2015;212(6):861–70.
[106] Schlaudecker EP, Ambroggio L, McNeal MM, Finkelman FD, Way SS. Declining responsiveness to influenza vaccination with progression of human pregnancy. Vaccine 2018;36(31):4734–41.
[107] Jackson LA, Patel SM, Swamy GK, Frey SE, Creech CB, Munoz FM, et al. Immunogenicity of an inactivated monovalent 2009 H1N1 influenza vaccine in pregnant women. J Infect Dis 2011;204(6):854–63.
[108] Tsatsaris V, Capitant C, Schmitz T, Chazallon C, Bulifon S, Riethmuller D, et al. Maternal immune response and neonatal seroprotection from a single dose of a monovalent nonadjuvanted 2009 influenza A(H1N1) vaccine: a single-group trial. Ann Intern Med 2011;155(11):733–41.
[109] Ohfuji S, Fukushima W, Deguchi M, Kawabata K, Yoshida H, Hatayama H, et al. Immunogenicity of a monovalent 2009 influenza A (H1N1) vaccine among pregnant women: lowered antibody response by prior seasonal vaccination. J Infect Dis 2011;203(9):1301–8.
[110] Horiya M, Hisano M, Iwasaki Y, Hanaoka M, Watanabe N, Ito Y, et al. Efficacy of double vaccination with the 2009 pandemic influenza A (H1N1) vaccine during pregnancy. Obstet Gynecol 2011;118(4):887–94.
[111] Zuccotti G, Pogliani L, Pariani E, Amendola A, Zanetti A. Transplacental antibody transfer following maternal immunization with a pandemic 2009 influenza A(H1N1) MF59-adjuvanted vaccine. JAMA 2010;304(21):2360–1.
[112] Kostinov MP, Cherdantsev AP, Akhmatova NK, Praulova DA, Kostinova AM, Akhmatova EA, et al. Immunogenicity and safety of subunit influenza vaccines in pregnant women. ERJ Open Res 2018;4(2).
[113] Blanchard-Rohner G, Meier S, Bel M, Combescure C, Othenin-Girard V, Swali RA, et al. Influenza vaccination given at least 2 weeks before delivery to pregnant women facilitates transmission of seroprotective influenza-specific antibodies to the newborn. Pediatr Infect Dis J 2013;32(12):1374–80.
[114] Malek A, Sager R, Kuhn P, Nicolaides KH, Schneider H. Evolution of maternofetal transport of immunoglobulins during human pregnancy. Am J Reprod Immunol 1996;36(5):248–55.
[115] Englund JA, Mbawuike IN, Hammill H, Holleman MC, Baxter BD, Glezen WP. Maternal immunization with influenza or tetanus toxoid vaccine for passive antibody protection in young infants. J Infect Dis 1993;168(3):647–56.
[116] Nunes MC, Cutland CL, Dighero B, Bate J, Jones S, Hugo A, et al. Kinetics of hemagglutination-inhibiting antibodies following maternal influenza vaccination among mothers with and those without HIV infection and their infants. J Infect Dis 2015;212(12):1976–87.
[117] Nunes MC, Cutland CL, Jones S, Hugo A, Madimabe R, Simoes EA, et al. Duration of infant protection against influenza illness conferred by maternal immunization: secondary analysis of a randomized clinical trial. JAMA Pediatr 2016;170(9):840–7.

[118] Munoz FM, Jackson LA, Swamy GK, Edwards KM, Frey SE, Stephens I, et al. Safety and immunogenicity of seasonal trivalent inactivated influenza vaccines in pregnant women. Vaccine 2018;36(52):8054–61.
[119] Omer SB, Richards JL, Madhi SA, Tapia MD, Steinhoff MC, Aqil AR, et al. Three randomized trials of maternal influenza immunization in Mali, Nepal, and South Africa: methods and expectations. Vaccine 2015;33(32):3801–12.
[120] Moro PL, Broder K, Zheteyeva Y, Revzina N, Tepper N, Kissin D, et al. Adverse events following administration to pregnant women of influenza A (H1N1) 2009 monovalent vaccine reported to the Vaccine Adverse Event Reporting System. Am J Obstet Gynecol 2011;205(5). 473 e1-9.
[121] Regan AK, Blyth CC, Mak DB, Richmond PC, Effler PV. Using SMS to monitor adverse events following trivalent influenza vaccination in pregnant women. Aust N Z J Obstet Gynaecol 2014;54(6):522–8.
[122] Mackenzie IS, MacDonald TM, Shakir S, Dryburgh M, Mantay BJ, McDonnell P, et al. Influenza H1N1 (swine flu) vaccination: a safety surveillance feasibility study using self-reporting of serious adverse events and pregnancy outcomes. Br J Clin Pharmacol 2012;73(5):801–11.
[123] Tavares F, Nazareth I, Monegal JS, Kolte I, Verstraeten T, Bauchau V. Pregnancy and safety outcomes in women vaccinated with an AS03-adjuvanted split virion H1N1 (2009) pandemic influenza vaccine during pregnancy: a prospective cohort study. Vaccine 2011;29(37):6358–65.
[124] Moro PL, Tepper NK, Grohskopf LA, Vellozzi C, Broder K. Safety of seasonal influenza and influenza A (H1N1) 2009 monovalent vaccines in pregnancy. Expert Rev Vaccines 2012;11(8):911–21.
[125] Bednarczyk RA, Adjaye-Gbewonyo D, Omer SB. Safety of influenza immunization during pregnancy for the fetus and the neonate. Am J Obstet Gynecol 2012;207(3 Suppl):S38–46.
[126] Munoz FM. Safety of influenza vaccines in pregnant women. Am J Obstet Gynecol 2012;207(3 Suppl):S33–7.
[127] Munoz FM, Greisinger AJ, Wehmanen OA, Mouzoon ME, Hoyle JC, Smith FA, et al. Safety of influenza vaccination during pregnancy. Am J Obstet Gynecol 2005;192(4):1098–106.
[128] Nordin JD, Kharbanda EO, Vazquez Benitez G, Lipkind H, Vellozzi C, Destefano F, et al. Maternal influenza vaccine and risks for preterm or small for gestational age birth. J Pediatr 2014;164(5). 1051-7 e2.
[129] Omer SB, Goodman D, Steinhoff MC, Rochat R, Klugman KP, Stoll BJ, et al. Maternal influenza immunization and reduced likelihood of prematurity and small for gestational age births: a retrospective cohort study. PLoS Med 2011;8(5):e1000441.
[130] Dodds L, Macdonald N, Scott J, Spencer A, Allen VM, McNeil S. The association between influenza vaccine in pregnancy and adverse neonatal outcomes. J Obstet Gynaecol Can 2012;34(8):714–20.
[131] Sheffield JS, Greer LG, Rogers VL, Roberts SW, Lytle H, McIntire DD, et al. Effect of influenza vaccination in the first trimester of pregnancy. Obstet Gynecol 2012;120(3):532–7.
[132] Irving SA, Kieke BA, Donahue JG, Mascola MA, Baggs J, DeStefano F, et al. Trivalent inactivated influenza vaccine and spontaneous abortion. Obstet Gynecol 2013;121(1):159–65.
[133] Regan AK, Moore HC, de Klerk N, Omer SB, Shellam G, Mak DB, et al. Seasonal trivalent influenza vaccination during pregnancy and the incidence of stillbirth: population-based retrospective cohort study. Clin Infect Dis 2016;62(10):1221–7.
[134] Polyzos KA, Konstantelias AA, Pitsa CE, Falagas ME. Maternal influenza vaccination and risk for congenital malformations: a systematic review and meta-analysis. Obstet Gynecol 2015;126(5):1075–84.

[135] Chambers CD, Johnson DL, Xu R, Luo YJ, Louik C, Mitchell AA, et al. Safety of the 2010-11, 2011-12, 2012-13, and 2013-14 seasonal influenza vaccines in pregnancy: birth defects, spontaneous abortion, preterm delivery, and small for gestational age infants, a study from the cohort arm of VAMPSS. Vaccine 2016;34(37):4443–9.
[136] Louik C, Kerr S, Van Bennekom CM, Chambers C, Jones KL, Schatz M, et al. Safety of the 2011-12, 2012-13, and 2013-14 seasonal influenza vaccines in pregnancy: preterm delivery and specific malformations, a study from the case-control arm of VAMPSS. Vaccine 2016;34(37):4450–9.
[137] Ludvigsson JF, Strom P, Lundholm C, Cnattingius S, Ekbom A, Ortqvist A, et al. Risk for congenital malformation with H1N1 influenza vaccine: a cohort study with sibling analysis. Ann Intern Med 2016;165(12):848–55.
[138] Kharbanda EO, Vazquez-Benitez G, Romitti PA, Naleway AL, Cheetham TC, Lipkind HS, et al. First trimester influenza vaccination and risks for major structural birth defects in offspring. J Pediatr 2017;187:234-9 e4.
[139] Donahue JG, Kieke BA, King JP, DeStefano F, Mascola MA, Irving SA, et al. Association of spontaneous abortion with receipt of inactivated influenza vaccine containing H1N1pdm09 in 2010-11 and 2011-12. Vaccine 2017;35(40):5314–22.
[140] Chambers CD, Johnson D, Xu R, Luo Y, Louik C, Mitchell AA, et al. Risks and safety of pandemic H1N1 influenza vaccine in pregnancy: birth defects, spontaneous abortion, preterm delivery, and small for gestational age infants. Vaccine 2013;31(44):5026–32.
[141] Heikkinen T, Young J, van Beek E, Franke H, Verstraeten T, Weil JG, et al. Safety of MF59-adjuvanted A/H1N1 influenza vaccine in pregnancy: a comparative cohort study. Am J Obstet Gynecol 2012;207(3). 177 e1-8.
[142] Pasternak B, Svanstrom H, Molgaard-Nielsen D, Krause TG, Emborg HD, Melbye M, et al. Vaccination against pandemic A/H1N1 2009 influenza in pregnancy and risk of fetal death: cohort study in Denmark. BMJ 2012;344:e2794.
[143] Bratton KN, Wardle MT, Orenstein WA, Omer SB. Maternal influenza immunization and birth outcomes of stillbirth and spontaneous abortion: a systematic review and meta-analysis. Clin Infect Dis 2015;60(5):e11–9.
[144] van der Maas N, Dijs-Elsinga J, Kemmeren J, van Lier A, Knol M, de Melker H. Safety of vaccination against influenza A (H1N1) during pregnancy in the Netherlands: results on pregnancy outcomes and infant's health: cross-sectional linkage study. BJOG 2016;123(5):709–17.
[145] Fell DB, Wilson K, Ducharme R, Hawken S, Sprague AE, Kwong JC, et al. Infant respiratory outcomes associated with prenatal exposure to maternal 2009 A/H1N1 influenza vaccination. PLoS ONE 2016;11(8):e0160342.
[146] Hviid A, Svanstrom H, Molgaard-Nielsen D, Lambach P. Association between pandemic influenza A(H1N1) vaccination in pregnancy and early childhood morbidity in offspring. JAMA Pediatr 2017;171(3):239–48.
[147] Walsh LK, Donelle J, Dodds L, Hawken S, Wilson K, Benchimol EI, et al. Health outcomes of young children born to mothers who received 2009 pandemic H1N1 influenza vaccination during pregnancy: retrospective cohort study. BMJ 2019;366:l4151.
[148] Ludvigsson JF, Strom P, Lundholm C, Cnattingius S, Ekbom A, Ortqvist A, et al. Maternal vaccination against H1N1 influenza and offspring mortality: population based cohort study and sibling design. BMJ 2015;351:h5585.
[149] Zerbo O, Qian Y, Yoshida C, Fireman BH, Klein NP, Croen LA. Association Between Influenza Infection and Vaccination During Pregnancy and Risk of Autism Spectrum Disorder. JAMA Pediatr 2017;171(1):e163609.
[150] Naleway AL, Irving SA, Henninger ML, Li DK, Shifflett P, Ball S, et al. Safety of influenza vaccination during pregnancy: a review of subsequent maternal obstetric events and findings from two recent cohort studies. Vaccine 2014;32(26):3122–7.

[151] Conlin AM, Bukowinski AT, Sevick CJ, DeScisciolo C, Crum-Cianflone NF. Safety of the pandemic H1N1 influenza vaccine among pregnant U.S. military women and their newborns. Obstet Gynecol 2013;121(3):511–8.
[152] Kallen B, Olausson PO. Vaccination against H1N1 influenza with Pandemrix((R)) during pregnancy and delivery outcome: a Swedish register study. BJOG 2012;119(13):1583–90.
[153] Oppermann M, Fritzsche J, Weber-Schoendorfer C, Keller-Stanislawski B, Allignol A, Meister R, et al. A(H1N1)v2009: a controlled observational prospective cohort study on vaccine safety in pregnancy. Vaccine 2012;30(30):4445–52.
[154] Rubinstein F, Micone P, Bonotti A, Wainer V, Schwarcz A, Augustovski F, et al. Influenza A/H1N1 MF59 adjuvanted vaccine in pregnant women and adverse perinatal outcomes: multicentre study. BMJ 2013;346:f393.
[155] Kharbanda EO, Vazquez-Benitez G, Lipkind H, Naleway A, Lee G, Nordin JD, et al. Inactivated influenza vaccine during pregnancy and risks for adverse obstetric events. Obstet Gynecol 2013;122(3):659–67.
[156] Nordin JD, Kharbanda EO, Vazquez-Benitez G, Lipkind H, Lee GM, Naleway AL. Monovalent H1N1 influenza vaccine safety in pregnant women, risks for acute adverse events. Vaccine 2014;32(39):4985–92.
[157] Steinhoff MC, Omer SB, Roy E, El Arifeen S, Raqib R, Dodd C, et al. Neonatal outcomes after influenza immunization during pregnancy: a randomized controlled trial. Can Med Assoc J 2012;184(6):645–53.
[158] Omer SB. Maternal Immunization. N Engl J Med 2017;376(25):2497.
[159] Savitz DA, Fell DB, Ortiz JR, Bhat N. Does influenza vaccination improve pregnancy outcome? Methodological issues and research needs. Vaccine 2015;33(47):6430–5.
[160] Fell DB, Bhutta ZA, Hutcheon JA, Karron RA, Knight M, Kramer MS, et al. Report of the WHO technical consultation on the effect of maternal influenza and influenza vaccination on the developing fetus: Montreal, Canada, September 30-October 1, 2015. Vaccine 2017;35(18):2279–87.
[161] Hutcheon JA, Fell DB, Jackson ML, Kramer MS, Ortiz JR, Savitz DA, et al. Detectable risks in studies of the fetal benefits of maternal influenza vaccination. Am J Epidemiol 2016;184(3):227–32.
[162] Olsen SJ, Mirza SA, Vonglokham P, Khanthamaly V, Chitry B, Pholsena V, et al. The effect of influenza vaccination on birth outcomes in a cohort of pregnant women in Lao PDR, 2014-2015. Clin Infect Dis 2016;63(4):487–94.
[163] Vazquez-Benitez G, Kharbanda EO, Naleway AL, Lipkind H, Sukumaran L, McCarthy NL, et al. Risk of preterm or small-for-gestational-age birth after influenza vaccination during pregnancy: caveats when conducting retrospective observational studies. Am J Epidemiol 2016;184(3):176–86.
[164] Hutcheon JA, Savitz DA. Invited commentary: influenza, influenza immunization, and pregnancy-it's about time. Am J Epidemiol 2016;184(3):187–91.
[165] Nunes MC, Aqil AR, Omer SB, Madhi SA. The effects of influenza vaccination during pregnancy on birth outcomes: a systematic review and meta-analysis. Am J Perinatol 2016;33(11):1104–14.
[166] Slogrove AL, Goetghebuer T, Cotton MF, Singer J, Bettinger JA. Pattern of infectious morbidity in hiv-exposed uninfected infants and children. Front Immunol 2016;7:164.
[167] Richardson K, Weinberg A. Reduced immunogenicity of influenza vaccines in HIV-infected compared with uninfected pregnant women is associated with regulatory T cells. AIDS 2011;25(5):595–602.
[168] Abzug MJ, Nachman SA, Muresan P, Handelsman E, Watts DH, Fenton T, et al. Safety and Immunogenicity of 2009 pH1N1 vaccination in HIV-infected pregnant women. Clin Infect Dis 2013;56(10):1488–97.

[169] Weinberg A, Muresan P, Richardson KM, Fenton T, Dominguez T, Bloom A, et al. Determinants of vaccine immunogenicity in HIV-infected pregnant women: analysis of B and T cell responses to pandemic H1N1 monovalent vaccine. PLoS ONE 2015;10(4):e0122431.
[170] Ludvigsson JF, Zugna D, Cnattingius S, Richiardi L, Ekbom A, Ortqvist A, et al. Influenza H1N1 vaccination and adverse pregnancy outcome. Eur J Epidemiol 2013;28(7):579–88.
[171] Engels G, Hierweger AM, Hoffmann J, Thieme R, Thiele S, Bertram S, et al. Pregnancy-related immune adaptation promotes the emergence of highly virulent H1N1 influenza virus strains in allogenically pregnant mice. Cell Host Microbe 2017;21(3):321–33.
[172] Donahue JG, Kieke BA, King JP, Mascola MA, Shimabukuro TT, DeStefano F, et al. Inactivated influenza vaccine and spontaneous abortion in the Vaccine Safety Datalink in 2012–13, 2013–14, and 2014–15. Vaccine 2019;37(44):6673–81.

CHAPTER

8

Pertussis

Kirsten Maertens[a], *Kathryn Edwards*[b], *Elke E. Leuridan*[a]

[a]Centre for the Evaluation of Vaccination, Vaccine & Infectious Diseases Institute, Faculty of Medicine and Health Sciences, University of Antwerp, Antwerp, Belgium, [b]Division of Infectious Diseases, Vanderbilt Vaccine Research Program, Department of Pediatrics, Vanderbilt University School of Medicine, Nashville, TN, United States

Introduction

Neonates are more susceptible to severe pertussis disease and death than any other age group. Despite the high vaccine coverage of pertussis-containing vaccines globally, pertussis incidence has increased, particularly in high-income countries. The increasing incidence does not only affect young infants, yet also older age categories.

Immunization of pregnant women with acellular pertussis (aP) vaccines for adult use has been implemented in a number of high-income and middle-income countries to reduce the burden of disease among the youngest age categories. This strategy provides protection to neonates through the transport of pertussis specific maternal antibodies from mother to fetus, until infant vaccination is provided. Data on the immunogenicity and safety of maternal immunization have been obtained in large cohorts and are reassuring. Effectiveness data from United Kingdom (UK), United States (US) and some other countries, indicate that maternal immunization with an aP containing vaccine is highly effective in preventing severe pertussis disease and mortality among young infants. However, suppression of infant responses to primary and booster immunization, the so-called blunting effect by maternal antibodies after maternal immunization, has been demonstrated, yet the impact on disease burden has not been shown. Future studies of different or more restricted pertussis antigens in maternal vaccines and different

https://doi.org/10.1016/B978-0-12-814582-1.00009-7

immunization schedules in infants are needed to see if the impact on infant immune responses might be reduced.

Clinical burden of disease and epidemiology

Clinical burden of pertussis in children and adults

Pertussis disease occurs at all ages and is endemic in all countries worldwide. However, the clinical presentation and the resulting complications of disease vary in different age groups as outlined in Table 1.

Young infants are the most vulnerable to severe disease and mortality. During a recent outbreak in California, risk factors for fatal pertussis in infants were compared between 53 fatal and 183 non-fatal hospitalized cases [1]. Fatal cases had significantly lower birth weight, younger gestational age, younger age of disease onset, higher peak lymphocyte counts, and were less likely to have been treated with macrolide antibiotics. To better understand the pathogenesis of fatal pertussis in young infants, an autopsy series of 15 infants demonstrated necrotizing bronchiolitis, intra-alveolar hemorrhage, and fibrinous edema in the lungs [2]. Although pulmonary co-infections with other viral or bacterial organisms were reported in some of these fatal cases, the predominant histopathologic lung findings did not differ with co-infections, leading one to infer that the pneumonia associated with pertussis was caused by the organism itself. In addition, there has been an increasing appreciation for the role of pulmonary hypertension as a complication of pertussis in young infants, often contributing to the fatal outcome [3]. Encephalopathy is also

TABLE 1 Clinical presentation and complications of pertussis per age group.

Pertussis clinical features	Pertussis among children, adolescents and adults	Pertussis complications in children
Incubation period 7–10 days (range 4–21 days) Insidious onset, similar to the common cold with nonspecific cough Fever usually minimal throughout course of illness Catarrhal stage: 1–2 weeks Paroxysmal cough stage: 1–6 weeks Convalescence: weeks to months	Presentation in infants may be with cough and apnoea. The typical 'whoop' is often absent Disease in older children and adults is often milder than in infants and young children Infection may be asymptomatic, or may present as classic pertussis Persons with mild disease may transmit the infection Older persons often source of infection for children	Secondary bacterial pneumonia – most common Neurologic complications – seizures, encephalopathy more common among infants Otitis media Anorexia Dehydration Pneumothorax Epistaxis Subdural hematomas Hernias Rectal prolapse

Adapted from Pink book CDC. https://www.cdc.gov/vaccines/pubs/pinkbook/pert.html.

reported in infants with pertussis, but usually results from hypoxia or intracranial bleeding associated with paroxysmal cough. Approximately one-third of children with pertussis encephalopathy succumb to the acute illness, one-third survive with permanent brain damage, and one-third recover without obvious neurologic sequelae [4].

Pertussis also occurs in older children, adolescents and adults. Serologic surveys indicate that it is more common than generally appreciated [5–7]. Symptoms of pertussis infections in this age group tend to be less severe than in young children but are often associated with persistent cough. Occasionally they may even be asymptomatic (Table 1). In pregnant women, pertussis infection is not associated with increased morbidity, abnormal fetal development or adverse pregnancy outcomes [8,9].

Adolescents and adults frequently serve as the source of infection for infants [10,11]. In addition, family studies of children with culture-confirmed pertussis disease and seroprevalence studies have shown that asymptomatic infections are common in adolescents and adults [12]. However, the highest incidence of pertussis disease can still be found in infants below six months of age, too young to be completely protected by the currently available infant vaccines and vaccination schedules [13].

During the last two decades, an increase in the number of cases of pertussis in adolescents and young adults has been reported. As early as 2002, the Global Pertussis Initiative recommended the routine use of Tdap boosters for adolescents and adults within the immunization programs of developed countries [14]. This has not been uniformly implemented and rapid waning immunity of the boosters has been shown [15].

One of the most exciting advances in understanding the pathogenesis of *Bordetella pertussis* has been the development of the infant baboon model (*Papio anubis*) since it appears to mirror infant disease. Infecting infant baboons with live *B. pertussis* organisms results in low-grade fever, persistent paroxysmal cough, lymphocytosis and protection from subsequent pertussis challenge [16]. This model also facilitates the study of colonization and transmission of pertussis [17]. If infant baboons were vaccinated with either an aP or whole cell pertussis (wP) containing vaccine and subsequently challenged with *B. pertussis* organisms, the challenged baboons were protected from clinical illness. However, neither aP or wP containing vaccines protected against colonization after challenge. Yet, animals vaccinated with wP containing vaccines cleared colonization twice as fast as aP containing vaccinated animals [18]. Also, it was shown that aP vaccinated infant baboons challenged with *B. pertussis* organisms were still able to transmit infection to susceptible cage mates. Therefore, it is very likely that asymptomatic humans immunized with an aP containing vaccines could similarly be capable of transmitting pertussis to susceptible infants and children [19]. Maternal and neonatal immunization has also been studied in this model and protection against symptomatic disease, but not colonization, is conferred by both vaccine approaches [20].

The baboon model has great appeal for the further investigations of the mechanisms of transmission and disease and for the testing of new vaccines and therapeutic drugs. In addition, carefully monitored human pertussis challenge studies also have been proposed to further the understanding of the pathogenesis of pertussis and the extent of protection afforded by administration of aP containing vaccines [21].

Epidemiology of pertussis

Pertussis epidemiology has a cyclical nature, peaking every 3–5 years. However, during recent years, some countries have experienced an increase in the incidence of pertussis [22]. The 2014 World Health Organization (WHO)-Strategic Advisory Group of Experts (SAGE) reports reviewed country-specific data from 19 countries for the express purpose of assessing the status of pertussis disease control globally. Overall, this report provided no evidence of a broad resurgence of pertussis at the global level. However, data from 5 out of 19 countries in this report, namely Australia, Chile, Portugal, United Kingdom (UK) and United States (US), indicated a true resurgence of disease in the last decade [23]. The reasons for this resurgence are likely multifactorial including; an increase in disease awareness due to strengthening of surveillance systems, an increase in overall laboratory testing and an enhanced sensitivity of the polymerase chain reaction (PCR) diagnostic methods, waning immunity both after natural infection and vaccination [24], the switch from wP to aP containing vaccines in many high-income countries resulting in a switch from a T-Helper-1 (Th1) and Th17 driven immune response to a Th2 driven immune response [25], pathogen adaptation with changes in the antigenic and genotypic characteristics of circulating *Bordetella pertussis* strains resulting in the emergence of Pertactin (Prn) deficient strains, particularly in countries using aP vaccines, and additional factors such as variable vaccine uptake and inadequate adult booster coverage [26]. The relative contribution of each of these individual factors is not entirely clear. Prn deficient *Bordetella pertussis* strains were first reported in France in 2007 [27] and now many of the strains in the United States are Prn deficient. Yet, controversy exists on the clinical relevance of infection with Prn negative strains since enhanced surveillance has not revealed more severe disease nor lower vaccine efficacy against these strains [28–30].

The majority of data on the burden of pertussis disease come from high-income countries. Two illustrative figures, one from the United Kingdom and another from the United States document the age distribution of cases in those regions (Fig. 1A and B).

The United Kingdom data demonstrate the large pertussis outbreak that occurred in 2012. Although pertussis disease was reported in all age groups, the highest incidence was noted in the youngest children,

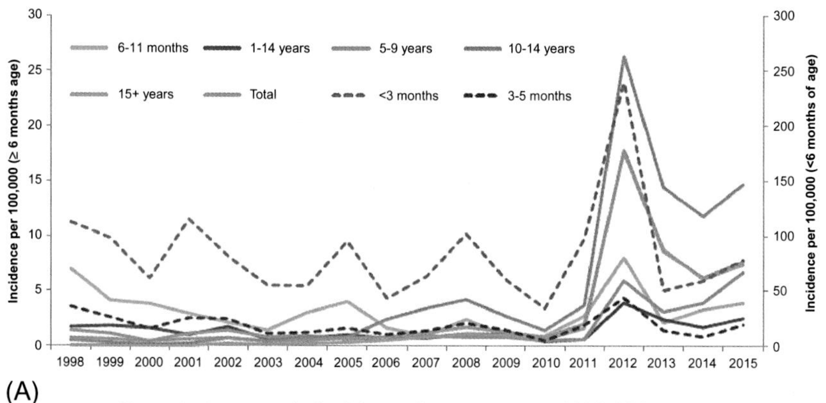

(A)

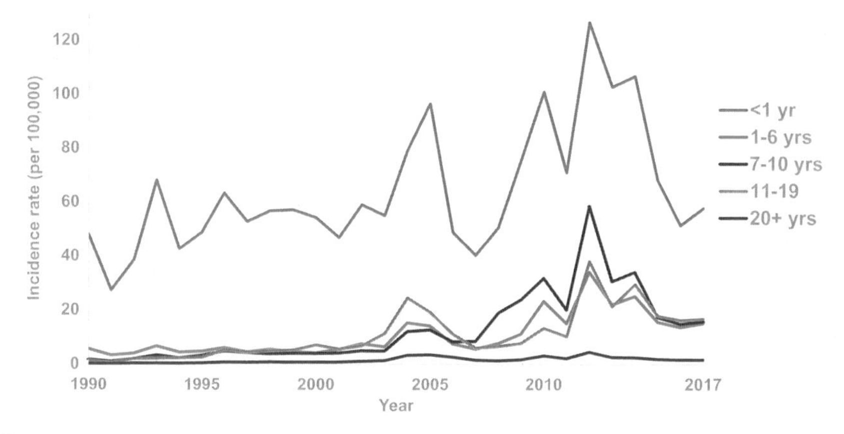

(B) SOURCE: CDC. National Notifiable Diseases Surveillance System, 2017.

FIG. 1 (A) Incidence of laboratory-confirmed pertussis cases by age group in England: 1998–2015. (B) Reported pertussis incidence by age group in the USA: 1990–2017. *Source: (A) Health Protection Report, vol. 10, No 16, May 6, 2016; (B) National Notifiable Diseases Surveillance System, 1990–2017, https://www.cdc.gov/pertussis/surv-reporting.html.*

particularly in those <3 months of age. Increased pertussis incidence has also been noted in the United States with the highest incidence reported in children under one year of age but with a marked increase in children between 7 and 10 years of age. This coincides with the reports on waning immunity in children receiving solely aP containing vaccines [31].

Assessing the global burden of pertussis is more challenging since comprehensive and reliable data from low- and middle-income countries are often lacking. WHO global figures for 2017 projected 143,963 pertussis cases, with an estimated 89,000 deaths [32]. A WHO modeling study was recently published using data from 2014 to estimate global pertussis cases and deaths. Pertussis cases were defined as a coughing illness lasting at least 2 weeks with paroxysms of coughing, inspiratory whooping, or post-tussive vomiting. Using this model, it was estimated that there were 24.1 million pertussis cases and 160,700 deaths from pertussis in children younger than 5 years during 2014, with the African region contributing the largest proportions [33]. Recently, Chow et al. systematically reviewed studies reporting pertussis mortality rates. In high-income countries, during a prevaccine observation period of more than 50 years, pertussis related mortality was reduced in both infants and 1–4 year-olds by >80%, concomitant with improvements in living conditions. After introduction of diphtheria, tetanus and pertussis (DTP) vaccines in high income countries (HIC), a further reduction of pertussis related deaths of over 98% occurred with few residual deaths in younger individuals [34]. Overall, the average annual pertussis mortality rates supplied by 15 countries to the WHO for the decade 2003–2012, mainly from high-income countries, ranged from 0.1 to 38.6 per 1 million births, with rates in most countries between 3.0 and 10.0 per 1 million births. These data are illustrated in Fig. 2 [35]. Gabutti and Rota also report a mean case fatality rate (CFR) of 0.2% in HIC and up to 4% in low- and middle-income countries (LMIC) in children below one year of age [36].

The epidemiology of pertussis in LMIC, where wP containing vaccines are more commonly used as part of national immunization programs, is less well documented, however most of the burden of pertussis is believed to occur in LMIC [37]. In 2016, a report from the Bill & Melinda Gates

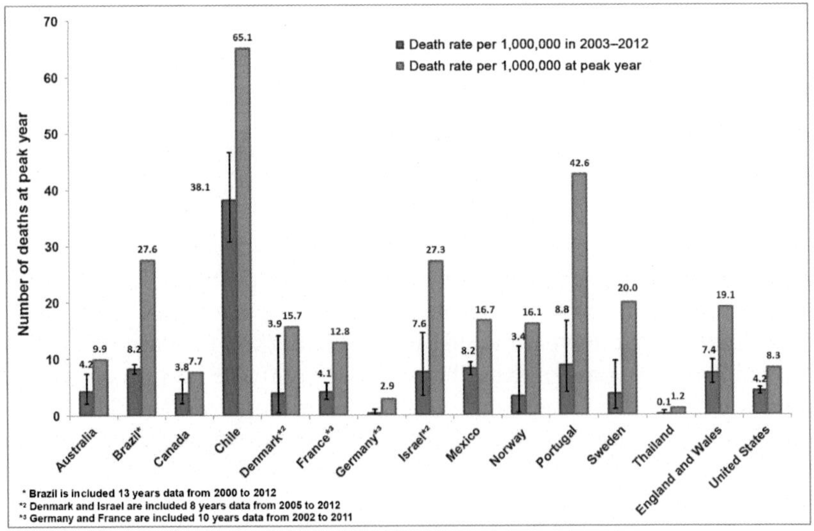

FIG. 2 Infant death rates per 1,000,000 per decade (2003–2012) and at peak year. *Source: SAGE report WHO.*

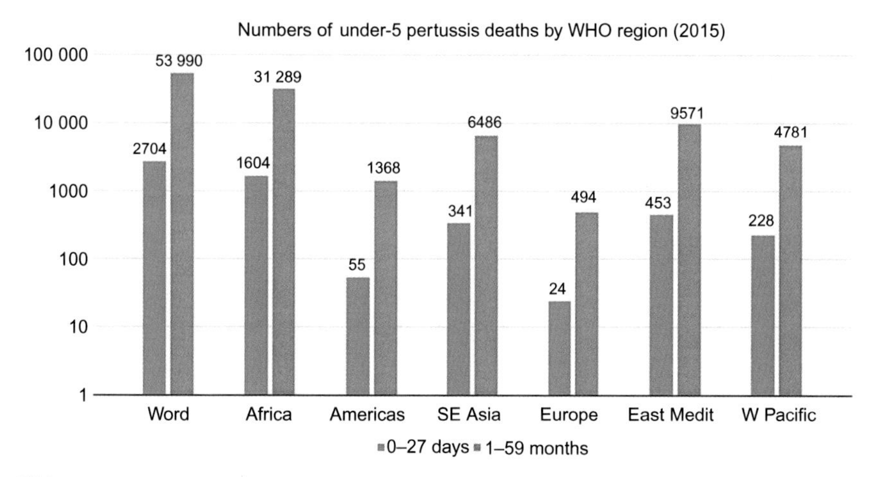

FIG. 3 Numbers of under-5 pertussis deaths by WHO region (2015).

Foundation Symposium reviewed the global burden of pertussis disease in LMIC and found major local differences [35]. Despite differences in study design, there is consistent evidence across studies for circulation of pertussis in all geographies. Overall, the pertussis burden is higher in Africa and Asia, especially when compared with Europe and the Americas (Fig. 3).

There are several difficulties with assessing and comparing the burden of pertussis disease between countries, for example, different case definitions are used worldwide, some categorizing cases as confirmed or probable and others based on clinical presentation alone [38], making comparisons problematic. In addition, there is a difference in access to diagnostic tools, lack of established surveillance systems, and underreporting or misdiagnosis due to lack of knowledge and awareness of the disease. For the African region, the lower childhood immunization coverage (74% for DTP3 in 2016) compared to the 95% global coverage rate, and a lack of general awareness and disease education, contributes to a higher disease burden but lower capture of the actual number of cases [39].

Not only do LMIC struggle with the reliability of the surveillance data, but problems also exist in HIC as well. Under-diagnosis can be due to a lack of appreciation for the disease, misdiagnosis, and underreporting. An excellent example is a recent capture-recapture analysis conducted in the Netherlands [40]. Pertussis hospitalization data were matched to reported pertussis cases for 2007–2014. The completeness of reporting in infants <2 years of age was about 50%, while in the older population it was only about 20%.

Several seroprevalence studies in the Netherlands, Belgium, The Gambia, the US and the Czech Republic compared reported pertussis incidence with seroprevalence studies and found a discrepancy between the number of individuals who had elevated antibody titers to pertussis antigens and to the actual reported pertussis incidence [6,7,41–44]. Also, in a Chinese study, 5% of the adult population between 20 and 39 years old had antibody concentrations indicative of a recent infection [45]. However, a recent report by Van der Lee et al. described the persistence of anti-PT antibodies after infection for a few years, suggesting that caution should be used in interpreting serologic data for the purpose of determining recent infection [46]. In contrast a number of earlier studies suggested that pertussis toxin antibody titers waned rapidly and elevated titers were indicative of recent disease [47]. Some of these studies have included pregnant women [48,49]. In addition, sero-surveys have documented the low levels of antibody in pregnant women who have not been recently immunized, and the lack of transplacental antibodies in women who were not recently immunized to their infants to protect them from disease [50–53].

Pertussis vaccines

Two general types of pertussis vaccines are currently available globally [54]. In many high income countries only aP vaccines combined with diphtheria and tetanus toxoids and other antigens are available, while in LMIC wP combination vaccines are largely used in the public sector with aP combination vaccines available in the private market. The wP vaccines were developed first and are suspensions of the inactivated whole *Bordetella pertussis* organisms combined with tetanus and diphtheria toxoids. These wP vaccines are only licensed for use in young children, with booster doses administered to adolescents and adults being all aP based vaccines. The pertussis components in the aP vaccines include one or more highly purified pertussis antigens. All contain pertussis toxin (PT), and in different combination, they also contain filamentous hemagglutinin (FHA), pertactin (Prn) or fimbriae (FIM) in addition to other non-pertussis antigens [55]. Studies using a genetically detoxified pertussis vaccine in adolescents have demonstrated significantly enhanced immunogenicity when compared with a standard Tdap vaccine. These results led to the licensure of a vaccine containing only genetically inactivated PT either with or without the diphtheria and tetanus toxoids in Thailand (BioNet Asia) [56].

Many high-income countries have replaced the infant primary series of DTwP with DTaP vaccines due to the improved reaction profile of the DTaP. However, the WHO currently recommends that countries using wP vaccines for their primary immunization schedules should continue to do so, since wP vaccines appear to provide more lasting protection [55]. Studies conducted in Australia by Sheridan et al. [57], confirm an increased risk of pertussis disease at the age of 10–13 years, in children receiving only aP vaccine in a primary and booster series when compared with children receiving one or more wP containing vaccines for priming with aP childhood booster doses.

More detailed immunologic studies have recently been published assessing the impact of wP and aP priming in infancy on responses to Tdap boosters in adolescents. It has been demonstrated that aP primed children had significantly lower antibody titers after booster doses, compared to wP primed infants who had a more protective Th1 type response and better immune memory [58]. Currently most pregnant women who receive Tdap in those countries where it is recommended have been primed with wP vaccines in infancy. However, in the future, pregnant women will be more likely to be aP primed. Thus, immune responses after repeated Tdap in pregnant women who have been primed in infancy with aP containing vaccines will need to be assessed [58].

There are several different pertussis-containing vaccines that may be used in adolescents and adults, that are included in national-level recommendations for use in pregnancy.

Adacel (marketed in Europe as Covaxis or Triaxis) is a five-component acellular pertussis vaccine. It contains PT (2.5 μg), FHA (5 μg), Prn (3 μg), FIM type 2 and 3 (5 μg), tetanus toxoid (5Lf [limit of flocculation]), diphtheria toxoid (2 Lf) and adjuvant (0.33 mg aluminum). Adacel is licensed for use in most countries for persons 4 years of age or older (in the United States, persons 10 to 64 years of age) and is supplied in thimerosal-free, single-dose vials. Combined with IPV, the product is marketed as Repevax [59].

Boostrix is a three-component acellular pertussis vaccine. It contains PT (8 μg), FHA (8 μg), Prn (2.5 μg), tetanus (5 Lf) toxoid and diphtheria toxoid (2.5 Lf), and tetanus toxoid (5Lf), and adjuvant (<0.39 mg aluminum). Boostrix is licensed in most countries for persons 4 years of age or older (in the United States, persons 10 years of age and older). It is also available combined with IPV as Boostrix-IPV [60].

Manufacturers in Asia have developed a variety of acellular pertussis vaccines for their local markets. For example, in China, at least 5 locally-produced acellular pertussis vaccines are licensed [61]. The aforementioned Thai manufactured vaccine contains genetically inactivated PT either with or without the diphtheria and tetanus toxoids [56].

Current recommendations for the use of pertussis vaccines in pregnant women

As a result of the resurgence of pertussis disease, particularly in infants, national advisory bodies (e.g. US (2011) [62], UK (2012) [63], Belgium (2013) [64] and others), have recommended immunization with tetanus, diphtheria and acellular pertussis (Tdap) vaccine for all pregnant women.

This strategy offers passive protection immediately at birth until the initiation of the primary immunization series in the first months of life, thus closing the neonatal susceptibility gap. There were, however, major gaps in the scientific knowledge at the time of the recommendation: on safety, immunogenicity, interference of maternal antibodies with aP or wP infant vaccine responses, and the impact on the breastmilk antibody composition.

In the US, Tdap vaccination has been recommended in the late second or third trimester of pregnancy by the Advisory Committee on Immunization practices (ACIP) (Centers for Disease Control and Prevention (CDC)) since August 2011 [9,65]. If the vaccination is not given during pregnancy, it should be administered in the immediate post-partum as part of the cocoon strategy [66]. In October 2012, this recommendation was updated stating that every woman should be vaccinated with a Tdap vaccine during every pregnancy [65].

In the UK, Tdap-Inactivated Polio Vaccine (IPV) vaccination during every pregnancy was also recommended by the Department of Health in October 2012. Originally, the recommendation was to vaccinate the pregnant women preferably between 28 and 32 weeks of pregnancy, but in April 2016, this time window was extended to 16–32 weeks of pregnancy to increase the opportunities for women to be vaccinated in pregnancy. Although the vaccination can be offered until 38 weeks of pregnancy [67].

In Belgium, maternal pertussis vaccination was recommended for every pregnancy between 24 and 32 weeks of gestation by the National Immunization Technical Advisory Group (NITAG) (Superior Health Council) in August 2013. If women are not vaccinated during pregnancy, they are recommended to receive Tdap vaccine in the immediate post-partum as part of the cocoon strategy [68]. The cocoon strategy is where all those in close contact with the infant receive pertussis vaccination, this is further explained in Chapter 2.

In addition, maternal vaccination is also recommended in a variety of other countries worldwide including Argentina [69], Australia [70], Brazil, Canada, Colombia, Czech Republic [71], El Salvador, Greece [72], Ireland [71], Israel [73], Italy [71], Mexico, New Zealand, Panama, Spain [74], Switzerland and The Netherlands [75]. Fig. 4 shows the countries recommending maternal Tdap vaccination, by March 2019, to the best of our knowledge.

Evidence of effectiveness of pertussis vaccination in pregnancy

The strategy of vaccinating pregnant women with pertussis-containing vaccines is not a new strategy. Already in the early 20th century, they vaccinated pregnant women with multiple doses of wP containing vaccines in the second or third trimester of pregnancy with a good safety, immunogenicity and effectiveness of the strategy [76–81] (Table 2).

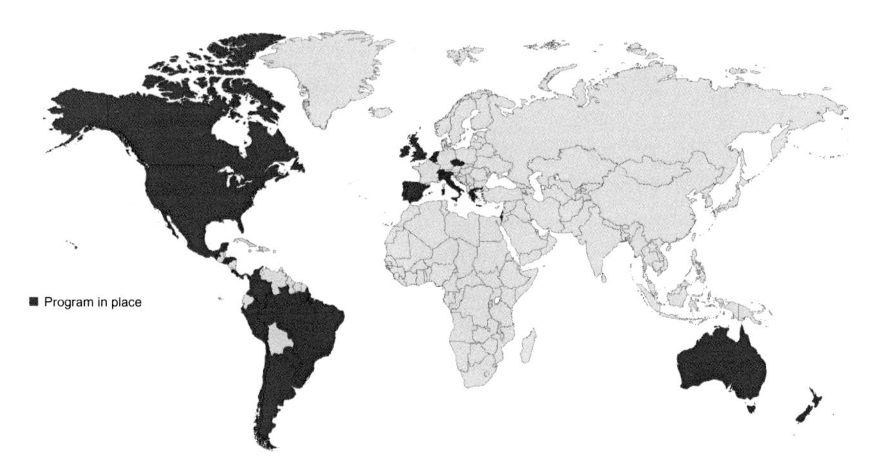

FIG. 4 Countries with existing maternal immunization programs (Status September 2019).

There is no serologic correlate of protection for pertussis, but higher antibody titers against pertussis toxin (PT) and to a lesser extent pertactin (Prn), have been associated with better protection [82]. Several studies confirmed the beneficial effect of maternal vaccination on the antibody titer in cord blood with documented higher titers of antibodies in neonates persisting until primary vaccination is started [52,83–85]. The maternal antibodies are transported in utero by the placental FcRn receptor, and the transport is most efficient during the last trimester of gestation [86]. Recent Swiss data indicate however, that maternal immunization in the second trimester is associated with higher levels of pertussis antibodies in cord blood compared to third trimester vaccination, both in term and preterm infants [87,88]. However, the optimal gestational age for Tdap vaccination is still subject for debate and continued study, but for any protection to be conferred to the infant, vaccination should occur at least 2 weeks prior to delivery [89].

It must be noted however, that high concentrations of maternal antibody are associated with a reduced infant antibody response to aP vaccines during their primary vaccination and to a lesser extent after booster vaccination [90], this is further explored in Chapter 4. For example, Voysey et al. reported that both maternal antibody concentrations and infant age at the time of the first primary vaccination influenced the magnitude of the infant antibody responses [91]. The suppressive effects of maternal antibody have been reported for nearly all vaccines contained in the primary immunization series, with suppression of some antibody responses, noted for periods as long as 24 months [92]. The interference effect is highly variable for different vaccines included in the infant vaccination schedule and even in different studies of the same vaccine. Blunting of infant responses to wP vaccination has been investigated in a study in Pakistan [93],

TABLE 2 Summary of historical maternal pertussis vaccination studies.

Study	Year	Number of vaccinated mothers	Vaccine type/doses administered during pregnancy	Timing of vaccination	Safety outcomes	Immunogenicity outcomes	Effectiveness outcomes
Lichty et al. [76]	1938	28	wP; 3 doses	Third trimester of pregnancy	Local adverse reactions, one systemic reaction	Neonatal antibody concentration influenced by maternal antibody level, history of pertussis and active immunization	Not reported
Mishulow et al. [77]	1942	29	wP; 3 doses	Third trimester of pregnancy	No adverse pregnancy outcomes	Increase in protective, agglutinating and complement-fixing antibodies in mothers	Not reported
Cohen and Scadron [78, 79]	1943, 1946	±170	wP; 6 doses	Second or third trimester of pregnancy	Arm pain, swelling, fever in 2% of cases; no adverse pregnancy outcomes	Transfer of protective antibodies to neonate; rapid decline of protective antibodies in infants	0/8 immunized and 3/6 unimmunized exposed infants developed pertussis
Kendrick et al. [80]	1945	57	wP; 3 doses	Not reported	Not reported	Increase in maternal antibody concentration; transfer of antibodies to infants	Not reported
Adams et al. [81]	1946	16	wP; 3 doses	Last trimester of pregnancy	Not reported	Good immune response to vaccination in mothers; rapid decline in antibodies in infants	Not reported

yet without maternal immunization. In contrast to Englund et al. [94], these authors report no blunting effect on wP responses in infants.

One must also be mindful that all the antigens included in the pertussis-containing vaccines recommended for use in pregnancy have the potential to suppress infant immune responses to primary immunization but the impact of this lower immune response on prevention of disease has not been demonstrated. As an example, blunting of the infant pneumococcal immune response after maternal Tdap vaccination is noted. However, despite this blunting, no effect on the seroprotection is seen [95, 96]. The administration of a monovalent pertussis vaccine consisting of a single pertussis antigen, pertussis toxin (PT), without the diphtheria and tetanus toxoids, to pregnant baboons recently demonstrated complete protection against respiratory pertussis challenge in their unborn infants [97]. The baboon-model provides evidence that transplacental antibody to PT alone was sufficient for newborn protection and that the avoidance of the other antigens in Tdap might ameliorate issues related to suppression of infant responses to routine vaccination.

Until recently, limited or no evidence regarding the effectiveness of maternal vaccination with aP containing vaccines on morbidity and mortality was available except from two UK studies conducted in the first months following the introduction of their national recommendation during their outbreak [98]. The first study, a case-control study with a relatively small sample size of 58 cases and 55 controls showed a vaccine effectiveness (VE) for maternal Tdap vaccination against laboratory confirmed pertussis disease of 93% (95% CI 81–97%) [99]. A comparable VE was found in another study (N=26,684) using the screening method, also known as the case-coverage method, with an overall VE of 91% (95% CI 84–95%) in infants below 3 months of age if mothers were vaccinated at least 7 days before delivery. A significant drop in VE was observed if the vaccine was given later in pregnancy with an effectiveness of only 38% in infants from mothers vaccinated between 6 and 13 days before delivery [89].

Longer term UK reports indicate that there were significantly fewer pertussis related deaths with a VE against infant death of 95% (95% CI 79–100%) after the maternal Tdap recommendation was implemented [100,101], with a sustained effect for 3 years after the program began. Additionally, the number of pertussis-related hospitalizations of young infants, was reduced significantly after the recommendation for maternal immunization was made, with a larger effect when prenatal vaccination was performed with longer interval before delivery. The fatality rate in the total population decreased also in Argentina (69.9%; 95% CI 50.4–81.8%) 2 years after introduction of the maternal vaccination strategy, with highest reduction rate among infants below 2 years of age (83.7%; 95% CI 63.9–92.6%) [69].

When looking at the effect of maternal Tdap on hospitalizations in preterm born infants in the UK, an overrepresentation of this population is seen with even a longer duration of hospitalization [102] indicating less benefit from the maternal pertussis vaccination strategy probably due to higher vulnerability, lesser time for transplacental transport and the reduced opportunity for maternal vaccination in this population.

In the last two years, more globally obtained data on the effectiveness of the maternal pertussis vaccination strategy have been published. A case-control study conducted in Spain showed similar results to the UK with an overall VE of 90.9% (95% CI 56.6–98.1%) [103]. A US study performed on a large cohort of infants (N=148,981) demonstrated a VE of 91.4% (95% CI 19.5–99.1%) in the first two months of life and of 69.0% (95% CI 43.6–82.9%) during the first year of life [104]. The effectiveness of each of the subsequent doses of the primary vaccination schedule was also checked and VE at each DTaP dose from infants born to mothers who received Tdap during pregnancy was similar in the sensitivity analysis. In addition, receipt of a Tdap vaccine within two years before pregnancy, which is in fact pre-pregnancy vaccination, provided protection against disease with a VE of 55.6% (95% CI 20.1–75.4%) in the first year of life. Post-partum vaccination on the other hand did not provide protection against pertussis in the first year of life [104]. Another study in California by Winter et al. also demonstrated that prenatal Tdap vaccination was 85% (95% CI 33–98%) more effective than post-partum vaccination in preventing pertussis in infants below 8 weeks of age [105]. If infants, despite maternal immunization, still contract pertussis disease, maternal vaccination is 58% (95% CI 15–80%) effective in preventing hospitalization and ICU admission [106]. In a case-control study conducted in the United States by Skoff et al. [107], no effectiveness was shown when the Tdap vaccine was administered during the post-partum period. When Tdap was administered at any time before pregnancy, an overall VE of 50.8% (95% CI 2.1–75.2%) was found with a VE of 83.0% (95% CI 49.6–94.3%) if the vaccine was administered within 2 years before pregnancy, higher than the pre-pregnancy Tdap VE reported by Baxter et al. In the study by Skoff et al., Tdap VE was 77.7% (95% CI 48.3–90.4%) for third trimester vaccination compared to a VE of 64.3% (95% CI 13.8-88.8%) for first or second trimester vaccination. In addition, when vaccinating pregnant women according the ACIP recommendation between 27 and 36 weeks of gestation, a VE of 78.4% (95% CI 49.8–90.7%) was reported. Finally, in an Australian case-control study, maternal Tdap vaccine had a VE of 69% (95% CI 13–89%) for the prevention of laboratory confirmed pertussis in infants <3 months of age and a VE against pertussis disease requiring hospitalization of 94% (95% CI 59–99%). Maternal Tdap vaccination is recommended between 28 and 32 weeks of gestation of every pregnancy in Australia [108].

Evidence of safety of pertussis vaccination in pregnancy

The first large prospective safety study of Tdap vaccination in pregnancy resulted from the implementation of universal maternal Tdap administration in the UK. In that report, 20,074 pregnant women vaccinated with Tdap during pregnancy were compared with a historical control group of unvaccinated women. These data showed no increased risk of stillbirth within 14 days after vaccination, no maternal and neonatal deaths, no increase in pre-eclampsia, and no other predefined adverse events related to pregnancy [109].

In the US, 123,494 singleton pregnancies were monitored for adverse reactions and within this cohort, 26,229 women received a Tdap vaccine during pregnancy. There was no association with maternal Tdap vaccination and increased risk for preterm delivery, small for gestational age or hypertensive disorders during pregnancy. Only a small increased relative risk for chorioamnionitis during pregnancy was noted. However, these data should be interpreted carefully since there was no increased risk for preterm birth, a direct consequence of chorioamnionitis [110]. To further assess this possible relation between maternal Tdap vaccination and chorioamnionitis, a review of the Vaccine Adverse Event Reporting System (VAERS) was carried out. All cases of chorioamnionitis after any vaccine in pregnancy between 1990 and 2014 were considered. Only a few cases of chorioamnionitis were found accounting for 1% of all pregnancy reports to the VAERS [111]. Also, two other small observational studies could not replicate this association [112,113]. In contrast, a recent study carried out in the US supports the results found by Kharbanda et al. with a small but positive association between maternal Tdap and chorioamnionitis, but no increased risk for associated indicators of infant morbidity [114]. To further clarify the possible relation between maternal Tdap and chorioamnionitis, future studies should use standardized case definitions for chorioamnionitis, as suggested by the Global Alignment of Immunization safety Assessment in pregnancy (GAIA) consortium [115].

Since recommendations support repeat booster doses at every pregnancy, Sukumaran et al. studied the possible adverse outcomes of repeat Tdap vaccines during subsequent pregnancies [116,117]. Using the Vaccine Safety Datalink network in the United States that included 29,155 pregnant women, they found no increased risk of large local reactions, premature delivery, small for gestational age, or other adverse events following immunization when repeated Tdap booster doses were administered with intervals of 2 years, between 2 and 5 years, or greater than 5 year intervals [116].

Future prospects

Given the increase in pertussis in industrialized countries, there are a number of new pertussis vaccine approaches that have been proposed during the last decade. These theoretical approaches include returning to the whole cell vaccine in infancy, the addition of new pertussis antigens to the current acellular vaccines, the addition of adjuvants to the current vaccines, the development of live-attenuated pertussis vaccines and the development of a low-cost PT vaccine. None of these options have yet led to a new licensed pertussis vaccine, and none have been studied in pregnant women.

Conclusion

Pertussis immunization in pregnancy has been recommended in many HIC countries to protect young infants from severe pertussis disease. The strategy is safe and effective, based on the currently available data. There are indications of a blunting effect by vaccine-induced maternal antibodies, on the infant immune responses to aP vaccines, yet the clinical meaning is unclear and the effect appears transient. Many gaps remain in our basic knowledge of maternal pertussis vaccination: what is the functionality of these vaccine induced maternal antibodies, is there possibly a minimal effective interval between repeat doses, could we identify a biomarker for correlate of protection (either antibodies or something else), are there other (new) pertussis vaccines or strategies that would confer longer lasting immunity, what is the effect on cellular immune responses in infants in the presence of high maternal antibody titers, what is the effect of maternal pertussis antibodies when infants receive wP vaccines, and what is the effect on breastmilk composition. Research opportunities are plenty, in this quickly evolving field.

The relevance of the ongoing research on pertussis is not only offering scientific background for the strategy, but also providing answers and tools to policy makers, vaccine manufacturers and field workers for better care of pregnant women and their offspring.

Conflict of interest statement

Authors do not have a commercial or other association that might pose a conflict of interest (e.g., pharmaceutical stock ownership, consultancy, pharmaceutical board membership, relevant patents, or research funding).

References

[1] Winter K, Zipprich J, Harriman K, Murray EL, Gornbein J, Hammer SJ, et al. Risk factors associated with infant deaths from pertussis: a case-control study. Clin Infect Dis 2015;61:1099–106.
[2] Paddock CD, Sanden GN, Cherry JD, Gal AA, Langston C, Tatti KM, et al. Pathology and pathogenesis of fatal *Bordetella pertussis* infection in infants. Clin Infect Dis 2008;47:328–38.
[3] Goulin GD, Kaya KM, Bradley JS. Severe pulmonary hypertension associated with shock and death in infants infected with *Bordetella pertussis*. Crit Care Med 1993;21:1791–4.
[4] Litvak AM, Gibel H, et al. Cerebral complications in pertussis. J Pediatr 1948;32:357–79.
[5] Centers for Disease Control and Prevention. Pertussis (whooping cough): Surveillance and reporting; 2018.
[6] Huygen K, Rodeghiero C, Govaerts D, Leroux-Roels I, Melin P, Reynders M, et al. *Bordetella pertussis* seroprevalence in Belgian adults aged 20-39 years, 2012. Epidemiol Infect 2014;142:724–8.
[7] de Melker HE, Versteegh FG, Schellekens JF, Teunis PF, Kretzschmar M. The incidence of *Bordetella pertussis* infections estimated in the population from a combination of serological surveys. J Infect 2006;53:106–13.
[8] Hubka TA, Wisner KP. Vaccinations recommended during pregnancy and breastfeeding. J Am Osteopath Assoc 2011;111:S23–30.
[9] Murphy TV, Slade BA, Broder KR, Kretsinger K, Tiwari T, Joyce PM, et al. Prevention of pertussis, tetanus, and diphtheria among pregnant and postpartum women and their infants recommendations of the Advisory Committee on Immunization Practices (ACIP). MMWR Recomm Rep 2008;57:1–51.
[10] Bisgard KM, Pascual FB, Ehresmann KR, Miller CA, Cianfrini C, Jennings CE, et al. Infant pertussis: who was the source? Pediatr Infect Dis J 2004;23:985–9.
[11] Wendelboe AM, Njamkepo E, Bourillon A, Floret DD, Gaudelus J, Gerber M, et al. Transmission of *Bordetella pertussis* to young infants. Pediatr Infect Dis J 2007;26:293–9.
[12] Long SS, Welkon CJ, Clark JL. Widespread silent transmission of pertussis in families: antibody correlates of infection and symptomatology. J Infect Dis 1990;161:480–6.
[13] Masseria C, Martin CK, Krishnarajah G, Becker LK, Buikema A, Tan TQ. Incidence and burden of pertussis among infants less than 1 year of age. Pediatr Infect Dis J 2017;36:e54–61.
[14] Forsyth KD, Wirsing von Konig CH, Tan T, Caro J, Plotkin S. Prevention of pertussis: recommendations derived from the second Global Pertussis Initiative roundtable meeting. Vaccine 2007;25:2634–42.
[15] Klein NP, Bartlett J, Rowhani-Rahbar A, Fireman B, Baxter R. Waning protection after fifth dose of acellular pertussis vaccine in children. N Engl J Med 2012;367:1012–9.
[16] Warfel JM, Beren J, Kelly VK, Lee G, Merkel TJ. Nonhuman primate model of pertussis. Infect Immun 2012;80:1530–6.
[17] Warfel JM, Beren J, Merkel TJ. Airborne transmission of *Bordetella pertussis*. J Infect Dis 2012;206:902–6.
[18] Warfel JM, Zimmerman LI, Merkel TJ. Acellular pertussis vaccines protect against disease but fail to prevent infection and transmission in a nonhuman primate model. Proc Natl Acad Sci U S A 2014;111:787–92.
[19] Edwards KM. Unraveling the challenges of pertussis. Proc Natl Acad Sci U S A 2014;111:575–6.
[20] Warfel JM, Papin JF, Wolf RF, Zimmerman LI, Merkel TJ. Maternal and neonatal vaccination protects newborn baboons from pertussis infection. J Infect Dis 2014;210:604–10.
[21] Merkel TJ, Halperin SA. Nonhuman primate and human challenge models of pertussis. J Infect Dis 2014;209(Suppl. 1):S20–3.

[22] Plotkin SA. The pertussis problem. Clin Infect Dis 2014;58:830–3.
[23] WHO. WHO SAGE pertussis working group: Background paper; 2014.
[24] Sheridan SL, Frith K, Snelling TL, Grimwood K, McIntyre PB, Lambert SB. Waning vaccine immunity in teenagers primed with whole cell and acellular pertussis vaccine: recent epidemiology. Expert Rev Vaccines 2014;13:1081–106.
[25] Warfel JM, Edwards KM. Pertussis vaccines and the challenge of inducing durable immunity. Curr Opin Immunol 2015;35:48–54.
[26] Libster R, Edwards KM. Re-emergence of pertussis: what are the solutions? Expert Rev Vaccines 2012;11:1331–46.
[27] Bouchez V, Brun D, Cantinelli T, Dore G, Njamkepo E, Guiso N. First report and detailed characterization of *B. pertussis* isolates not expressing Pertussis Toxin or Pertactin. Vaccine 2009;27:6034–41.
[28] Bodilis H, Guiso N. Virulence of pertactin-negative *Bordetella pertussis* isolates from infants, France. Emerg Infect Dis 2013;19:471–4.
[29] Vodzak J, Queenan AM, Souder E, Evangelista AT, Long SS. Clinical manifestations and molecular characterization of pertactin-deficient and pertactin-producing *Bordetella pertussis* in children, Philadelphia 2007-2014. Clin Infect Dis 2017;64:60–6.
[30] Breakwell L, Kelso P, Finley C, Schoenfeld S, Goode B, Misegades LK, et al. Pertussis vaccine effectiveness in the setting of pertactin-deficient pertussis. Pediatrics 2016;137.
[31] Klein NP, Zerbo O. Use of acellular pertussis vaccines in the United States: can we do better? Expert Rev Vaccines 2017;16:1175–9.
[32] WHO. Immunization, vaccines and biologicals: Pertussis; 2018.
[33] Yeung KHT, Duclos P, Nelson EAS, Hutubessy RCW. An update of the global burden of pertussis in children younger than 5 years: a modelling study. Lancet Infect Dis 2017,17(9):974–980.
[34] Chow MY, Khandaker G, McIntyre P. Global childhood deaths from pertussis: a historical review. Clin Infect Dis 2016;63:S134–41.
[35] Sobanjo-Ter Meulen A, Duclos P, McIntyre P, Lewis KD, Van Damme P, O'Brien KL, et al. Assessing the evidence for maternal pertussis immunization: a report from the Bill & Melinda Gates Foundation Symposium on pertussis infant disease burden in low- and lower-middle-income countries. Clin Infect Dis 2016;63:S123–33.
[36] Gabutti G, Rota MC. Pertussis: a review of disease epidemiology worldwide and in Italy. Int J Environ Res Public Health 2012;9:4626–38.
[37] Tan T, Dalby T, Forsyth K, Halperin SA, Heininger U, Hozbor D, et al. Pertussis across the globe: recent epidemiologic trends from 2000 to 2013. Pediatr Infect Dis J 2015;34:e222–32.
[38] Hartzell JD, Blaylock JM. Whooping cough in 2014 and beyond: an update and review. Chest 2014;146:205–14.
[39] Muloiwa R, Wolter N, Mupere E, Tan T, Chitkara AJ, Forsyth KD, et al. Pertussis in Africa: findings and recommendations of the Global Pertussis Initiative (GPI). Vaccine 2018;36:2385–93.
[40] van der Maas NAT, Hoes J, Sanders EAM, de Melker HE. Severe underestimation of pertussis related hospitalizations and deaths in the Netherlands: a capture-recapture analysis. Vaccine 2017;35:4162–6.
[41] Scott S, van der Sande M, Faye-Joof T, Mendy M, Sanneh B, Barry Jallow F, et al. Seroprevalence of pertussis in the Gambia: evidence for continued circulation of *Bordetella pertussis* despite high vaccination rates. Pediatr Infect Dis J 2015;34:333–8.
[42] Chlibek R, Smetana J, Sosovickova R, Fabianova K, Zavadilova J, Dite P, et al. Seroepidemiology of whooping cough in the Czech Republic: estimates of incidence of infection in adults. Public Health 2017;150:77–83.
[43] Sutter RW, Cochi SL. Pertussis hospitalizations and mortality in the United States, 1985-1988. Evaluation of the completeness of national reporting. JAMA 1992;267:386–91.

[44] Heininger U. Pertussis vaccines. In: Vesikari T, Van Damme P, editors. Pediatric vaccines and vaccinations: A European textbook. Springer; 2017. p. 161–9.
[45] Chen Z, Zhang J, Cao L, Zhang N, Zhu J, Ping G, et al. Seroprevalence of pertussis among adults in China where whole cell vaccines have been used for 50 years. J Infect 2016;73:38–44.
[46] van der Lee S, Stoof SP, van Ravenhorst MB, van Gageldonk PGM, van der Maas NAT, Sanders EAM, et al. Enhanced *Bordetella pertussis* acquisition rate in adolescents during the 2012 epidemic in the Netherlands and evidence for prolonged antibody persistence after infection. Euro Surveill 2017;22 (47). https://doi.org/10.2807/1560-7917.ES.2017.22.47.17-00011.
[47] Le T, Cherry JD, Chang SJ, Knoll MD, Lee ML, Barenkamp S, et al. Immune responses and antibody decay after immunization of adolescents and adults with an acellular pertussis vaccine: the APERT Study. J Infect Dis 2004;190:535–44.
[48] Huygen K, Cabore RN, Maertens K, Van Damme P, Leuridan E. Humoral and cell mediated immune responses to a pertussis containing vaccine in pregnant and non-pregnant women. Vaccine 2015;33:4117–23.
[49] Abu Raya B, Srugo I, Kessel A, Peterman M, Vaknin A, Bamberger E. The decline of pertussis-specific antibodies after tetanus, diphtheria, and acellular pertussis immunization in late pregnancy. J Infect Dis 2015;212:1869–73.
[50] Healy CM, Munoz FM, Rench MA, Halasa NB, Edwards KM, Baker CJ. Prevalence of pertussis antibodies in maternal delivery, cord, and infant serum. J Infect Dis 2004;190:335–40.
[51] Hughes MM, Englund JA, Edwards K, Yoder S, Tielsch JM, Steinhoff M, et al. Pertussis seroepidemiology in women and their infants in Sarlahi District, Nepal. Vaccine 2017;35:6766–73.
[52] Hoang HT, Leuridan E, Maertens K, Nguyen TD, Hens N, Vu NH, et al. Pertussis vaccination during pregnancy in Vietnam: results of a randomized controlled trial. Vaccine 2016;34:151–9.
[53] Wanlapakorn N, Thongmee T, Vichaiwattana P, Leuridan E, Vongpunsawad S, Poovorawan Y. Antibodies to *Bordetella pertussis* antigens in maternal and cord blood pairs: a Thai cohort study. PeerJ 2017;5:e4043.
[54] Edwards K, Decker M. Pertussis vaccine. In: Plotkin vaccines; 2018. p. 711–61.
[55] WHO. Pertussis vaccines: WHO position paper—August 2015. Wkly Epidemiol Rec 2015;90:433–58.
[56] Sricharoenchai S, Sirivichayakul C, Chokephaibulkit K, Pitisuttithum P, Dhitavat J, Pitisuthitham A, et al. A genetically inactivated two-component acellular pertussis vaccine, alone or combined with tetanus and reduced-dose diphtheria vaccines, in adolescents: a phase 2/3, randomised controlled non-inferiority trial. Lancet Infect Dis 2018;18:58–67.
[57] Sheridan SL, Ware RS, Grimwood K, Lambert SB. Reduced risk of pertussis in whole-cell compared to acellular vaccine recipients is not confounded by age or receipt of booster-doses. Vaccine 2015;33:5027–30.
[58] van der Lee S, Hendrikx LH, Sanders EAM, Berbers GAM, Buisman AM. Whole-cell or acellular pertussis primary immunizations in infancy determines adolescent cellular immune profiles. Front Immunol 2018;9:51.
[59] Sanofi Pasteur Limited Company. Summary of product characteristics. Adacel; 2011.
[60] GlaxoSmithKline Inc. Product monograph. Boostrix; 2013.
[61] Wang L, Lei D, Zhang S. Acellular pertussis vaccines in China. Vaccine 2012;30:7174–8.
[62] ACIP. ACIP provisional recommendations for pregnant women on use of tetanus toxoid. In: Reduced diphtheria toxoid and acellular pertussis vaccine (Tdap); 2011.
[63] Department of Health and Social Care and Public Health England. Whooping cough vaccination programme for pregnant women: Extension to 2014; 2013.
[64] Hoge Gezondheidsraad B. Pertussis vaccination; 2013.

[65] Centers for Disease Control and Prevention (CDC). Updated recommendations for use of tetanus toxoid, reduced diphtheria toxoid, and acellular pertussis vaccine (Tdap) in pregnant women—Advisory Committee on Immunization Practices (ACIP), 2012. MMWR Morb Mortal Wkly Rep 2013;62:131–5.
[66] Centers for Disease Control and Prevention. Updated recommendations for use of tetanus toxoid, reduced diphtheria toxoid and acellular pertussis vaccine (Tdap) in pregnant women and persons who have or anticipate having close contact with an infant aged <12 months—Advisory Committee on Immunization Practices (ACIP), 2011. MMWR Morb Mortal Wkly Rep 2011; 60(41): 1424–1426.
[67] Public Health England. Vaccination against pertussis (whooping cough) for pregnant women—2016: Information for healthcare professionals; 2016.
[68] Superior Health Council Belgium. Pertussis vaccination. Update 2013. Superior Health Council; 2013.
[69] Vizzotti C, Neyro S, Katz N, Juarez MV, Perez Carrega ME, Aquino A, et al. Maternal immunization in Argentina: a storyline from the prospective of a middle income country. Vaccine 2015;33:6413–9.
[70] Government of South Australia. Whooping cough vaccine in pregnancy. Dent Prog 2017.
[71] ECDC. Vaccine schedule; 2016.
[72] Gkentzi D, Katsakiori P, Marangos M, Hsia Y, Amirthalingam G, Heath PT, et al. Maternal vaccination against pertussis: a systematic review of the recent literature. Arch Dis Child Fetal Neonatal Ed 2017.
[73] State of Israel: Ministry of Health. Whooping cough vaccination in pregnant women; 2017.
[74] Moreno-Perez D, Alvarez Garcia FJ, Aristegui Fernandez J, Cilleruelo Ortega MJ, Corretger Rauet JM, Garcia Sanchez N, et al. Immunisation schedule of the Spanish Association of Paediatrics: 2016 recommendations. An Pediatr 2016;84:60. e1–13.
[75] Rijksinstituut voor Volksgezondheid en Milieu. Kinkhoestvaccinatie bij zwangere vrouwen; 2016.
[76] Lichty JA, Slavin B, Bradford WL. An attempt to increase resistance to pertussis in newborn infants by immunizing their mothers during pregnancy. J Clin Invest 1938;17:613–21.
[77] Mishulow L, Leifer L, Sherwood C, Schlesinger SL, Berkey SR. Pertussis antibodies in pregnant women. Protective, agglutinating and complement-fixing antibodies before and after vaccination. Am J Dis Child 1942;64:608–17.
[78] Cohen P, Scadron SJ. The effects of active immunization of the mother upon the offspring. J Pediatr 1946;29:609–19.
[79] Cohen P, Scadron SJ. The placental transmission of protective antibodies against whooping cough. JAMA 1943;121:656–62.
[80] Kendrick PL, Gottshall RY, Anderson HD, Volk VK, Bunney WE, Top FH. Pertussis agglutinins in adults. Public Health Rep 1969;84:9–15.
[81] Adams JM, Kimball AC, Adams FH. Early immunization against pertussis. Am J Dis Child 1947;74:10–8.
[82] Taranger J, Trollfors B, Lagergard T, Sundh V, Bryla DA, Schneerson R, et al. Correlation between pertussis toxin IgG antibodies in postvaccination sera and subsequent protection against pertussis. J Infect Dis 2000;181:1010–3.
[83] Munoz FM, Bond NH, Maccato M, Pinell P, Hammill HA, Swamy GK, et al. Safety and immunogenicity of tetanus diphtheria and acellular pertussis (Tdap) immunization during pregnancy in mothers and infants: a randomized clinical trial. JAMA 2014;311:1760–9.
[84] Maertens K, Cabore RN, Huygen K, Hens N, Van Damme P, Leuridan E. Pertussis vaccination during pregnancy in Belgium: results of a prospective controlled cohort study. Vaccine 2016;34:142–50.

[85] Vilajeliu A, Gonce A, Lopez M, Costa J, Rocamora L, Rios J, et al. Combined tetanus-diphtheria and pertussis vaccine during pregnancy: transfer of maternal pertussis antibodies to the newborn. Vaccine 2015;33:1056–62.
[86] Faucette AN, Unger BL, Gonik B, Chen K. Maternal vaccination: moving the science forward. Hum Reprod Update 2015;21:119–35.
[87] Eberhardt CS, Blanchard-Rohner G, Lemaitre B, Boukrid M, Combescure C, Othenin-Girard V, et al. Maternal immunization earlier in pregnancy maximizes antibody transfer and expected infant seropositivity against pertussis. Clin Infect Dis 2016;62:829–36.
[88] Eberhardt CS, Blanchard-Rohner G, Lemaitre B, Combescure C, Othenin-Girard V, Chilin A, et al. Pertussis antibody transfer to preterm neonates after second-versus third-trimester maternal immunization. Clin Infect Dis 2017;64:1129–32.
[89] Amirthalingam G, Andrews N, Campbell H, Ribeiro S, Kara E, Donegan K, et al. Effectiveness of maternal pertussis vaccination in England: an observational study. Lancet 2014;384:1521–8.
[90] Edwards KM. Maternal antibodies and infant immune responses to vaccines. Vaccine 2015;33:6469–72.
[91] Voysey M, Kelly DF, Fanshawe TR, Sadarangani M, O'Brien KL, Perera R, et al. The influence of maternally derived antibody and infant age at vaccination on infant vaccine responses: an individual participant meta-analysis. JAMA Pediatr 2017;171:637–46.
[92] Gans H, Yasukawa L, Rinki M, DeHovitz R, Forghani B, Beeler J, et al. Immune responses to measles and mumps vaccination of infants at 6, 9, and 12 months. J Infect Dis 2001;184:817–26.
[93] Ibrahim R, Ali SA, Kazi AM, Rizvi A, Guterman LB, Bednarczyk RA, et al. Impact of maternally derived pertussis antibody titers on infant whole-cell pertussis vaccine response in a low income setting. Vaccine 2018;36:7048–53.
[94] Englund JA, Anderson EL, Reed GF, Decker MD, Edwards KM, Pichichero ME, et al. The effect of maternal antibody on the serologic response and the incidence of adverse reactions after primary immunization with acellular and whole-cell pertussis vaccines combined with diphtheria and tetanus toxoids. Pediatrics 1995;96:580–4.
[95] Ladhani SN, Andrews NJ, Southern J, Jones CE, Amirthalingam G, Waight PA, et al. Antibody responses after primary immunization in infants born to women receiving a pertussis-containing vaccine during pregnancy: single arm observational study with a historical comparator. Clin Infect Dis 2015;61:1637–44.
[96] Maertens K, Burbidge P, Van Damme P, Goldblatt D, Leuridan E. Pneumococcal immune response in infants whose mothers received Tdap vaccination during pregnancy. Pediatr Infect Dis J 2017.
[97] Kapil P, Papin JF, Wolf RF, Zimmerman LI, Wagner LD, Merkel TJ. Maternal vaccination with a monocomponent pertussis toxoid vaccine is sufficient to protect infants in a baboon model of whooping cough. J Infect Dis 2018;217:1231–6.
[98] Department of Health England. Whooping cough vaccination programme for pregnant women: Guidance; 2012.
[99] Dabrera G, Amirthalingam G, Andrews N, Campbell H, Ribeiro S, Kara E, et al. A case-control study to estimate the effectiveness of maternal pertussis vaccination in protecting newborn infants in England and Wales, 2012-2013. Clin Infect Dis 2015;60:333–7.
[100] Amirthalingam G, Letley L, Campbell H, Green D, Yarwood J, Ramsay M. Lessons learnt from the implementation of maternal immunization programs in England. Hum Vaccin Immunother 2016;12:2934–9.
[101] Amirthalingam G, Campbell H, Ribeiro S, Fry NK, Ramsay M, Miller E, et al. Sustained effectiveness of the maternal pertussis immunization program in England 3 years following introduction. Clin Infect Dis 2016;63:S236–43.
[102] Byrne L, Campbell H, Andrews N, Ribeiro S, Amirthalingam G. Hospitalisation of preterm infants with pertussis in the context of a maternal vaccination programme in England. Arch Dis Child 2018;103:224–9.

[103] Bellido-Blasco J, Guiral-Rodrigo S, Miguez-Santiyan A, Salazar-Cifre A, Gonzalez-Moran F. A case-control study to assess the effectiveness of pertussis vaccination during pregnancy on newborns, Valencian community, Spain, 1 March 2015 to 29 February 2016. Euro Surveill 2017;22(22):30545.
[104] Baxter R, Bartlett J, Fireman B, Lewis E, Klein NP. Effectiveness of vaccination during pregnancy to prevent infant pertussis. Pediatrics 2017;139(5).
[105] Winter K, Nickell S, Powell M, Harriman K. Effectiveness of prenatal versus postpartum tetanus, diphtheria, and acellular pertussis vaccination in preventing infant pertussis. Clin Infect Dis 2017;64:3–8.
[106] Winter K, Cherry JD, Harriman K. Effectiveness of prenatal tetanus, diphtheria, and acellular pertussis vaccination on pertussis severity in infants. Clin Infect Dis 2017;64:9–14.
[107] Skoff TH, Blain AE, Watt J, Scherzinger K, McMahon M, Zansky SM, et al. Impact of the US maternal tetanus, diphtheria, and acellular pertussis vaccination program on preventing pertussis in infants <2 months of age: a case-control evaluation. Clin Infect Dis 2017;65:1977–83.
[108] Saul N, Wang K, Bag S, Baldwin H, Alexander K, Chandra M, et al. Effectiveness of maternal pertussis vaccination in preventing infection and disease in infants: the NSW Public Health Network case-control study. Vaccine 2018;36:1887–92.
[109] Donegan K, King B, Bryan P. Safety of pertussis vaccination in pregnant women in UK: observational study. BMJ 2014;349:g4219.
[110] Kharbanda EO, Vazquez-Benitez G, Lipkind HS, Klein NP, Cheetham TC, Naleway A, et al. Evaluation of the association of maternal pertussis vaccination with obstetric events and birth outcomes. JAMA 2014;312:1897–904.
[111] Datwani H, Moro PL, Harrington T, Broder KR. Chorioamnionitis following vaccination in the Vaccine Adverse Event Reporting System. Vaccine 2015;33:3110–3.
[112] Berenson AB, Hirth JM, Rahman M, Laz TH, Rupp RE, Sarpong KO. Maternal and infant outcomes among women vaccinated against pertussis during pregnancy. Hum Vaccin Immunother 2016;12:1965–71.
[113] Morgan JL, Baggari SR, McIntire DD, Sheffield JS. Pregnancy outcomes after antepartum tetanus, diphtheria, and acellular pertussis vaccination. Obstet Gynecol 2015;125:1433–8.
[114] DeSilva M, Vazquez-Benitez G, Nordin JD, Lipkind HS, Klein NP, Cheetham TC, et al. Maternal Tdap vaccination and risk of infant morbidity. Vaccine 2017;35:3655–60.
[115] Bonhoeffer J, Kochhar S, Hirschfeld S, Heath PT, Jones CE, Bauwens J, et al. Global alignment of immunization safety assessment in pregnancy—The GAIA project. Vaccine 2016;34:(49):5993–5997.
[116] Sukumaran L, McCarthy NL, Kharbanda EO, McNeil MM, Naleway AL, Klein NP, et al. Association of Tdap vaccination with acute events and adverse birth outcomes among pregnant women with prior tetanus-containing immunizations. JAMA 2015;314:1581–7.
[117] Sukumaran L, Omer SB. Repeat Tdap vaccination and adverse birth outcomes—Reply. JAMA 2016;315:1286.

CHAPTER

9

Vaccination in pregnancy in specific circumstances

Anna Calvert[a], *Miranda Quenby*[b], *Paul T. Heath*[a]

[a]Vaccine Institute & Paediatric Infectious Diseases Research Group, St Georges, University of London, London, United Kingdom, [b]Monash University, Melbourne, VIC, Australia

Introduction

Vaccination in pregnancy is a strategy that has been shown to protect pregnant women and their infants against a number of pathogens and which is offered as part of routine care in many countries. Current national and international recommendations exist for tetanus, influenza and pertussis. There is however still some uncertainty from pregnant women and health care providers about receiving and offering vaccines in pregnancy, an uncertainty which can be greater for those vaccines which are not part of national recommendations and about which there is often less evidence. Whenever a medicinal product is used in pregnancy it is important to consider carefully the risks and benefits, both to the woman and to the developing fetus. We consider vaccines that may be considered for use in pregnancy: vaccines for *Streptococcus pneumoniae* and *Haemophilus influenzae* type b (Hib); vaccines which might be recommended during an outbreak; travel vaccines; vaccines for those at increased risk of a particular infection and vaccines after exposure to infection.

Vaccines for *Streptococcus pneumoniae* and *Haemophilus influenzae* type b

Streptococcus pneumoniae and *Haemophilus influenzae* type b are both significant pathogens which continue to be responsible for considerable

Maternal Immunization
https://doi.org/10.1016/B978-0-12-814582-1.00010-3

morbidity and mortality across the world. Vaccines which provide protection against these pathogens are widely and effectively used in infant vaccination programs, including in many low- and middle-income countries (LMIC). Neither of these vaccinations are currently included in national vaccination programs for pregnant women.

Pneumococcal vaccines

Streptococcus pneumoniae is a Gram positive bacteria with numerous distinct capsular polysaccharide serotypes [1]. Pneumococcal infection causes a range of clinical disease, including pneumonia, meningitis, sepsis and otitis media. Pneumococcal infections are defined as invasive, in which *Streptococcus pneumoniae* is isolated from a normally sterile site, or non-invasive. Pneumococcal conjugate vaccines (PCV including 7, 10 or 13 serotypes) are recommended for use in infancy by the World Health Organisation (WHO) as part of the Expanded Programme on Immunization (EPI) according to a 3+0 or a 2+1 schedule and pneumococcal vaccination is becoming more established globally, including in LMIC [2]. Infants remain vulnerable to infection until about two weeks after completion of the primary schedule at which point an adequate immune response has been elicited. Despite high vaccine coverage in infants, pneumococcal infection still remains an important cause of morbidity and mortality in some countries, in part because of serotype replacement in which there is an increase in the incidence of pneumococcal disease caused by serotypes not covered by the available vaccines. There were an estimated 294,000 pneumococcal deaths in HIV uninfected children aged 1–59 months in 2015 and an additional 23,300 deaths associated with pneumococcus in HIV infected children [3]. In addition, *S. pneumoniae* remains an important cause of neonatal sepsis [2].

There is a substantial body of evidence which suggests that pneumococcal polysaccharide vaccination in pregnancy is safe [2]. Studies in the Philippines [4–6], Papua New Guinea [7], the Gambia [8], Bangladesh [9], Brazil [10] and the United States (US) [11] have also shown that vaccination in pregnancy is immunogenic in women and results in high levels of anti-polysaccharide antibody in their infants.

As pneumococcal colonization is an important part of the pathogenesis of invasive pneumococcal disease, the impact of pneumococcal vaccination in pregnancy on rates of colonization is also of interest. Studies of pneumococcal polysaccharide vaccine from Bangladesh [9], Brazil [12,13] and in Human Immunodeficiency Virus (HIV) infected mothers [14] showed no difference in the rates of colonization in infants up to six months of age between those born to vaccinated and unvaccinated mothers and an Australian study showed that pneumococcal polysaccharide vaccine in pregnancy did not have a significant impact on colonization at 7 months of age [15]. A US study did show a reduction in the rates of colonization in infants born to vaccinated mothers, although this was only significant at 16 months ($p = 0.021$) [11].

Few investigators have reported disease outcomes and most have found no difference in the rate of disease between infants born to vaccinated and unvaccinated mothers, possibly explained by small sample sizes [7,8,11]. A recent Australian study did not show a significant difference in the rates of middle ear disease at seven months in infants born to women vaccinated in pregnancy and those not vaccinated in pregnancy [15].

There is some evidence that vaccination with a pneumococcal polysaccharide vaccine in pregnancy increases pneumococcal-specific immunity in breastmilk. Studies from the Gambia [16], the US [11], Bangladesh [9] and Australia [15] have all shown an elevation in serotype-specific antibodies in the breastmilk of mothers vaccinated in pregnancy compared with those not vaccinated in pregnancy. In another small study, the colostrum from women who had received the pneumococcal polysaccharide vaccine was shown to have an enhanced ability to inhibit adherence of pneumococci to pharyngeal epithelial cells [17].

The vast majority of research in this area has been on polysaccharide vaccines. These have a theoretical disadvantage when compared with conjugate vaccines as the antibody response to polysaccharide vaccines tends to be predominantly immunoglobulin G2 (IgG2) which is less efficiently transported across the placenta. Also, conjugate vaccines are known to reduce colonization with serotypes included in the vaccines in a way which is not seen with polysaccharide vaccines. It is possible therefore that whilst use of polysaccharide vaccines in pregnancy was not associated with differences in colonization in the infants, this might be seen if a pneumococcal conjugate vaccine was used. There is one published study using a conjugate vaccine. In this study 153 women were randomised to receive a 9-valent pneumococcal conjugate vaccine (PCV-9) or placebo between 30 and 35 weeks and the primary outcome was incidence of Acute Otitis Media (AOM) in early infancy. Infants born to vaccinated mothers had an AOM free rate in the first six months of 74% compared with 89% in infants born to unvaccinated mothers ($p = 0.03$). The investigators hypothesized that this might be due to a reduced response to infant vaccination in those infants who had higher maternal antibody levels following vaccination in pregnancy [18].

There is evidence that pneumococcal polysaccharide vaccination in pregnancy is safe and can increase the anti-polysaccharide antibody in infants. There is an absence of high quality evidence however, that this leads to a reduction in disease in infants. In 2015 a Cochrane review again reported that there is insufficient evidence that pneumococcal vaccination in pregnancy reduces infant infections [19]. Further work needs to be done with larger studies powered to adequately investigate the relatively rare endpoint of invasive infection in the infant. These future studies must adequately explore conjugate vaccines as these have the potential to generate higher antibody concentrations in the infant [2]. There are several studies in progress investigating pneumococcal vaccination in pregnancy. The PROPEL trial, which is planned to finish at the end of 2019, is investigating the effect of a maternal or neonatal dose of pneumococcal conjugate vaccine on pneumococcal colonization in the nose (NCT02628886, clinicaltrials.gov). Another study is investigating various aspects of vaccination with PCV-10 or polysaccharide vaccine in pregnant women with HIV (NCT02717494, clinicaltrials.gov).

Pneumococcal vaccines are safe to be administered in pregnancy, if protection of the pregnant woman is necessary for any reason.

Haemophilus influenzae vaccine

Haemophilus influenzae type b (Hib) is an important cause of bacterial meningitis and bacterial pneumonia in young children in countries where Hib conjugate vaccines are not routinely given to infants. In countries with routine infant programs, infants are protected through a combination of direct and indirect (herd) immunity.

Trials of Hib conjugate and polysaccharide vaccines given in the third trimester of pregnancy have shown vaccination to be safe and immunogenic [20,21]. Conjugate vaccine results in higher infant antibody concentrations at birth and 2 months of age than polysaccharide vaccines and are therefore preferred [21]. Interference by maternal Hib vaccination on the infant response to Hib conjugate vaccine has not been demonstrated [21].

Maternal immunization appears to be safe and could potentially be used to protect young infants under 6 months of age where there is no routine infant program and ongoing evidence of invasive Hib disease occurring in young infants. However, as noted in a Cochrane review, there is currently limited evidence on the effectiveness of this approach and further trials are required [22]. Additionally, in places where infant vaccination is performed and working well, maternal vaccination against Hib is probably not required.

Vaccines in an outbreak situation

Meningococcal vaccines

Meningococcal disease is a devastating illness which can cause meningitis and sepsis [23]. If left untreated, it can lead to the death of previously healthy individuals within hours of symptom onset [23]. Infants aged less than one year make up the majority of cases due to their immature immunity and limited response to vaccination, and thus rely heavily on maternal antibodies to protect them in the first few months of life [24].

There are 12 meningococcal serogroups, of which six cause most disease [24]. Meningococcal A (MenA) causes regular epidemics in the 'Meningitis belt' of sub-Saharan Africa, where the majority of the world's cases occur. Typically, Europe, Australia and the US, are affected by meningococcal B (MenB) strains, while meningococcal serogroups C, Y and W also cause smaller outbreaks [25–27].

There are five vaccines which are currently used most widely to protect against meningococcal disease: three conjugate vaccines – monovalent conjugate vaccines against serogroups A and C and a quadrivalent conjugate vaccine against serogroups A, C, W and Y and two protein vaccines against Men B [27].

Meningococcal vaccines are not routinely recommended during pregnancy except in women considered high risk, for example those living in endemic regions, military recruits, patients with terminal complement component deficiencies or people with functional or anatomical asplenia [25,26].

The data supporting the use of meningococcal vaccines in pregnancy are limited and focussed on polysaccharide vaccines. This literature shows that polysaccharide vaccines when given in pregnancy are safe [28–32], immunogenic and result in increased antibody levels in the infant when compared to the infants of unvaccinated mothers [28–31]. Shahid and colleagues also assessed breastmilk antibody and found that polysaccharide vaccination in pregnancy resulted in significantly higher IgA in breastmilk up to six months of age [29].

A significant limitation of these studies is that they considered IgG concentrations only, which do not correlate well with protection [24,33]. As protection against invasive meningococcal disease is predominantly determined by antibody based complement dependent bacteriolysis, a better assessment of protection following vaccination can be obtained through use of serum bactericidal antibody (SBA) assays [24] which were not used in these studies.

There is some evidence around the safety of meningococcal conjugate vaccines in pregnancy. In a review of reports to the Vaccine Adverse Event Reporting System (VAERS) following vaccination with a conjugate quadrivalent vaccine, 14 women were identified who had received the vaccine in pregnancy. Of these, one had reported an adverse event of fever and muscle soreness, but delivery information was available for only three of the women. None had reported any problems [34]. In a review of reports to VAERS following vaccination with a different conjugate quadrivalent vaccine, 103 women were identified who had received the vaccine whilst pregnant. This study identified no safety concerns [35]. An observational comparative cohort study in Ghana reviewed 1730 pregnant women who had received a serogroup A conjugate vaccine in pregnancy and found no safety concerns [36]. There is no evidence regarding the immunogenicity or effectiveness of conjugate vaccines.

The current meningococcal conjugate vaccines appear to be safe to use in pregnancy. If a woman is at high risk of developing meningococcal disease, pregnancy does not contraindicate vaccination, and it is routinely recommended during an outbreak for the protection of both the woman and her infant.

Travel vaccines

Rabies vaccine

Rabies is a zoonosis caused by the rabies virus which is of the genus *Lyssavirus* and is transmitted primarily from animals to humans [37]. The incubation period of rabies in humans is variable and rabies results in severe neurological disease, it is classified either as furious or paralytic and is usually followed by death within 7 days of symptom onset for furious disease and within 21 days for paralytic disease [38].

Rabies threatens 3.3 billion people worldwide and is estimated to cause 60,000 deaths a year although these figures are likely to be significant underestimates [37]. There is uncertainty about whether the rabies virus can cross the placenta and cause disease in the infant [37].

Rabies vaccination plays an important role in the prevention of rabies disease. There are two strategies for the use of rabies vaccination: pre-exposure and post exposure vaccination. Pre-exposure vaccination aims to provide individuals at risk of infection with the ability to produce sufficient neutralizing antibody if they are exposed to the virus. This means that if they are exposed they need fewer booster doses of rabies vaccine and allows more time to seek help. Post-exposure vaccination aims to intervene in the incubation period to neutralize the virus before it gains entry to the central nervous system. This can be combined with rabies immunoglobulin (RIG) in those who have never received rabies vaccination in order to provide neutralizing antibody cover until the recipient is able to make their own antibody response [39]. According to the American Advisory Committee on Immunization Practices (ACIP), pregnancy is not considered a contraindication to post-exposure prophylaxis because of the extremely high case fatality rate and 'if the risk of exposure to rabies is substantial, pre-exposure prophylaxis also might be indicated' [40].

Post-exposure rabies immunization

There are several studies including around 350 women on the use of rabies vaccine in pregnant women following suspected exposure, which show that there is no increase in risk of adverse pregnancy outcome in those women who received post-exposure vaccination compared to the background rate of adverse pregnancy outcomes in the population [41–48]. In two studies in India the antibody response to vaccination in pregnancy has been shown to be adequate [41,42].

Pre-exposure rabies vaccination

Although most of the published studies have focussed on the rabies vaccine being given in a post-exposure setting, the evidence presented above [41–48] demonstrates that the vaccine is safe when given in pregnancy and therefore should be considered in pregnant women who are at risk.

Typhoid vaccine

Typhoid fever is caused by *Salmonella enterica* serotype typhi and although it has become a rare disease in high income countries, it remains a significant pathogen in much of the world. In 2000 there were an estimated 21.7 million cases and 217,000 deaths from typhoid fever [49]. Pregnancy is considered to be a high-risk situation for the development of typhoid fever because of the combination of a reduction in peristaltic force and frequency within the gastrointestinal tract and the frequent use of antacid medication for gastro-oesophageal reflux disease [50] and a retrospective study in Pakistan of 181 adult women admitted with blood culture confirmed typhoid fever found pregnancy to be a risk factor for typhoid disease [51]. Interestingly the latter study also found that the clinical presentation in pregnant women was somewhat different than in non-pregnant women, with cough being more frequent and nausea and vomiting less in the pregnant women [51]. This study also showed no significant difference in pregnancy outcomes between those pregnant women who had typhoid fever and age matched pregnant controls [51].

There are two vaccines available against typhoid – the oral live attenuated Ty21a vaccine and the parenteral Vi polysaccharide vaccine [52]. There are no specific data on the use of typhoid vaccines in pregnancy, but the live attenuated vaccine is contraindicated. ACIP guidelines recommend the use of inactivated typhoid vaccine when the benefits are likely to outweigh the risks as, although the safety has not been determined, the theoretical risk from an inactivated vaccine is low [40].

Japanese encephalitis vaccine

Japanese encephalitis virus is a flavivirus which is transmitted to humans by the *Culex* species of mosquito. Japanese encephalitis virus is the main cause of viral encephalitis in many countries of Asia. Three billion people are at risk of infection and there are an estimated 68,000 cases each year with 13,600–20,400 deaths [53]. Most human infections are asymptomatic, but of those who develop clinical symptoms, acute encephalitis is the most common. The case fatality rate is 20–30% and sequelae are common in survivors [54].

Japanese encephalitis virus is able to cross the placenta in humans [55] and infection within the first 22 weeks gestation is associated with pregnancy loss [55,56]. There are 4 types of vaccine available: inactivated mouse brain-derived vaccines, inactivated vero cell-derived vaccines, live attenuated vaccines and live recombinant vaccines. There are no available data regarding the use of Japanese encephalitis virus vaccines in pregnancy.

Vaccination with a live JEV vaccine is contraindicated in pregnancy. Vaccination with an inactivated JEV vaccine may be considered in pregnancy if women are travelling to an endemic area where they are likely to experience significant exposure [57], however there is no evidence on which to base this recommendation.

Yellow fever vaccine

The Yellow fever virus is a flavivirus which is transmitted to humans by the *Aedes* and *Haemogogus* species of mosquito [58]. Most infection is asymptomatic, but symptomatic infection causes fever, headache, myalgia, photophobia, back pain, vomiting, restlessness and anorexia and about 15% of patients experience severe disease with jaundice and in some cases haemorrhagic symptoms and multi-organ complications. In severe disease with hepatorenal dysfunction, the case fatality rate is 20–50% [58].

Yellow fever vaccination has been available for more than 60 years and is very effective, but as it is a live vaccine there are concerns about its use in pregnancy, particularly regarding the possibility of central nervous system complications. The ACIP guidelines point out that whilst most live vaccines are contraindicated in pregnancy, yellow fever vaccination should be considered with caution and recommends that if the risks of yellow fever exposure outweigh the risks of vaccination it should be given. If the risks are not felt to outweigh the benefits, but travel is necessary to an area which requires vaccination, a medical waiver can be issued [40]. Pregnant women are usually excluded from large yellow fever vaccination campaigns, but they are not infrequently inadvertently vaccinated, particularly early in their pregnancy. A multicentre European study of 74 women showed no increase in risk of miscarriage or fetal anomaly [59] and a Nigerian study also reported no increased risk of adverse pregnancy outcomes in vaccinated mothers, although in the latter study the immunogenicity of the vaccine was shown to be reduced in pregnancy [60]. In contrast, a Brazilian study of 441 women showed good immunogenicity of the vaccine with a seroconversion rate of 98.2% and again no increase in risk of adverse pregnancy outcomes [61]. A report from a Swiss travel clinic included 6 women who received yellow fever vaccination, two in the first and four in the second trimesters and they reported no adverse events associated with this [62].

A Brazilian study found a two-fold increased risk of miscarriage in vaccinated mothers, but this was not statistically significant and the sample size was small [63]. A retrospective study in Trinidad reported one case of congenital infection in 41 infants born to vaccinated mothers and no serious adverse events [64]. In a Brazilian study looking at the relationship between the yellow fever vaccination in pregnancy and structural malformation in the newborn, investigators found no association for major malformations, and the increase in cases of minor malformations in those infants exposed to the vaccine in utero was hypothesized to be a result of evaluation bias [65].

Although there is a theoretical risk of using a live vaccine in pregnancy and there is the possibility of causing congenital infection, this should be weighed against the risks to the pregnant woman if she is visiting an area of high risk.

Yellow fever vaccination during lactation has been linked to neonatal Yellow fever disease, caused by the vaccine strain, see Chapter 2.

Hepatitis A vaccine

Hepatitis A virus is a non-enveloped RNA virus which is transmitted from person to person and is more common in environments with poor sanitation. The clinical presentation is variable with infections being inapparent, subclinical, anicteric or icteric. [66]. When symptoms do occur they include fever, myalgia, anorexia, malaise, nausea and vomiting followed a few days later by dark urine, light coloured stools, jaundice and scleral icterus [66]. Globally there are an estimated 1.4 million cases of hepatitis A infection each year and unlike hepatitis B and C it does not cause chronic liver disease [67].

Hepatitis A infection in the second and third trimester of pregnancy is associated with complications including preterm labor [68]. Vertical transmission is rare although there are some case reports describing this [69–71] and two case reports reporting intrauterine infection with hepatitis A complicated by fetal ascites and bowel perforation diagnosed after birth [72,73].

There are both inactivated and live attenuated vaccines available against hepatitis A, although the latter are used primarily in China and occasionally privately in India [74]. The ACIP guidelines recommend that the inactivated hepatitis A vaccine should be used in pregnancy after an assessment of the likely risks of exposure to hepatitis A virus [40]. In a US study looking at reports submitted through the VAERS over an 18-year period there was no increase in adverse pregnancy specific outcomes in those receiving an inactivated hepatitis A or combined hepatitis A and B vaccination in pregnancy [75]. In a study of vaccines given in pregnancy in a Swiss travel clinic, 29 women received inactivated hepatitis A vaccination, 16 in the first trimester, 12 in the second trimester and 1 in the third trimester. There was no increase in adverse events associated with vaccination [62].

Cholera vaccine

Cholera is a diarrhoeal illness caused by *Vibrio cholerae* and is mainly spread through contaminated water [76]. Epidemic cholera is caused by the toxigenic serogroups of 01 and 0139 [77]. This toxin is responsible for the hypersecretion of electrolytes and water which causes significant diarrhea. Pregnant women have been shown to respond to cholera infection in a similar manner to non-pregnant women despite the changes in immunological status during pregnancy [78]. Cholera infection can cause serious complications in pregnancy, including fetal loss in 2–36% of cases [79].

In 2016, 132,121 cases were reported to WHO with 2420 deaths [76] although this is likely to be a significant underestimate. There are 1.3 billion people at risk of cholera in cholera endemic countries and a further 99 million people at risk in non-endemic countries [80].

In 2010 the WHO recommended that oral cholera vaccine should be used preventatively as well as reactively in cholera outbreaks [81].

There are currently four licenced, killed, whole-cell oral cholera vaccines (OCV), all of which require two doses for full protection. The product information for these vaccines recommend caution in pregnancy and pregnant women have historically been excluded from campaigns with these vaccines. In three retrospective studies including a total of nearly 3000 women who had been inadvertently vaccinated in pregnancy in Guinea [82], Zanzibar [83] and Bangladesh [84] there was no statistically significant increase in the rate of any adverse pregnancy outcomes. In an observational cohort study in Malawi [80] there was no statistically significant increase in the risk of pregnancy loss or of neonatal death.

These findings support observations made by several authors that it is reasonable to consider that OCVs will be safe in pregnancy as the bacteria contained within them are killed and cannot replicate, the vaccine antigens act locally on the GI mucosa and are unlikely to cause systemic toxicity and there is experience of the safety of other killed vaccines in pregnancy [80,85].

The findings from the published studies, our knowledge of the mechanism of action of OCVs and the risk that cholera poses to pregnancy would support the administration for those who are at high risk for disease.

Tick borne encephalitis virus (TBEV) vaccine

Tick borne encephalitis virus is a flavivirus which is transmitted through the bite of an infected *Ixodes* spp. tick [58]. There are three subtypes:

European (TBEV-Eur), Siberian (TBEV-Sib) and Far Eastern (TBEV-FE) [86]. Clinical presentation is usually biphasic with symptoms of the first phase including fever, fatigue, general malaise, headache and body pain and those of the second phase including a wide spectrum of presentations from mild meningitis to severe encephalitis [86].

There are two inactivated vaccines against TBEV available in Europe both based on TBEV-Eur, both of which provide protection against all three TBEV subtypes [58]. There are no studies of TBEV vaccinations in pregnancy and no official recommendations about their use.

Vaccines for those at increased risk of infection

Hepatitis B vaccine

The focus of hepatitis B vaccination in pregnancy is to prevent vertical transmission from a mother with chronic hepatitis B infection to her newborn infant. However, a Cochrane review did not identify any randomised controlled trials assessing hepatitis B vaccine compared with placebo or no treatment during pregnancy for preventing infant infection [87]. One study in HBsAg negative pregnant women showed that a 3-dose schedule resulted in higher antibody concentrations in the mother and the baby than a 2-dose schedule [88] and a clinical trial of an accelerated hepatitis B vaccination schedule, given at 0, 1, and 4 months to pregnant women considered to be at high risk of infection (n=168) was found to be feasible, well tolerated and to achieve similar seroconversion rates to a standard schedule: 90% (95% CI, 85–94%) of women seroconverted after completing three doses, comparable to that expected after a standard 0, 1, 6 month schedule. There were no serious adverse events and injection site discomfort was the most prevalent complaint (10.5%) [89]. Other studies have shown that the passive immunity conferred on the infants is of short duration [90] and therefore infants at an increased risk of infection should still be vaccinated early.

An analysis of 192 reports in VAERS relating to hepatitis B vaccination in pregnancy did not find any unusual or unexpected adverse events [91].

Q fever vaccine

Q fever is a zoonosis caused by *Coxiella burnetii* with a presentation ranging from asymptomatic through to possibly fatal chronic Q fever. Transmission occurs primarily through inhalation of aerosols from contaminated soil or animal waste or ingestion of contaminated milk or milk products. Q fever may result in adverse pregnancy outcomes, including spontaneous abortion, intrauterine growth retardation, oligohydramnios, intrauterine fetal death, and premature delivery [92].

The preferred means of infection control for *C. burnetii* is vaccination of at-risk populations. The current vaccine, Q-VAX® (Commonwealth Serum Laboratories Ltd.), is approved for human use only in Australia and requires pre-screening as vaccination can elicit severe reactogenic responses in individuals previously exposed to the bacterium [93]. There are no data on the safety and efficacy of this or other candidate vaccines in pregnant women and no recommendations about the use of Q fever vaccination in pregnancy.

Vaccines after exposure to infection

Anthrax vaccine

Although it is not clear whether pregnancy puts women at higher risk of acquiring *B. anthracis* infection or of developing more severe disease, it is known that anthrax is associated with maternal and fetal deaths [94,95]. Given the severity of anthrax, pregnant women should receive the same post-exposure prophylaxis (PEP) and treatment regimens as non-pregnant adults. These regimens would include antimicrobial drug treatment and vaccination for those who have had direct exposure to anthrax spores. Limited data exist to guide the use of BioThrax Anthrax Vaccine Adsorbed (AVA; Emergent BioSolutions, Rockville, Maryland) in pregnant women. However, where risk for anthrax exposure is low, vaccination of pregnant women is not recommended. It is recommended that pregnant women at risk for inhalation anthrax should receive Anthrax Vaccine Adsorbed vaccine and antimicrobial drug therapy regardless of gestation [94].

Anthrax vaccine adsorbed vaccination is compulsory for United States military service members with operational indicators. After adjusting for other potential risk factors, no association has been shown between inadvertent Anthrax Vaccine Adsorbed vaccination during pregnancy and risk of birth defects [96,97].

Conclusion

We have provided an overview of vaccines which may be considered in particular circumstances in pregnancy outside of national programs (see Table 1 below).

For some vaccines, there is limited evidence and few official recommendations available and so careful consideration of the potential risks and benefits is necessary before using these vaccines in pregnancy.

TABLE 1 Summary of vaccines which may be considered in particular circumstances in pregnancy.

Vaccine	Comments
Streptococcus pneumoniae (pneumococcal)	**Polysaccharide vaccines:** Safe and increase anti-polysaccharide antibody in infants. No evidence that polysaccharide vaccines reduce disease in infants. **Conjugate vaccines:** Limited evidence. Only study showed increased disease in infants of vaccinated mothers.
Haemophilus influenzae type b (Hib)	**Polysaccharide and conjugate vaccines:** Safe, immunogenic and result in increased antibody concentrations in the infant – these are higher following conjugate vaccines. No evidence for effectiveness in reducing disease incidence in infants.
Vaccines in outbreak situations	
Neisseria meningitidis (meningococcal)	**Polysaccharide vaccines:** Safe, immunogenic and result in higher antibody concentrations in the infant. **Conjugate vaccines:** Safe. No evidence about immunogenicity or effectiveness when given in pregnancy.
Travel vaccines	
Rabies	No evidence that rabies vaccination in pregnancy is associated with worse pregnancy outcomes and given the high case fatality rate in rabies pregnancy should not be considered a contraindication to post-exposure prophylaxis and may be considered for pre-exposure prophylaxis for women at risk.
Typhoid	**Live attenuated vaccine:** Contraindicated **Inactivated:** Safety has not been determined but theoretical risk is low so may be considered when benefits are likely to outweigh risks.
Japanese encephalitis virus (JEV)	**Live vaccine:** Contraindicated **Inactivated:** No evidence about use of JEV vaccine in pregnancy. May be considered if travelling to an endemic area where likely to experience significant exposure.
Yellow fever	Only a live vaccine is available for yellow fever. There is some evidence that yellow fever vaccination in pregnancy is not associated with an increased incidence of adverse pregnancy outcomes, although congenital infection is possible. Use of this live vaccine can be considered if it is thought that the risks of infection outweigh the possible risks of vaccination. If the risks of vaccination are considered to outweigh the risks of yellow fever but travel is required to an area which requires vaccination, a medical waiver can be issued.
Hepatitis A	**Inactivated vaccine:** No evidence of increase in adverse pregnancy outcomes following hepatitis A vaccination in pregnancy. May be used after consideration of the likely risks of exposure.

Continued

TABLE 1 Summary of vaccines which may be considered in particular circumstances in pregnancy—cont'd

Vaccine	Comments
Cholera	Theoretically safe as bacteria within the vaccine are killed and cannot replicate and the vaccine antigens act locally on GI mucosa and are unlikely to cause systemic toxicity. No increase in pregnancy adverse outcomes in those women who inadvertently received cholera vaccination in pregnancy. May be considered for women who are at high risk for disease.
Tick borne encephalitis virus (TBEV)	There are no studies of TBEV vaccines in pregnancy and no official recommendations about their use.
Vaccines for those at increased risk of infection	
Hepatitis B	No evidence that vaccination in pregnancy prevents infant infection. Hepatitis B vaccination in pregnancy is not associated with an increase in adverse pregnancy outcomes.
Q fever	There are no studies of Q fever vaccines in pregnancy and no official recommendations about their use.
Anthrax	No association has been shown between inadvertent anthrax vaccination in pregnancy and risk of birth defects. Because of the severity of anthrax infection it is recommended that pregnant women should receive the same post exposure prophylaxis as non-pregnant adults, including vaccination. If women are at risk of inhalational anthrax they should receive anthrax vaccine regardless of gestation.

References

[1] Ampofo K, Byington CL. Streptococcus pneumoniae. In: Long S, Prober C, Fischer M, editors. Principles and practice of pediatric infectious diseases. Philadelphia: Elsevier; 2018. p. 737–45.

[2] Clarke E, Kampmann B, Goldblatt D. Maternal and neonatal pneumococcal vaccination—where are we now? Expert Rev Vaccines 2016, Oct;15(10):1305–17.

[3] Wahl B, O'Brien KL, Greenbaum A, Majumber A, Liu L, Chu Y, et al. Burden of *Streptococcus pneumoniae* and *Haemophilus influenzae* type b disease in children in the era of conjugate vaccines: global, regional, and national estimates for 2000-15. Lancet Glob Health 2018, Jul, 1;6(7):e744–57.

[4] Quiambo BP, Nohynek H, Kayhty H, Ollgren J, Gozum L, Gepanayao CP, et al. Maternal immunization with pneumococcal polysaccharide vaccine in the Philippines. Vaccine 2003, Jul, 28;21(24):3451–4.

[5] Quiambo BP, Nohynek H, Kayhty H, Ollgren J, Gozum L, Gepanayao CP, et al. Immunogenicity and reactogenicity of 23-valent pneumococcal polysaccharide vaccine among pregnant Filipino women and placental transfer of antibodies. Vaccine 2007, May, 30;25(22):4470–7.

[6] Holmlund E, Nohynek H, Quiambao B, Ollgren J, Kayhty H. Mother-infant vaccination with pneumococcal polysaccharide vaccine: persistence of maternal antibodies and responses of infants to vaccination. Vaccine 2011, Jun, 20;29(28):4565–75.

[7] Lehmann D, Pomat WS, Combs B, Dyke T, Alpers MP. Maternal immunization with pneumococcal polysaccharide vaccine in the highlands of Papua New Guinea. Vaccine 2002, Mar, 15;20(13–14):1837–45.

[8] O'Dempsey TJD, McArdle T, Ceesay SJ, Banya WA, Demba E, Secka O, et al. Immunization with a pneumococcal capsular polysaccharide vaccine during pregnancy. Vaccine 1996, Jul;14(10):963–70.

[9] Shahid NS, Steinhoff MC, Hogue SS, Begum T, Thompson C, Siber GR. Serum, breastmilk, and infant antibody after maternal immunisation with pneumococcal vaccine. Lancet 1995, Nov, 11;346(8985):1252–7.

[10] Berezin EN, Lopes CC, Cardoso MRA. Maternal immunization with pneumococcal polysaccharide vaccine: persistence of maternal antibodies in infants. J Trop Pediatr 2017, Apr, 1;63(2):118–23.

[11] Munoz FM, Englund JA, Cheesman CC, Maccato ML, Pinell PM, Nahm MH, et al. Maternal immunization with pneumococcal polysaccharide vaccine in the third trimester of gestation. Vaccine 2001, Dec, 12;20(5–6):826–37.

[12] Lopes CC, Berezin EN, Scheffer D, Huziwara R, Sliva MI, Brandao A, et al. Pneumococcal nasopharyngeal carriage in infants of mothers immunized with 23V non-conjugate pneumococcal polysaccharide vaccine. J Trop Pediatr 2012, Oct, 1;58(5):348–52.

[13] Lopes CR, Berezin EN, Ching TH, de Souza Canuto J, Costa VO, Klering EM. Ineffectiveness for infants of immunization of mothers with pneumococcal capsular polysaccharide vaccine during pregnancy. Braz J Infect Dis 2009, Apr;13(2):104–6.

[14] de Almeida V, Negrini B, Cervi M, de Isaac M, Mussi-Pinhata M. Pneumococcal nasopharyngeal carriage among infants born to human immunodeficiency virus-infected mothers immunized with pneumococcal polysaccharide vaccine during gestation. Pediatr Infect Dis J 2011, Jun;30(6):466–70.

[15] Binks MJ, Moberley SA, Balloch A, Leach AJ, Nelson S, Hare KM, et al. PneuMum: impact from a randomised controlled trial of maternal 23-valent pneumococcal polysaccharide vaccination on middle ear disease amongst Indigenous infants, Northern Territory, Australia. Vaccine 2015, Nov, 27;33(48):6579–87.

[16] Obaro SK, Deubzer HE, Newman VO, Adegbola RA, Greenwood BM, Henderson DC. Serotype-specific pneumococcal antibodies in breastmilk of Gambian women immunized with a pneumococcal polysaccharide vaccine during pregnancy. Pediatr Infect Dis J 2004, Nov;23(11):1023–9.

[17] Deubzer HE, Obaro SK, Newman VO, Adegbola RA, Greenwood BM, Henderson DC. Colostrum obtained from women vaccinated with pneumococcal vaccine during pregnancy inhibits epithelial adhesion of *Streptococcus pneumoniae*. J Infect Dis 2004, Nov, 15;190(10):1758–61.
[18] Daly KA, Giebink S, Lindgren BR, Knox J, Haggerty BJ, Nordin J, et al. Maternal immunization with pneumococcal 9-valent conjugate vaccine and early otitis media. Vaccine 2014, Dec, 5;32(51):6948–55.
[19] Chaithongwongwatthana S, Yamasmit W, Limpongsanurak S, Lumbiganon P, Tolosa JE. Pneumococcal vaccination during pregnancy for preventing infant infection. Cochrane Database Syst Rev 2015;1:CD004903.
[20] Mulholland K, Suara RO, Siber G, Roberton D, Jaffar S, N'Jie J, et al. Maternal immunization with *Haemophilus influenzae* type b polysaccharide-tetanus protein conjugate vaccine in The Gambia. JAMA 1996, Apr, 17;275(15):1182–8.
[21] Englund JA, Glezen WP, Thompson C, Anwaruddin R, Turner CS, Siber GR. *Haemophilus influenzae* type b-specific antibody in infants after maternal immunization. Pediatr Infect Dis J 1997, Dec;16(12):1122–30.
[22] Salam RA, Das JK, Dojo Soeandy C, Lassi ZS, Bhutta ZA. Impact of *Haemophilus influenzae* type B (Hib) and viral influenza vaccinations in pregnancy for improving maternal, neonatal and infant health outcomes. Cochrane Database Syst Rev 2015;6:CD009982.
[23] Thompson MJ, Ninis N, Perera R, Mayon-White R, Phillips C, Bailey L, et al. Clinical recognition of meningococcal disease in children and adolescents. Lancet 2006, Feb, 4;367(9508):397–403.
[24] Abu Raya B, Sadrangani M. Meningococcal vaccination in pregnancy. Hum Vaccin Immunother 2018, May, 4;14(5):1188–96.
[25] Australian Department of Health. The Australian immunisation handbook. Available from: http://www.immunise.health.gov.au/internet/immunise/publishing.nsf/Content/Handbook10-home~handbook10part4~handbook10-4-10.
[26] Centers for Disease Control and Prevention. Meningococcal disease. Available from: https://www.cdc.gov/meningococcal/index.html.
[27] Public Health England. Guidance for the public health management of meningococcal disease in the UK. Available from: https://www.gov.uk/government/publications/meningococcal-disease-guidance-on-public-health-management.
[28] de Andrade CA, Giampaglia CM, Kimura H, de Pereira OA, Farhat CK, Neves JC, et al. Maternal and infant antibody response to meningococcal vaccination in pregnancy. Lancet 1977, Oct, 15;2(8042):809–11.
[29] Shahid NS, Steinhoff MC, Roy E, Begum T, Thompson CM, Siber GR. Placental and breast transfer of antibodies after maternal immunization with polysaccharide meningococcal vaccine: a randomized, controlled evaluation. Vaccine 2002, May, 22;20(17–18):2404–9.
[30] O'Dempsey TJ, McArdle T, Ceesay SJ, Secka O, Demba E, Banya WA, et al. Meningococcal antibody titres in infants of women immunised with meningococcal polysaccharide vaccine during pregnancy. Arch Dis Child Fetal Neonatal Ed 1996, Jan;74(1):F43–6.
[31] McCormick JB, Gusmao HH, Nakamura S, Freire JB, Veras J, Gorman G, et al. Antibody response to serogroup A and C meningococcal polysaccharide vaccines in infants born of mothers vaccinated during pregnancy. J Clin Invest 1980, May;65(5):1141–4.
[32] Letson GW, Little JR, Ottman J, Miller GL. Meningococcal vaccine in pregnancy: an assessment of infant risk. Pediatr Infect Dis J 1998, Mar;17(3):261–3.
[33] de Voer RM, van der Klis FR, Nooitgedagt JE, Versteegh FG, van Huisseling JC, van Rooijen DM, et al. Seroprevalence and placental transportation of maternal antibodies specific for *Neisseria meningitidis* serogroup C, *Haemophilus influenzae* type B, diphtheria, tetanus, and pertussis. Clin Infect Dis 2009, Jul, 1;49(1):58–64.

[34] Myers TR, McNeil MM, Ng CS, Li R, Lewis PW, Cano MV. Adverse events following quadrivalent meningococcal CRM-conjugate vaccine (Menveo(R)) reported to the Vaccine Adverse Event Reporting system (VAERS), 2010-2015. Vaccine 2017, Mar, 27;35(14):1758–63.
[35] Zheteyeva Y, Moro PL, Yue X, Broder K. Safety of meningococcal polysaccharide-protein conjugate vaccine in pregnancy: a review of the Vaccine Adverse Event Reporting System. Am J Obstet Gynecol 2013, Jun;208(6). 478 e1–6.
[36] Wak G, Williams J, Oduro A, Maure C, Zuber PLF, Black S. The safety of PsA-TT in pregnancy: an assessment performed within the Navrongo Health and Demographic Surveillance Site in Ghana. Clin Infect Dis 2015, Nov, 15;61(Suppl. 5):S489–92.
[37] Giesen A, Gniel D, Malerczyk C. 30 years of rabies vaccination with Rabipur: a summary of clinical data and global experience. Expert Rev Vaccines 2015, Mar;14(3):351–67.
[38] Willoughby Jr RE. Rabies virus. In: Long S, Prober C, Fischer M, editors. Principles and practice of pediatric infectious diseases. Philadelphia: Elsevier; 2017. p. 1176–80.
[39] Toovey S. Preventing rabies with the Verorab vaccine: 1985-2005. Twenty years of clinical experience. Travel Med Infect Dis 2007, Nov;5(6):327–48.
[40] CDC. Guidelines for vaccinating pregnant women. Available from: https://www.cdc.gov/vaccines/pregnancy/hcp/guidelines.html.
[41] Sudarshan MK, Madhusudana SN, Mahendra BJ. Post-exposure prophylaxis with purified vero cell rabies vaccine during pregnancy—safety and immunogenicity. J Commun Dis 1999, Dec;31(4):229–36.
[42] Sudarshan MK, Madhusudana SN, Mahendra BJ, Ashwathnarayana DH, Jayakumary M, Gangaboriah. Post exposure prophylaxis with Purified Verocell Rabies Vaccine: a study of immunoresponse in pregnant women and their matched controls. Indian J Public Health 1999, Apr-Jun;43(2):76–8.
[43] Sudarshan MK, Giri MS, Mahendra BJ, Venkatesh GM, Sanjay TV, Narayana DH, et al. Assessing the safety of post-exposure rabies immunization in pregnancy. Hum Vaccin 2007, May-Jun;3(3):87–9.
[44] Chutivongse S, Wilde H. Postexposure rabies vaccination during pregnancy: experience with 21 patients. Vaccine 1989, Dec;7(6):546–8.
[45] Chutivongse S, Wilde H, Benjavongkulchai M, Chomchey P, Punthawong S. Postexposure rabies vaccination during pregnancy: effect on 202 women and their infants. Clin Infect Dis 1995, Apr;20(4):818–20.
[46] Fayaz A, Simani S, Fallahian V, Eslamifar A, Hazrati M, Farahtaj F, et al. Rabies antibody levels in pregnant women and their newborns after rabies post-exposure prophylaxis. Iran J Reprod Med 2012, Mar;10(2):161–3.
[47] Huang G, Liu H, Cao Q, Liu B, Pan H, Fu C. Safety of post-exposure rabies prophylaxis during pregnancy. Hum Vaccin Immunother 2013, Jan, 1;9(1):177–83.
[48] Fescharek R, Quat U, Dechert G. Postexposure rabies vaccination during pregnancy: experience from post-marketing surveillance with 16 patients. Vaccine 1990, Aug;8(4):409.
[49] Crump JA, Luby SP, Mintz ED. The global burden of typhoid fever. Bull World Health Organ 2004, May;82(5):346–53.
[50] Touchan F, Hall JD, Lee RV. Typhoid fever during pregnancy: case report and review. Obstet Med 2009, Dec;2(4):161–3.
[51] Sulaiman K, Sarwari AR. Culture-confirmed typhoid fever and pregnancy. Int J Infect Dis 2007, Jul;11(4):337–41.
[52] Milligan R, Paul M, Richardson M, Neuberger A. Vaccines for preventing typhoid fever. Cochrane Database Syst Rev 2018;5:CD001261.
[53] WHO. Japanese encephalitis. Available from: http://www.who.int/news-room/fact-sheets/detail/japanese-encephalitis.

[54] Yun SI, Lee YM. Japanese encephalitis—the virus and vaccines. Hum Vaccin Immunother 2014, Feb, 1;10(2):263–79.
[55] Chaturvedi UC, Mathur A, Chandra A, Das SK, Tandon HO, Singh UK. Transplacental infection with Japanese encephalitis virus. J Infect Dis 1980, Jun;141(6):712–5.
[56] Charlier C, Beaudoin MC, Couderc T, Lortholary O, Lecuit M. Arboviruses and pregnancy: maternal, fetal, and neonatal effects. Lancet Child Adolesc Health 2017, Oct;1(2):134–46.
[57] Fischer M, Lindsey N, Staples E, Hills S. Japanese encephalitis vaccines: recommendations of the Advisory Committee on Immunization Practices (ACIP). MMWR Recomm Rep 2010, March, 12;59(1):1–27.
[58] Hills SL, Fischer ML. Flaviviruses. In: Long S, Prober C, Fischer M, editors. Principles and practice of pediatric infectious diseases. Philadelphia: Elsevier; 2018. p. 1128–31.
[59] Robert E, Vial T, Schaefer C, Arnon J, Reuvers M. Exposure to yellow fever vaccine in early pregnancy. Vaccine 1999, Jan, 21;17(3):283–5.
[60] Nasidi A, Monath TP, Vandenberg J, Tomori O, Calisher CH, Hurtgen X, et al. Yellow fever vaccination and pregnancy: a four-year prospective study. Trans R Soc Trop Med Hyg 1993, May-Jun;87(3):337–9.
[61] Suzano CE, Amarai E, Sato HK, Papaiordanou PM. The effects of yellow fever immunization (17DD) inadvertently used in early pregnancy during a mass campaign in Brazil. Vaccine 2006, Feb, 27;24(9):1421–6.
[62] D'Acremont V, Tremblay S, Genton B. Impact of vaccines given during pregnancy on the offspring of women consulting a travel clinic: a longitudinal study. J Travel Med 2008, Mar-Apr;15(2):77–81.
[63] Nishioka SA, Nunes-Araujo FR, Pires WP, Silva FA, Costa HL. Yellow fever vaccination during pregnancy and spontaneous abortion: a case-control study. Trop Med Int Health 1998, Jan;3(1):29–33.
[64] Tsai TF, Paul R, Lynberg MC, Letson GW. Congenital yellow fever virus infection after immunization in pregnancy. J Infect Dis 1993, Dec;168(6):1520–3.
[65] Cavalcanti DP, Salomao MA, Lopez-Camelo J, Pessoto MA. Early exposure to yellow fever vaccine during pregnancy. Trop Med Int Health 2007, Jul;12(7):833–7.
[66] Collier MG, Nelson NP. Hepatitis A virus. In: Long S, Prober C, Fischer M, editors. Principles and practice of pediatric infectious diseases. Philadelphia: Elsevier; 2017. p. 1214–8.
[67] WHO. Hepatitis A. Available from: http://www.who.int/immunization/diseases/hepatitisA/en/.
[68] Elinav E, Ben-Dov I, Shapira Y, Daudi N, Adler R, Shouval D, Ackerman Z. Acute hepatitis A infection in pregnancy is associated with high rates of gestational complications and preterm labor. Gastroenterology 2006, Apr;130(4):1129–34.
[69] Renge RL, Dani VS, Chitambar SD, Arankalle VA. Vertical transmission of hepatitis A. Indian J Pediatr 2002, Jun;69(6):535–6.
[70] Tanaka I, Shima M, Kubota Y, Takahashi Y, Kawamata O, Yoshioka A. Vertical transmission of hepatitis A virus. Lancet 1995, Feb, 11;345(8946):397.
[71] Erkan T, Kutlu T, Cullu F, Tumay GT. A case of vertical transmission of hepatitis A infection. Acta Paediatr 1998, Sep;87(9):1008–9.
[72] Leikin E, Lysikiewicz A, Garry D, Tejani N. Intrauterine transmission of hepatitis A virus. Obstet Gynecol 1996, Oct;88(4):690–1.
[73] McDuffie Jr RS, Bader T. Fetal meconium peritonitis after maternal hepatitis A. Am J Obstet Gynecol 1999, Apr;180(4):1031–2.
[74] WHO. Hepatitis A. Available from: http://www.who.int/ith/vaccines/hepatitisA/en/.
[75] Moro PL, Museru OI, Niu M, Lewis P, Broder K. Reports to the Vaccine Adverse Event Reporting System after hepatitis A and hepatitis AB vaccines in pregnant women. Am J Obstet Gynecol 2014, Jun;210(6). 561e1–6.

[76] WHO. Cholera. Available from: http://www.who.int/news-room/fact-sheets/detail/cholera.
[77] Routh JA, Matanock A, Mintz ED. In: Long S, Prober C, Fischer M, editors. Principles and practice of pediatric infectious diseases. Philadelphia: Elsevier; 2018. p. 874–8.
[78] Khan AI, Chowdhury F, Leung DT, Larocque RC, Harris JB, Ryan ET, et al. Cholera in pregnancy: clinical and immunological aspects. Int J Infect Dis 2015, Oct;39:20–4.
[79] Moro PL, Sukumaran L. Cholera vaccination: pregnant women excluded no more. Lancet 2017, May, 1;17(5):469–70.
[80] Ali M, Nelson A, Luquero FJ, Azman AS, Debes AK, M'bang'ombe MM, et al. Safety of a killed oral cholera vaccine (Shancol) in pregnant women in Malawi: an observational cohort study. Lancet Infect Dis 2017, May;17(5):538–44.
[81] WHO. Meeting of the Strategic Advisory Group of Experts on immunization, October 2009—conclusions and recommendations. Wkly Epidemiol Rec 2009, Oct;84(50):517–32.
[82] Grout L, Martinez-Pino I, Ciglenecki I, Keita S, Diallo AA, Traore B, et al. Pregnancy outcomes after a mass vaccination campaign with an oral cholera vaccine in Guinea: a retrospective cohort study. PLoS Negl Trop Dis 2015, Dec, 29;9(12):e0004274.
[83] Hashim R, Khatib AM, Enwere G, Park JK, Reyburn R, Ali M, et al. Safety of the recombinant cholera toxin B subunit, killed whole-cell (rBS-WC) oral cholera vaccine in pregnancy. PLoS Negl Trop Dis 2012;6(7):e1743.
[84] Khan AI, Ali M, Chowdhury F, Saha A, Khan IA, Khan A, et al. Safety of the oral cholera vaccine in pregnancy: retrospective findings from a subgroup following mass vaccination campaign in Dhaka, Bangladesh. Vaccine 2017, Mar, 13;35(11):1538–43.
[85] Stop Cholera. Available from: https://www.stopcholera.org/sites/cholera/files/cholera_and_the_use_of_ocv_in_pregant_women_0.pdf.
[86] Lindquist L, Vapalahti O. Tick-borne encephalitis. Lancet 2008, May, 31;371(9627):1861–71.
[87] Sangkomkamhang US, Lumbiganon P, Laopaiboon M. Hepatitis B vaccination during pregnancy for preventing infant infection. Cochrane Database Syst Rev 2014;11:CD007879.
[88] Gupta I, Ratho RK. Immunogenicity and safety of two schedules of Hepatitis B vaccination during pregnancy. J Obstet Gynaecol Res 2003, Apr;29(2):84–6.
[89] Sheffield JS, Hickman A, Tang J, Moss K, Kourosh A, Crawford NM, et al. Efficacy of an accelerated hepatitis B vaccination program during pregnancy. Obstet Gynecol 2011, May;117(5):1130–5.
[90] Ayoola EA, Johnson AO. Hepatitis B vaccine in pregnancy: immunogenicity, safety and transfer of antibodies to infants. Int J Gynaecol Obstet 1987, Aug;25(4):297–301.
[91] Moro PL, Zheteyeva Y, Barash F, Lewis P, Cano M. Assessing the safety of hepatitis B vaccination during pregnancy in the Vaccine Adverse Event Reporting System (VAERS), 1990–2016. Vaccine 2018, Jan, 2;36(1):50–4.
[92] Nielsen SY, Molbak K, Henriksen TB, Krogfelt KA, Larsen CS, Villumsen S. Adverse pregnancy outcomes and *Coxiella burnetii* antibodies in pregnant women, Denmark. Emerg Infect Dis 2014, Jun;20(6):925–31.
[93] Reeves PM, Paul SR, Sluder AE, Brauns TA, Poznansky MC. Q-vaxcelerate: a distributed development approach for a new *Coxiella burnetii* vaccine. Hum Vaccin Immunother 2017, Dec, 2;13(12):2977–81.
[94] Meaney-Delman D, Zotti ME, Creanga AA, Misegades LK, Wako E, Treadwell TA, et al. Special considerations for prophylaxis for and treatment of anthrax in pregnant and postpartum women. Emerg Infect Dis 2014, Feb;20(2):e130611.
[95] Meaney-Delman D, Zotti ME, Rasmussen SA, Strasser S, Shadomy S, Turcios-Ruiz RM, et al. Anthrax cases in pregnant and postpartum women: a systematic review. Obstet Gynecol 2012, Dec;120(6):1439–49.

[96] Conlin AM, Bukowinski AT, Gumbs GR. Analysis of pregnancy and infant health outcomes among women in the National Smallpox Vaccine in Pregnancy Registry who received Anthrax Vaccine Adsorbed. Vaccine 2015, Aug, 26;33(36):4387–90.
[97] Conlin AM, Sevick CJ, Gumbs GR, Khodr ZG, Bukowinski AT. Safety of inadvertent anthrax vaccination during pregnancy: an analysis of birth defects in the U.S. military population, 2003-2010. Vaccine 2017, Aug, 3;35(34):4414–20.

PART III

Future vaccines for use in pregnancy

CHAPTER

10

Respiratory syncytial virus

Flor M. Munoz[a], *Janet A. Englund*[b]

[a]Department of Pediatrics, Section of Infectious Diseases, Baylor College of Medicine, Houston, TX, United States, [b]Department of Pediatrics, Pediatric Infectious Diseases, Seattle Children's Hospital, University of Washington, Seattle, WA, United States

Introduction

The impact of respiratory syncytial virus (RSV) on global health is becoming increasingly appreciated. Advances in RSV diagnostic methods, with increasing availability of rapid sensitive and specific diagnostic tests, as well as enhanced understanding of the molecular structure and function of the virus, have accelerated our understanding of RSV biology and epidemiology over the past decade. Serious RSV disease in infants, elderly, and immunocompromised hosts is now well recognized, and development of potential RSV preventative measures are underway. Although no RSV vaccine is currently available to prevent RSV, the World Health Organization (WHO) estimates that an RSV vaccine will be in clinical use within the next 5–10 years [1]. Many strategies for the prevention of RSV in infants, children, and adults are under consideration. The potential of maternal immunization to prevent serious disease in infants will be discussed in this chapter.

Clinical burden of disease and epidemiology

Clinical burden of RSV in children

Respiratory syncytial virus is the primary cause of viral lower respiratory tract infection (LRTI) in infants worldwide. RSV is associated with high rates of hospitalization, particularly in infants less than 6 months of

https://doi.org/10.1016/B978-0-12-814582-1.00011-5

age [2], with an estimated global burden of 33 million infections and 60,000 deaths annually in children under five [3]. In the United States (US) and many industrialized countries, RSV is the most common infection resulting in hospital admission among infants [2–4]. In temperate climates, RSV produces annual midwinter epidemics clinically characterized by bronchiolitis in infants. RSV outbreaks are less clearly delineated in tropical areas, where year-round infection may occur with or without epidemics [5]. An estimated 99% of deaths due to RSV in children occur in resource-limited settings, making the prevention of RSV-associated mortality in regions of the world with limited access to health care a high priority [3, 4, 6, 7].

Clinical manifestations of RSV infection in young children include nasal congestion, difficulty breathing potentially resulting in respiratory distress, bronchiolitis and pneumonia with or without fever, as well as apnea in young infants. While RSV associated illness is common in children less than 5 years of age, moderate to severe disease is seen primarily in infants. RSV disease in infants under 6 months of age accounts for about half of documented RSV cases in many studies. The vast majority of RSV-related hospitalizations occur in young children during the first two years of life. In one prospective cohort study in Arizona, US, almost 30% of infants with a medically attended illness in the first year of life was due to RSV, which was usually diagnosed as bronchiolitis or pneumonia [8] and similar rates have been observed in developing countries [9]. At least 2% of all US infants are hospitalized with RSV disease, with the peak occurrence in the second month of life. Hospitalization rates are higher in risk groups including premature infants and those with underlying cardiac or pulmonary diseases. Infants in aboriginal populations such as Native Americans or Alaska Natives may have hospitalization rates 3–5 times higher than that of the general US population [2, 10, 11]. To date, RSV treatment is mainly supportive and consists of respiratory support and fluid management. Support to treat apnea associated with RSV infection in preterm and very young infants may also be required. Despite the relatively minimal care typically administered to RSV-infected children in industrialized countries, RSV-related mortality is substantially higher in low and middle income countries [3, 4].

Clinical burden of RSV in pregnant women

RSV is a common cause of symptomatic respiratory illness in healthy adult populations between 18 and 60 years of age. Compared to influenza, RSV infection symptoms generally appear to last longer with less fever. In 2001 in a US study, 26% of adults with RSV had lower respiratory tract symptoms, defined as tracheobronchitis, bronchitis, or wheezing [12]. In a more recent study of pneumonia in hospitalized patients over the age of 18 years, RSV was responsible for between 0.2 and 5 cases of pneumonia per 10,000 individuals, with increasing hospitalization rates associated with increasing age [13]. However, rates and severity of RSV infection or RSV pneumonia in pregnant women remains poorly characterized, reflecting a major gap in RSV epidemiology [14].

Most studies of RSV infection or disease during pregnancy are based on case reports or secondary analyses of influenza vaccine trials [15–17]. A recent prospective longitudinal community based study conducted in Houston, Texas, provides additional information regarding the epidemiology of RSV and other respiratory viruses in healthy pregnant women receiving prenatal care at a community obstetrical clinic [18]. This study evaluated both symptomatic and asymptomatic women in an obstetric population that reflected the contemporary make-up of pregnant women in the US, with a median age of 31 years and a broad representation of racial and ethnic backgrounds. The most significant finding in the Houston study was that over one-third of women with acute respiratory symptoms had disease consistent with LRTI, which the authors defined as difficulty breathing or shortness of breath, wheezing, or cyanosis. Morbidity was associated with multiple respiratory viruses including RSV, with considerable symptoms noted for all viruses. Common viruses detected in this study included rhinovirus (27%), coronavirus (17%), and RSV. The overall attack rate of RSV among ambulatory pregnant women in this population was 10% based on laboratory-confirmed RSV infection documented by polymerase chain reaction (PCR) testing of respiratory secretions, and up to 13% when serologic diagnosis was added [19].

Data from influenza vaccine trials has been commonly used to determine the incidence of RSV and other respiratory viruses in order to estimate RSV epidemiology in pregnant women. This approach permits the evaluation of many pregnant women in prospective studies but has potential issues, with the main limitations being the use of influenza-like-illness (ILI) criteria requiring the presence of fever to guide which samples are obtained for the diagnosis of respiratory illness. Also, surveillance has often been ongoing only during influenza season. ILI criteria are typically based on the combination of fever and cough and/or sore throat, but the majority of cases of RSV in adults are afebrile [12]. Additionally, influenza and RSV seasons may not overlap, likely leading to underestimation of RSV burden if sampling is performed only during periods of influenza virus circulation [20]. Two large maternal influenza vaccine studies sponsored by the Bill & Melinda Gates Foundation evaluated the incidence of maternal respiratory disease due to RSV during pregnancy [21, 22]. In Nepal, where nasal swabs were collected prospectively from women with a fever and respiratory symptoms year round, RSV prevalence was 0.2%, with an incidence of 3.9/1000 person-years overall. In South Africa, RSV prevalence during the maternal influenza trial where specimens were obtained based on the presence of any respiratory symptoms was much higher at 2%, or an incidence of 14.4–48.0 cases per 1000 person-years overall [23]. Importantly, no maternal RSV case in the South Africa trial had a fever documented. RSV was not associated with an increased risk of low birth weight or preterm birth in either the South African or Nepal studies, although numbers of RSV-infected women in both studies were small.

More recently, the Pregnancy Influenza Vaccine Effectiveness Network (PREVENT) of the US Centers for Disease Control and Prevention and sites in four countries evaluated the impact of RSV in hospitalization among pregnant women in high income countries – the US, Canada, Israel and Australia [24]. Among a total population of 1,604,2016 pregnant women hospitalized over various seasons from 2010 to 2016, 15,287 (15%) had at least one hospitalization associated with a diagnosis of acute respiratory or febrile illness, but only 6% of those admitted were tested for viruses, including RSV, highlighting the low level of awareness and testing for RSV by obstetric providers. Importantly, 2.5% (range 1.9–3.1%, 21 RSV-positive cases in total) of women tested were positive for RSV by PCR. Two-thirds of the tests and diagnoses of RSV occurred in the third trimester of pregnancy. A pre-existing health condition was documented in 38% of women, with asthma being the most common. Importantly, a diagnosis of pneumonia was more frequent among RSV-positive compared to RSV negative women (38% vs. 19%, $p=0.046$), and nearly half of RSV-positive women required admission for ≥3 days. A significant association was documented between admission for RSV confirmed infection and subsequent preterm birth, which occurred in 29% of RSV-positive women compared to 15% RSV-negative women ($p=0.034$). There were, however, no differences in the frequency of overall low birth weight and small-for-gestational age infants between RSV-positive and negative women. This study suggests that RSV is an important cause of LRTI in hospitalized pregnant women, particularly those with underlying medical conditions, however given the small proportion of pregnant women tested for RSV during a hospital admission with acute respiratory illness, further data is required to fully appreciate the burden in this population.

Protection against RSV

Maternally-derived antibody

RSV was first identified by Chanock in 1957 [25] and identified as a pathogen in young infants several years later [26]. Reasons for disease predilection in young children have been studied since that time, and the impact of maternal antibody on the subsequent development of RSV disease in young children has been under investigation for years. Originally, concern for potential augmentation of RSV disease in the presence of maternally-derived antibody in young infants was considered a possibility, but subsequent studies utilizing different study designs and laboratory analysis of RSV-specific antibodies dispelled that theory, and actually demonstrated that infants with RSV disease have a 2–4-lower concentration of RSV-specific antibodies compared to infants with no disease [27–32]. In a landmark study conducted by Glezen et al. in Houston, Texas, RSV-specific serum neutralizing antibodies were shown to be efficiently transferred from the mother to the newborn and high levels of neutralizing antibodies acquired transplacentally by the neonate protected the infant against LRTI during the first few months of life [27]. A later study from Denmark calculated a 26% reduction in hospitalization during the first 6 months of life for every two-fold increase in cord blood neutralizing antibody [33]. Efficient transplacental transfer of maternal RSV antibody has been demonstrated in multiple studies in industrialized countries as well as in Africa and Asia [34–37]. In healthy full-term maternal/cord blood pairs, maternal: cord RSV-specific antibody ratios in these studies demonstrated more antibody in infants than in the mothers, with transplacental transfer ratios in both vaccinated and non-vaccinated women to be approximately 1.02–1.03 [27–32]. The decline of virus-specific immunity provided by maternal RSV antibodies closely mirrors the half-life of immunoglobulin G1 (IgG1) (approximately 30–40 days), the principal IgG subclass antibody to RSV that is transplacentally transferred in preterm and term neonates. Currently, both human and animal data from mice, cotton rats, lambs, calves, and non-human primates support serum neutralizing antibody as a good correlate of immunity against RSV disease of the lower respiratory tract but not necessarily of the upper respiratory tract. Standardization of RSV neutralization assays and work toward developing an international standard are underway [38].

The structure of RSV and immunity

RSV is a member of the genus pneumovirus, one of two Paramixoviruses in this group, with human metapneumovirus, and has two antigenically distinct subgroups, RSV-A and RSV-B. RSV has a negative sense non-segmented RNA genome that encodes 11 proteins (Table 1). Two glycoproteins in the virion membrane, the fusion (F) and attachment or binding (G) proteins, carry the antigenic determinants that elicit neutralizing antibodies against RSV (Table 1). While immunity against RSV relies primarily on the development of neutralizing antibodies to

TABLE 1 The structure of respiratory syncytial virus and antigenic sites [39].

RSV structure and protein function	
Non Structural Proteins NS1 and NS2 M2-2	Counteract interferon responses
Internal Structural Proteins N, P, M, M2-1, L	
Surface Structural Proteins Fusion (F) Attachment (G) Small hydrophobic (SH)	F – determines viral entry to host cells G – determines binding to host cells F and G proteins have the antigenic determinants that elicit neutralizing antibodies to RSV
Main antigenic sites of RSV F-protein listed by order of ability to elicit neutralizing antibodies	
Site Φ Site V	Sites Φ and V are pre-F specific epitopes, present at the apex of the pre-F protein and therefore highly neutralizing sensitive. Antibodies to these sites have the highest neutralizing potency
Site III Site IV Site II	Sites III, IV and II are present in both pre-F and post-F conformation. Palivizumab binds to site II.
Site I	Site I is mostly present in post-F conformation

these surface proteins, non-neutralizing antibodies to the F, G and Small Hydrophobic (SH) surface proteins can inhibit infection by complement mediated neutralization or antibody dependent cell mediated toxicity, and all viral antigens can induce protection by T cell mediated immunity. Given that the F-protein plays a critical role in viral entry to the host cell, and that it is highly conserved within RSV-A and RSV-B subtypes, it is the preferred target of RSV vaccines and monoclonal antibodies. The F-protein contains 6 antigenic sites that elicit the production of more than 90% of the high potency neutralizing antibodies against RSV. The F-protein presents two structural conformations, a pre-fusion (pre-F) and a post-fusion (post-F) form, in the course of infecting host cells. Antibodies directed to antigenic sites present in the pre-F conformation are more efficient at neutralizing RSV than those directed to antigenic sites present in the more stable post-F confirmation. Although it contains a central conserved domain available for neutralizing antibody binding, the G-protein is not a primary target for vaccine development because it is mostly covered in glycans, and it is not as well conserved within RSV types as the F-protein [39, 40].

Monoclonal antibody prophylaxis to protect young infants against RSV

Early studies targeting antibody prophylaxis to prevent RSV infection in high-risk infants utilized high-titer human immunoglobulin products obtained from screened volunteers administered to infants intravenously [41]. In one study, where high-risk children were administered monthly doses of intravenous immunoglobulin containing high levels of RSV neutralizing antibody (RSV-IVIG) had reduced rates of hospitalization from RSV infection. This human immunoglobulin product, Respigam (MedImmune, Gaithersburg, MD), was licensed in 1996. Subsequently, a humanized monoclonal antibody specific for the F-protein of RSV, Palivizumab (Synagis; MedImmune and AstraZeneca, Cambridge, UK), was evaluated and approved in 1998 by the US Food and Drug Administration (FDA) for use by intramuscular injection in high-risk children. The studies definitively demonstrated that RSV-specific antibody alone may prevent or reduce RSV disease in infants [42]. Currently, Palivizumab is the only FDA- and European Medicines Agency-approved therapy utilized for the prophylaxis of RSV in infants and young children who are at increased risk of hospitalization. Palivizumab was originally approved for preterm infants <35 weeks gestational age, infants with chronic lung disease of prematurity, and those with hemodynamically significant congenital heart disease. This preventive approach, requiring the intramuscular administration of 15 mg/kg of Palivizumab at 4-week intervals during the RSV season, has been shown to be safe and effective in preventing serious RSV illness and hospitalization in high-risk young children [43]. Further refinements of recommendations for Palivizumab prophylaxis have changed over the past decades due in part to considerations of cost effectiveness [44, 45]. Impacts of a more restrictive policy on Palivizumab use in the US which limits its use among preterm infants to those less than 29 weeks of gestation at birth have been described, with increasing cases of RSV hospitalizations and morbidity in young infants 29–34 weeks gestation reported in some studies [46, 47]. Surveillance systems in place and other studies, including birth cohorts, will further inform policy on the use of Palivizumab for the prevention of severe RSV in high-risk infants in the US. Specific recommendations for pediatric populations that should receive Palivizumab therapy currently varies by country.

RSV F-protein and novel monoclonal antibodies

Major advances in the understanding of the structure and function of the highly conserved RSV F-protein, the target of most antibodies and current vaccine candidates, have been made this century. For example, it is known that activation of the RSV F-protein from the pre-fusion state to the post-fusion state requires a structural change [48]. This conformational change exposes various antigenic sites, which each elicit the production of neutralizing antibodies that vary in potency depending on the site. The RSV F-protein has been crystallized and its pre-fusion form stabilized [40, 49]. A substantial proportion of the neutralizing antibody response to RSV has been shown to be directed against the pre-fusion conformation of the F protein. Pre-F antibody has been postulated to serve as a more accurate correlate of protection allowing for the identification of novel antigenic sites that may serve as potential antigens for monoclonal antibody or vaccine development [40, 48, 50].

Next generation RSV F-protein monoclonal antibodies that target antigenic sites on the pre-fusion protein that have enhanced neutralizing activity are under active study. Motavizumab (MedImmune), a derivative of Palivizumab with higher affinity and a longer half-life, was effective in reducing RSV hospitalization in high-risk full-term infants in the US, but was not licensed due to safety concerns (allergic reactions) [51]. Another monoclonal antibody product, Suptavumab (Regeneron Pharmaceuticals, Tarrytown, NY) failed to meet the primary outcome of preventing medically attended RSV infection in preterm infants born at >29 weeks of gestation in a 2017 Phase III infant clinical trial (http://investor.regeneron.com/releaseDetail.cfm?releaseid1037184). A novel investigational monoclonal antibody, MEDI8897 (MedImmune and AztraZeneca LLC), is a recombinant human immunoglobulin G1 monoclonal antibody that targets the pre-fusion conformation of the RSV F-protein [52]. This monoclonal antibody binds a highly conserved epitope on RSV F and neutralizes a diverse group of RSV A and B strains with over 50-fold higher activity than Palivizumab. MEDI8897 has an extended half-life compared to Palivizumab due to mutations in the Fc-domain (YTE). This antibody has been studied for safety, tolerability and pharmacokinetics in adults and healthy preterm infants [53, 54] and is currently undergoing clinical efficacy studies in preterm young infants who do not qualify for Palivizumab in the US (Clinical Trials. Gov ID: NCT02325791 and NCT02290340). The US FDA granted Breakthrough Therapy Designation of this product in February 2019, indicating this product has received an expedited development and regulatory review (website accessed May 14, 2019: www.astrazeneca.com/media-centre/press-releases/2019/us-fda-grants-breakthrough-therapy-designation-for-potential-next-generation-rsv-medicine-medi8897.htmlww).

RSV vaccines

The development of vaccines for the prevention of RSV has been ongoing since its initial isolation in infants with severe LRTI in the 1960s. The first vaccine evaluated in a clinical trial in infants was a formalin-inactivated vaccine, which resulted in inadequate neutralizing antibody responses and augmentation of disease after subsequent wild-type infection in vaccine recipients [50]. This experience led to the careful consideration of the potential strategies for RSV prevention and the development of various RSV vaccine candidates which are currently undergoing evaluation, that include live attenuated and chimeric vaccines, subunit vaccines, particle based vaccines, including virus-like particle and nanoparticles, nucleic acid and recombinant vector vaccines. While no vaccine is yet licensed for the prevention of RSV, there are more than 20 vaccines candidates in various phases of preclinical and clinical development, targeting specific populations at risk, including infants and young children, the elderly, and pregnant women who would receive vaccines to protect their newborns against severe RSV disease (Table 2 and https://www.path.org/resources/rsv-vaccine-and-mab-snapshot/). Vaccines specifically tailored for each of these populations are necessary to establish an effective multi-prong approach to reduce the burden of RSV, particularly in young children. For example, live attenuated RSV vaccines may be administered directly into the nasal mucosa in young children to induce a robust local and systemic immune response, given that responses to vaccination under 6 months of age may be limited by interference from maternal antibody and overall lower immunogenicity, requiring repeated doses of vaccines before lasting protection can be achieved [55]. A particular challenge is that infants with low RSV antibody concentrations are at greater risk to acquire infection earlier in life, and natural RSV infection does not confer long lasting immunity, therefore reinfections in childhood and through life are common. Furthermore, the highest morbidity and mortality from RSV in both healthy term infants and those with underlying conditions that increase the risk for complications, such as prematurity and cardiopulmonary disease, occur in the first 3–6 months of life, too early for active immunization to be effective. Consequently, passive antibody administration has been the major strategy for RSV prevention in high-risk infants until now. However, passive immunization is limited by the need to restrict this costly intervention to the groups with highest risk, leaving the majority of infants, those born at term, susceptible to RSV infection and its complications.

TABLE 2 RSV vaccines in development, by vaccine type, phase of development and target population. Number of vaccine products in development is shown in parenthesis.

RSV vaccine type	Preclinical	Clinical phase I	Clinical phase II	Clinical phase III
Live attenuated/Chimeric	RSV (3) RSV/PIV-1-3 (1)	RSV-ΔG (1) Pediatric RSV ΔNS2/Δ1313/I11314L (1) Pediatric RSV 6120/ΔNS2/1030s (1) Pediatric RSV Δ46/NS2/N/ΔM2-2-*Hin*dIII (1) Pediatric SeV/RSV (1) Pediatric BCG/RSV (1) Pediatric	–	–
Whole virus inactivated	RSV (1)	–	–	–
Particle based	VLP (7) Peptide microparticle (1)	RSF F Nanoparticle (1) Pediatric	RSV F Nanoparticle (1) Elderly	RSV F Nanoparticle (1) **Maternal**[a]
Subunit	RSV F-protein (2) RSV G-protein (2)	RSV F-protein (2) **Maternal**[b] (3) Elderly RSV G-protein (1) Pediatric and Elderly DPX-RSV-SM Protein (1) Elderly	RSV F-Protein (1) **Maternal**[c] and Elderly	–

Nucleic acid	RNA (1) DNA (1)	–	–	–
Recombinant vector	Adenovirus (1)	Adenovirus (1) Elderly	Adenovirus (2) Pediatric (1) Elderly MVA (1) Elderly	–

[a] *RSV F nanoparticle (Novavax, Rockville, MD) Post-F "prefusogenic", adjuvanted with* $ALPO_4$ *or Matrix M.*
[b] *DS-Cav 1 (NIH) Pre-F subunit, Alum adjuvanted vaccine and GSK RSV F (GlaxoSmithKline, Brentford, UK) Pre-F with and without adjuvant.*
[c] *RSV F (Pfizer, New York, NY), Pre-F with and without adjuvant.*
Bold highlights represent maternal vaccines.
Source: Adapted from PATH https://www.path.org/resources/rsv-vaccine-and-mab-snapshot/.

Maternal immunization for the prevention of RSV in infants

Maternal vaccination appears to be the most direct and optimal strategy to prevent early RSV disease in most neonates and young infants. This strategy is supported by the following observations:

- Maternal RSV infection prior to delivery correlates with decreased incidence of RSV bronchiolitis in infants.
- Higher concentrations of RSV-specific maternally-derived antibody at birth are associated with lower incidence and later onset of RSV disease in neonates during the first months of life.
- Passive anti-F IgG (e.g. Palivizumab) administration reduces the incidence of severe RSV disease.
- Pregnant women are primed for RSV from previous infections and vaccination will boost their antibodies, achieving higher concentrations at a time when active antibody transfer mechanisms in the placenta favor the passage of antibodies to the fetus.
- Transplacental transfer of RSV-specific IgG from mothers to neonates is highly efficient.
- In addition to transplacental antibody transfer, there is also a potential for protection of lactating infants from RSV-specific breastmilk antibodies (natural and vaccine induced).
- There is a precedent of success of maternal immunization as a strategy for the prevention of maternal and infant tetanus, pertussis, and influenza.

The goals of a maternal immunization program for the prevention of RSV in infants include the prevention of infant death and hospitalization, the prevention and/or reduction of severe RSV LRTI, and the delay in the onset of the first episode of RSV infection. This delay in onset is important given that infants may be more capable to resist and overcome the initial RSV illness when it occurs later in the first or even second year of life. Indirect benefits from breastmilk antibodies prolonging the duration of protection of transplacentally derived antibodies, and reduction of household transmission of infection through maternal and infant protection have been documented [56]. Protection of mothers during pregnancy against obstetric and illness complications associated with RSV infection in the third trimester of gestation may be additional potential benefits.

An ideal RSV vaccine for maternal immunization is one that is not a live virus vaccine, can be administered as a single dose during pregnancy, preferably in the second or early third trimester of gestation to allow for sufficient time for transplacental antibody transfer, is safe for mother and infant, and is able to induce protective antibodies that are efficiently transferred to the infant via the placenta to achieve high concentration of neutralizing antibodies to protect the infant long enough in the first months of life. The natural decay of passively acquired maternal antibodies allows for the highest protection in the infant to be in the first 2–3 months of life, declining afterwards. Therefore, maternal immunization will likely need to be followed by active infant immunization starting as early as the second month after birth, to reduce the impact of RSV in the first year of life. This strategy, maternal followed by infant immunization, would be most effective for infants born at term, for whom passive antibody administration might not be an option. Infants of vaccinated mothers who receive active immunization would less likely acquire RSV infection in the first 3 months of life, the period of greatest vulnerability, or experience acquired infections that are modified or rendered less severe by the presence of maternally derived and vaccine induced antibodies. Maternal immunization is not likely to be an effective method to prevent early disease among premature infants who would not benefit from a sufficient duration of gestation to achieve optimal transplacental transfer and high antibody concentrations at birth. Therefore, administration of passive RSV antibodies (e.g. Palivizumab) followed by active infant immunization would be the most adequate strategy for prevention of RSV in preterm infants. The duration of protection of maternally-derived antibodies in all infants is limited by the concentration of antibodies present at birth, and the relatively short antibody half-life, which results in a rapid natural antibody decay in the first 6 months of life. Therefore, both passive and active immunization will have an important role in the prevention of RSV in infants, even if routine maternal immunization or administration of monoclonal antibodies after birth become available. One remaining challenge in the development of RSV vaccines is the lack of a known serologic correlate of protection to evaluate vaccine efficacy and effectiveness. Until a correlate of protection is identified, large clinical trials are required to assess the efficacy of candidate RSV vaccines for maternal or infant immunization. Another important area of research in the field of RSV addresses the need to better understand the mechanisms of protection beyond antibodies, such as innate and cellular immunity. Definitive progress in the design and development of RSV vaccines was driven by a better understanding of the conformation depending immunogenicity of the F-protein and the need to target neutralization sensitive epitopes on the functional forms of F-protein [40, 57]. All RSV vaccines for maternal immunization developed to date have been based on the F-protein of RSV. There are currently four vaccines in development for maternal immunization, two in phase I clinical trials, and one each in phase II and III clinical trials (Table 2).

A non-adjuvanted purified F-protein (PFP-2) vaccine for RSV was the first vaccine evaluated in pregnancy nearly 20 years ago, in the early 2000s [36]. This was a small proof of concept, randomized, placebo-controlled study in 35 healthy women who were vaccinated at 30–34 weeks of gestation, and in whom the RSV PFP-2 vaccine was well tolerated, though it did not significantly increase neutralizing antibody titers to RSV, and was not further developed. Nevertheless, maternal immunization with this experimental vaccine was safe in mothers and infants, no obstetric adverse events were associated with vaccination, efficient transplacental passage of vaccine induced antibodies was demonstrated, the half-life of these antibodies was estimated to be between 30 and 40 days, and anti-F IgA and IgG were present in breastmilk of vaccine recipients at higher concentrations than placebo recipients in the first 6 months after delivery. Importantly, after close follow up of these infants for natural infection in two consecutive viral seasons, no enhanced RSV disease was observed.

A purified recombinant pre-F protein vaccine prepared in Chinese Hamster Ovary Cells (GlaxoSmithKline [GSK] Investigational Vaccines) [58], has been evaluated in approximately 600 non-pregnant women to assess the reactogenicity and immunogenicity of various dosages of adjuvanted and non-adjuvanted formulations. All formulations of this vaccine boosted pre-existing antibodies in these 19–45 year old women, with comparable immunogenicity, and the safety profile was similar to that of tetanus, diphtheria, and pertussis (Tdap) vaccine, which is administered routinely during pregnancy in the US and various countries [58]. This vaccine construct is planned to be evaluated in pregnant women in a global phase I/II study expected to begin in 2019–20.

Lastly, an aluminum-adjuvanted RSV F-protein nanoparticle vaccine (Novavax, Inc., Gaithersburg, Maryland, US) was evaluated in women of childbearing age [59] and in a phase II study in pregnant women prior to being the first vaccine evaluated in a phase III clinical trial, seeking an indication for administration in pregnancy. The phase II study demonstrated that the vaccine was safe and well tolerated, and induced significantly higher concentrations of both anti F-IgG, but also neutralizing antibodies and Palivizumab competing antibodies [60]. This study also found that maternal antibody responses peaked at 14 days after vaccination, and that an interval of ≥30 days from maternal vaccination to delivery maximized transplacental antibody transfer for all measured antibodies [60]. This study allowed the progression to a global phase III efficacy clinical trial to evaluate this maternal vaccine for the protection of RSV in infants. This clinical trial in healthy pregnant women was conducted in 11 countries worldwide from 2015 to 2019. The primary objective of the study was to determine the efficacy of maternal immunization against medically significant RSV LRTI in infants through various time points including the first 90 days of life, and up to 180 days, when mothers were vaccinated with a single dose of vaccine at 28–36 weeks of gestation. The clinical definition of medically significant RSV LRTI included the presence of at least two clinical symptoms of illness, laboratory confirmation of RSV infection by PCR, and objective assessment of tachypnea and hypoxemia (O_2 saturation < 95%). Secondary endpoints included the reduction of RSV LRTI hospitalization and RSV LRTI with severe hypoxemia (O_2 saturation < 92%). Preliminary data shared to date by the manufacturer indicates that among 4636 women who were randomized 2:1 to receive vaccine or placebo, the vaccine was well tolerated and was not associated with any vaccine-related maternal, infant or obstetric adverse event. This study failed to meet its primary objective of efficacy against medically significant RSV LRTI in infants, which was estimated to be 39.4% (97.5% CI, −1 to 63.7) at 0–90 days of life, and the secondary outcome of RSV LRTI with severe hypoxemia was also not met. However, maternal vaccination was effective in protecting infants against RSV-associated LRTI hospitalization in the first 90 days and up to 180 days after birth (Table 3) (http://ir.novavax.com/news-releases/news-release-details/novavax-announces-topline-results-phase-3-preparetm-trial#). While vaccine licensure based on this study is unlikely in the US given the observed geographic variability in vaccine efficacy and overall lower numbers of cases than expected, the results are encouraging. When pre-specified analyses were conducted to evaluate these results utilizing expanded data (which increased the number of evaluable subjects), the vaccine efficacy achieved in the first 90 days of life varied between 40% and nearly 60% for these primary and secondary outcomes (Table 3), which is considered a result worthy of continuing the development and evaluation of this vaccine. This study was pivotal in generating data to better understand the potential for maternal RSV immunization to prevent the most severe outcomes of RSV in infants, and to answer specific programmatic questions, such as the impact of the timing of immunization during pregnancy and the interval between vaccination and

TABLE 3 Preliminary efficacy of the maternal RSV F-protein nanoparticle vaccine (Novavax) against severe RSV infection in infants.

Endpoints	Per-protocol analysis	Pre-specified expanded data analysis which includes data from hospitalization records
Primary		
Medically significant RSV LRTI	39% (97.5%CI, −1% to 64%)	41% (95%CI, 16% to 58%)
Secondary		
RSV LRTI Hospitalization	44% (95%CI, 20% to 62%)	42% (95%CI, 17% to 59%)
RSV LRTI with severe hypoxemia	48% (95%CI, −8% to 75%)	60% (95%CI, 32% to 76%)

Source: Novavax press release, available at http://ir.novavax.com/news-releases/news-release-details/novavax-announces-topline-results-phase-3-preparetm-trial#.

delivery to optimize the impact of maternal immunization in protecting the infant. The observation that geographic, seasonal, and potentially, population based differences in vaccine efficacy can occur will help inform the design of future clinical trials. While these and other results from this study continue to be evaluated, the experience of this trial has provided further impetus for the ongoing development and evaluation of RSV vaccines for maternal immunization, with an important contribution to defining the clinical endpoints to which clinical trials should be powered to.

Other experimental vaccines in advanced development for potential administration during pregnancy include an aluminum hydroxide adjuvanted RSV stabilized pre-fusion F subunit vaccine, which began enrollment in a phase IIB placebo controlled randomized clinical trial in pregnant women in the US in August 2019 (Pfizer. NCT04032093 – https://clinicaltrials.gov/ct2/show/NCT04032093), and an unadjuvanted pre-F subunit vaccine that will begin phase II evaluation in pregnant women in 2020 after having completed dose ranging immunogenicity and safety evaluation in non-pregnant healthy women (GSK Biologicals, GSK3888550A, NCT03674177, https://clinicaltrials.gov/ct2/show/NCT03674177) (Table 2).

In parallel, the research field is active in establishing standardized methods for the assessment of RSV infection and immunity, the clinical assessment of severity of disease, and the determination of correlates of protection against RSV. Preferred product characteristics for RSV vaccines have been developed by the WHO [61] and a clear road map for the research and development of technology associated with RSV vaccines is in place [62]. Various vaccine candidates remain in preclinical and clinical phases of development, while work to identify the key necessary elements for the implementation of maternal RSV vaccines worldwide continues with the support of multiple stakeholders [63–66]. Implementation of RSV vaccines for maternal immunization should be guided by accurate estimates of RSV disease burden in infants to determine their potential impact. Establishing robust surveillance systems to assess vaccine safety and efficacy outcomes after implementation is key, as is the support from key policy makers, national and local maternal and child health systems and other funding and administrative organizations (Fig. 1). A successful maternal immunization delivery platform would include integrated antenatal and neonatal care services and immunization programs, informed by existing vaccination (e.g. tetanus and influenza) and routine antenatal care interventions (e.g. HIV diagnosis and prevention), which would likely need strengthening, financial and policy support. Lastly, effective communication plans and advocacy planning are important elements to ensure vaccine uptake by educating providers, pregnant women, parents, and the public about the impact of

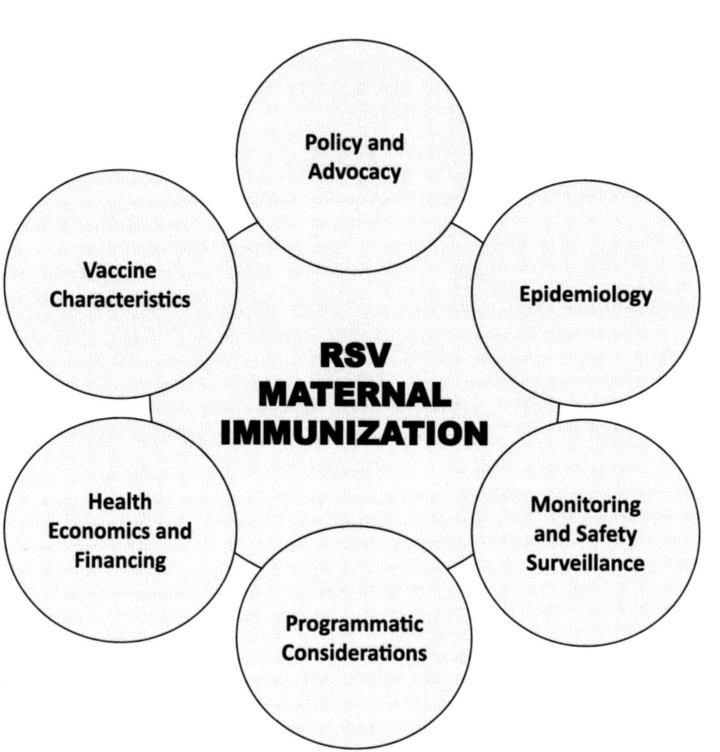

FIG. 1 Key elements for the implementation of RSV maternal immunization.Source: Adapted from PATH, 2018. Advancing RSV maternal immunization: A gap analysis report. PATH, Seattle, WA.

RSV and the role of maternal immunization in the prevention of infant disease and its associated mortality.

Conclusion

The global burden of morbidity and mortality associated with RSV disease in infants is substantial, particularly in low and low-middle income countries. Maternal immunization is a feasible strategy that has the potential to significantly reduce the impact of RSV disease in young infants worldwide. Vaccines for maternal immunization are in active preclinical and clinical development, with encouraging results. Ongoing research and implementation planning activities support the concept of RSV prevention through maternal immunization and the possibility to achieve tangible progress in the control of RSV disease in young infants in the near future.

References

[1] Modjarrad K, Giersing B, Kaslow DC, Smith PG, Moorthy VS, Group WRVCE. WHO consultation on respiratory syncytial virus vaccine development report from a World Health Organization meeting held on 23-24 March 2015. Vaccine 2016;34(2):190–7. https://doi.org/10.1016/j.vaccine.2015.05.093. 26100926.

[2] Hall CB, Weinberg GA, Iwane MK, Blumkin AK, Edwards KM, Staat MA, et al. The burden of respiratory syncytial virus infection in young children. N Engl J Med 2009;360(6):588–98. https://doi.org/10.1056/NEJMoa0804877. 19196675. PMCPMC4829966.

[3] Shi T, McAllister DA, O'Brien KL, Simoes EAF, Madhi SA, Gessner BD, et al. Global, regional, and national disease burden estimates of acute lower respiratory infections due to respiratory syncytial virus in young children in 2015: a systematic review and modelling study. Lancet 2017;390(10098):946–58. https://doi.org/10.1016/S0140-6736(17)30938-8. 28689664. PMCPMC5592248.

[4] Scheltema NM, Gentile A, Lucion F, Nokes DJ, Munywoki PK, Madhi SA, et al. Global respiratory syncytial virus-associated mortality in young children (RSV GOLD): a retrospective case series. Lancet Glob Health 2017;5(10):e984–91. https://doi.org/10.1016/S2214-109X(17)30344-3. 28911764. PMCPMC5599304.

[5] Glezen P, Denny FW. Epidemiology of acute lower respiratory disease in children. N Engl J Med 1973;288(10):498–505. https://doi.org/10.1056/NEJM197303082881005. 4346164.

[6] Nair H, Nokes DJ, Gessner BD, Dherani M, Madhi SA, Singleton RJ, et al. Global burden of acute lower respiratory infections due to respiratory syncytial virus in young children: a systematic review and meta-analysis. Lancet 2010;375(9725):1545–55. https://doi.org/10.1016/S0140-6736(10)60206-1. 20399493. PMCPMC2864404.

[7] Langley GF, Anderson LJ. Epidemiology and prevention of respiratory syncytial virus infections among infants and young children. Pediatr Infect Dis J 2011;30(6):510–7. https://doi.org/10.1097/INF.0b013e3182184ae7. 21487331.

[8] Wright AL, Taussig LM, Ray CG, Harrison HR, Holberg CJ. The Tucson Children's Respiratory Study. II. Lower respiratory tract illness in the first year of life. Am J Epidemiol 1989;129(6):1232–46. https://doi.org/10.1093/oxfordjournals.aje.a115243. 2729259.

[9] Nokes DJ, Okiro EA, Ngama M, Ochola R, White LJ, Scott PD, et al. Respiratory syncytial virus infection and disease in infants and young children observed from birth in Kilifi District, Kenya. Clin Infect Dis 2008;46(1):50–7. https://doi.org/10.1086/524019. 18171213. PMCPMC2358944.

[10] Bruden DJ, Singleton R, Hawk CS, Bulkow LR, Bentley S, Anderson LJ, et al. Eighteen years of respiratory syncytial virus surveillance: changes in seasonality and hospitalization rates in southwestern Alaska native children. Pediatr Infect Dis J 2015;34(9):945–50. https://doi.org/10.1097/INF.0000000000000772. 26065863.

[11] Shay DK, Holman RC, Newman RD, Liu LL, Stout JW, Anderson LJ. Bronchiolitis-associated hospitalizations among US children, 1980-1996. JAMA 1999;282(15):1440–6. 10535434.

[12] Hall CB, Long CE, Schnabel KC. Respiratory syncytial virus infections in previously healthy working adults. Clin Infect Dis 2001;33(6):792–6. https://doi.org/10.1086/322657. 11512084.

[13] Jain S, Self WH, Wunderink RG, Fakhran S, Balk R, Bramley AM, et al. Community-acquired pneumonia requiring hospitalization among U.S. adults. N Engl J Med 2015;373(5):415–27. https://doi.org/10.1056/NEJMoa1500245. 26172429. PMCPMC4728150.

[14] Kim L, Rha B, Abramson JS, Anderson LJ, Byington CL, Chen GL, et al. Identifying gaps in respiratory syncytial virus disease epidemiology in the United States prior to the introduction of vaccines. Clin Infect Dis 2017;65(6):1020–5. https://doi.org/10.1093/cid/cix432. 28903503. PMCPMC5850021.

[15] Chu HY, Katz J, Tielsch J, Khatry SK, Shrestha L, LeClerq SC, et al. Clinical presentation and birth outcomes associated with respiratory syncytial virus infection in pregnancy. PLoS One 2016;11(3):e0152015. Epub 2016/04/01, https://doi.org/10.1371/journal.pone.015201527031702. PMCPMC4816499.

[16] Wheeler SM, Dotters-Katz S, Heine RP, Grotegut CA, Swamy GK. Maternal effects of respiratory syncytial virus infection during pregnancy. Emerg Infect Dis 2015;21(11):1951–5. https://doi.org/10.3201/eid2111.150497. 26485575. PMCPMC4622246.

[17] Chaw L, Kamigaki T, Burmaa A, Urtnasan C, Od I, Nyamaa G, et al. Burden of influenza and respiratory syncytial virus infection in pregnant women and infants under 6 months in mongolia: a prospective cohort study. PLoS One 2016;11(2):e0148421. https://doi.org/10.1371/journal.pone.0148421. 26849042. PMCPMC4746066.

[18] Hause AM, Avadhanula V, Maccato ML, Pinell PM, Bond N, Santarcangelo P, et al. A cross-sectional surveillance study of the frequency and etiology of acute respiratory illness among pregnant women. J Infect Dis 2018;218(4):528–35. https://doi.org/10.1093/infdis/jiy167. 29741642.

[19] Hause AM, Avadhanula V, Maccato ML, Pinell PM, Bond N, Santarcangelo P, et al. Clinical characteristics and outcomes of respiratory syncytial virus infection in pregnant women. Vaccine 2019;37(26):3464–71. https://doi.org/10.1016/j.vaccine.2019.04.098. 31085002.

[20] Datta S, Walsh EE, Peterson DR, Falsey AR. Can analysis of routine viral testing provide accurate estimates of respiratory syncytial virus disease burden in adults? J Infect Dis 2017;215(11):1706–10. https://doi.org/10.1093/infdis/jix196. 28863444.

[21] Steinhoff MC, Katz J, Englund JA, Khatry SK, Shrestha L, Kuypers J, et al. Year-round influenza immunisation during pregnancy in Nepal: a phase 4, randomised, placebo-controlled trial. Lancet Infect Dis 2017. https://doi.org/10.1016/S1473-3099(17)30252-9. 28522338.

[22] Madhi SA, Cutland CL, Kuwanda L, Weinberg A, Hugo A, Jones S, et al. Influenza vaccination of pregnant women and protection of their infants. N Engl J Med 2014;371(10):918–31. https://doi.org/10.1056/NEJMoa1401480. 25184864.

[23] Madhi SA, Cutland CL, Downs S, Jones S, van Niekerk N, Simoes EAF, et al. Burden of respiratory syncytial virus infection in South African human immunodeficiency virus (HIV)-infected and HIV-uninfected pregnant and postpartum women: a longitudinal cohort study. Clin Infect Dis 2018;66(11):1658–65. https://doi.org/10.1093/cid/cix1088. 29253090. PMCPMC5961360.

[24] Regan AK, Klein NP, Langley G, Drews SJ, Buchan S, Ball S, et al. Respiratory syncytial virus hospitalization during pregnancy in 4 high-income countries, 2010-2016. Clin Infect Dis 2018;67(12):1915–8. https://doi.org/10.1093/cid/ciy439. 29800089.

[25] Chanock R, Roizman B, Myers R. Recovery from infants with respiratory illness of a virus related to chimpanzee coryza agent (CCA). I. Isolation, properties and characterization. Am J Hyg 1957;66(3):281–90. 13478578.

[26] Adams JM, Imagawa DT, Zike K. Epidemic bronchiolitis and pneumonitis related to respiratory syncytial virus. JAMA 1961;176:1037–9. 13681401.

[27] Glezen WP, Paredes A, Allison JE, Taber LH, Frank AL. Risk of respiratory syncytial virus infection for infants from low-income families in relationship to age, sex, ethnic group, and maternal antibody level. J Pediatr 1981;98(5):708–15. 7229749.

[28] Ogilvie MM, Vathenen AS, Radford M, Codd J, Key S. Maternal antibody and respiratory syncytial virus infection in infancy. J Med Virol 1981;7(4):263–71. 7038043.

[29] Roca A, Abacassamo F, Loscertales MP, Quinto L, Gomez-Olive X, Fenwick F, et al. Prevalence of respiratory syncytial virus IgG antibodies in infants living in a rural area of Mozambique. J Med Virol 2002;67(4):616–23. https://doi.org/10.1002/jmv.10148. 12116014.

[30] Piedra PA, Jewell AM, Cron SG, Atmar RL, Glezen WP. Correlates of immunity to respiratory syncytial virus (RSV) associated-hospitalization: establishment of minimum protective threshold levels of serum neutralizing antibodies. Vaccine 2003;21(24):3479–82. 12850364.

[31] Ochola R, Sande C, Fegan G, Scott PD, Medley GF, Cane PA, et al. The level and duration of RSV-specific maternal IgG in infants in Kilifi Kenya. PLoS One 2009;4(12):e8088. https://doi.org/10.1371/journal.pone.0008088. 19956576. PMCPMC2779853.

[32] Eick A, Karron R, Shaw J, Thumar B, Reid R, Santosham M, et al. The role of neutralizing antibodies in protection of American Indian infants against respiratory syncytial virus disease. Pediatr Infect Dis J 2008;27(3):207–12. https://doi.org/10.1097/INF.0b013e31815ac585. 18277934.

[33] Stensballe LG, Ravn H, Kristensen K, Agerskov K, Meakins T, Aaby P, et al. Respiratory syncytial virus neutralizing antibodies in cord blood, respiratory syncytial virus hospitalization, and recurrent wheeze. J Allergy Clin Immunol 2009;123(2):398–403. https://doi.org/10.1016/j.jaci.2008.10.043. 19101023.

[34] Chu HY, Steinhoff MC, Magaret A, Zaman K, Roy E, Langdon G, et al. Respiratory syncytial virus transplacental antibody transfer and kinetics in mother-infant pairs in Bangladesh. J Infect Dis 2014;210(10):1582–9. https://doi.org/10.1093/infdis/jiu316. 24903663. PMCPMC4334795.

[35] Chu HY, Tielsch J, Katz J, Magaret AS, Khatry S, LeClerq SC, et al. Transplacental transfer of maternal respiratory syncytial virus (RSV) antibody and protection against RSV disease in infants in rural Nepal. J Clin Virol 2017;95:90–5. https://doi.org/10.1016/j.jcv.2017.08.017. 28903080. PMCPMC5625849.

[36] Munoz FM, Piedra PA, Glezen WP. Safety and immunogenicity of respiratory syncytial virus purified fusion protein-2 vaccine in pregnant women. Vaccine 2003;21(24):3465–7. 12850361.

[37] Suara RO, Piedra PA, Glezen WP, Adegbola RA, Weber M, Mulholland EK, et al. Prevalence of neutralizing antibody to respiratory syncytial virus in sera from mothers and newborns residing in the Gambia and in The United States. Clin Diagn Lab Immunol 1996;3(4):477–9. 8807217. PMCPMC170373.

[38] Hosken N, Plikaytis B, Trujillo C, Mahmood K, Higgins D, Participating Laboratories Working Group. A multi-laboratory study of diverse RSV neutralization assays indicates feasibility for harmonization with an international standard. Vaccine 2017;35(23):3082–8. https://doi.org/10.1016/j.vaccine.2017.04.053. 28476625. PMCPMC5439532.

[39] Crank MC, Ruckwardt TJ, Chen M, Morabito KM, Phung E, Costner PJ, et al. A proof of concept for structure-based vaccine design targeting RSV in humans. Science 2019;365(6452):505–9. https://doi.org/10.1126/science.aav9033. 31371616.

[40] Graham BS, Gilman MSA, McLellan JS. Structure-based vaccine antigen design. Annu Rev Med 2019;70:91–104. https://doi.org/10.1146/annurev-med-121217-094234. 30691364.

[41] Reduction of respiratory syncytial virus hospitalization among premature infants and infants with bronchopulmonary dysplasia using respiratory syncytial virus immune globulin prophylaxis. The PREVENT Study Group. Pediatrics 1997;99(1):93–9. https://doi.org/10.1542/peds.99.1.93; . 8989345; .

[42] Mejias A, Ramilo O. Review of palivizumab in the prophylaxis of respiratory syncytial virus (RSV) in high-risk infants. Biologics 2008;2(3):433–9. 19707374. PMCPMC2721379.

[43] Palivizumab, a humanized respiratory syncytial virus monoclonal antibody, reduces hospitalization from respiratory syncytial virus infection in high-risk infants. The IMpact-RSV Study Group. Pediatrics. 1998;102(3 Pt 1):531–7. 9738173; .

[44] Sanchez-Luna M, Burgos-Pol R, Oyaguez I, Figueras-Aloy J, Sanchez-Solis M, Martinon-Torres F, et al. Cost-utility analysis of Palivizumab for Respiratory Syncytial Virus infection prophylaxis in preterm infants: update based on the clinical evidence in Spain. BMC Infect Dis 2017;17(1):687. https://doi.org/10.1186/s12879-017-2803-0. 29041909. PMCPMC5645982.

[45] Resch B, Sommer C, Nuijten MJ, Seidinger S, Walter E, Schoellbauer V, et al. Cost-effectiveness of palivizumab for respiratory syncytial virus infection in high-risk children, based on long-term epidemiologic data from Austria. Pediatr Infect Dis J 2012;31(1):e1–8. https://doi.org/10.1097/INF.0b013e318235455b. 21960187.

[46] Rajah B, Sanchez PJ, Garcia-Maurino C, Leber A, Ramilo O, Mejias A. Impact of the updated guidance for palivizumab prophylaxis against respiratory syncytial virus infection: a single center experience. J Pediatr 2017;181:183–8. e1 https://doi.org/10.1016/j.jpeds.2016.10.074. 27855996.

[47] Anderson EJ, Carosone-Link P, Yogev R, Yi J, Simoes EAF. Effectiveness of palivizumab in high-risk infants and children: a propensity score weighted regression analysis. Pediatr Infect Dis J 2017;36(8):699–704. https://doi.org/10.1097/INF.0000000000001533. 28709160. PMCPMC5516669.

[48] Phung E, Chang LA, Morabito KM, Kanekiyo M, Chen M, Nair D, et al. Epitope-specific serological assays for RSV: conformation matters. Vaccines (Basel) 2019;7(1). https://doi.org/10.3390/vaccines7010023. 30813394. PMCPMC6466065.

[49] McLellan JS, Chen M, Joyce MG, Sastry M, Stewart-Jones GB, Yang Y, et al. Structure-based design of a fusion glycoprotein vaccine for respiratory syncytial virus. Science 2013;342(6158):592–8. https://doi.org/10.1126/science.1243283. 24179220. PMCPMC4461862.

[50] Anderson LJ, Dormitzer PR, Nokes DJ, Rappuoli R, Roca A, Graham BS. Strategic priorities for respiratory syncytial virus (RSV) vaccine development. Vaccine 2013;31(Suppl. 2):B209–15. https://doi.org/10.1016/j.vaccine.2012.11.106. 23598484. PMCPMC3919153.

[51] O'Brien KL, Chandran A, Weatherholtz R, Jafri HS, Griffin MP, Bellamy T, et al. Efficacy of motavizumab for the prevention of respiratory syncytial virus disease in healthy Native American infants: a phase 3 randomised double-blind placebo-controlled trial. Lancet Infect Dis 2015;15(12):1398–408. https://doi.org/10.1016/S1473-3099(15)00247-9. 26511956.

[52] Zhu Q, McLellan JS, Kallewaard NL, Ulbrandt ND, Palaszynski S, Zhang J, et al. A highly potent extended half-life antibody as a potential RSV vaccine surrogate for all infants. Sci Transl Med 2017;9(388). https://doi.org/10.1126/scitranslmed.aaj1928. 28469033.

[53] Griffin MP, Khan AA, Esser MT, Jensen K, Takas T, Kankam MK, et al. Safety, tolerability, and pharmacokinetics of MEDI8897, the respiratory syncytial virus prefusion F-targeting monoclonal antibody with an extended half-life, in healthy adults. Antimicrob Agents Chemother 2017;61(3):https://doi.org/10.1128/AAC.01714-16. 27956428. PMCPMC5328523.

[54] Domachowske JB, Khan AA, Esser MT, Jensen K, Takas T, Villafana T, et al. Safety, tolerability and pharmacokinetics of MEDI8897, an extended half-life single-dose respiratory syncytial virus prefusion F-targeting monoclonal antibody administered as a single dose to healthy preterm infants. Pediatr Infect Dis J 2018;37(9):886–92. https://doi.org/10.1097/INF.0000000000001916. 29373476. PMCPMC6133204.

[55] Karron RA, Buchholz UJ, Collins PL. Live-attenuated respiratory syncytial virus vaccines. Curr Top Microbiol Immunol 2013;372:259–84. https://doi.org/10.1007/978-3-642-38919-1_13. 24362694. PMCPMC4794267.

[56] Mazur NI, Horsley NM, Englund JA, Nederend M, Magaret A, Kumar A, et al. Breastmilk prefusion F immunoglobulin G as a correlate of protection against respiratory syncytial virus acute respiratory illness. J Infect Dis 2019;219(1):59–67. https://doi.org/10.1093/infdis/jiy477. 30107412. PMCPMC6284547.

[57] Gilbert BE, Patel N, Lu H, Liu Y, Guebre-Xabier M, Piedra PA, et al. Respiratory syncytial virus fusion nanoparticle vaccine immune responses target multiple neutralizing epitopes that contribute to protection against wild-type and palivizumab-resistant mutant virus challenge. Vaccine 2018;36(52):8069–78. https://doi.org/10.1016/j.vaccine.2018.10.073. 30389195.

[58] Beran J, Lickliter JD, Schwarz TF, Johnson C, Chu L, Domachowske JB, et al. Safety and immunogenicity of 3 formulations of an investigational respiratory syncytial virus vaccine in nonpregnant women: results from 2 phase 2 trials. J Infect Dis 2018;217(10):1616–25. https://doi.org/10.1093/infdis/jiy065. 29401325. PMCPMC5913599.

[59] August A, Glenn GM, Kpamegan E, Hickman SP, Jani D, Lu H, et al. A Phase 2 randomized, observer-blind, placebo-controlled, dose-ranging trial of aluminum-adjuvanted respiratory syncytial virus F particle vaccine formulations in healthy women of childbearing age. Vaccine 2017;35(30):3749–59. https://doi.org/10.1016/j.vaccine.2017.05.045. 28579233.

[60] Munoz FM, Swamy GK, Hickman SP, Agrawal S, Piedra PA, Glenn GM, Patel N, August AM, Cho I, Fries F. Safety and immunogenicity of a respiratory syncytial visus fusion (F) protein nanoparticle vaccine in healthy third trimester pregnant women and their infants. J infect Dis 2019;1–13. https://doi.org/10.1093/infdis/jiz390. Aug 12 [Epub ahead of print]. PMID: 31402384.
[61] WHO preferred product characteristics for respiratory syncytial virus (RSV) vaccines. Geneva: World Health Organization; 2017. Contract No.: WHO/IVB/17.11.
[62] RSV vaccine research and development technology roadmap: Priority activities for development, testing, licensure and global use of RSV vaccines, with a specific focus on the medical need for young children in low- and middle-income countries. Geneva: World Health Organization; 2017. Contract No.: WHO/IVB/17.12.
[63] Sobanjo-Ter Meulen A, Munoz FM, Kaslow DC, Klugman KP, Omer SB, Vora P, et al. Maternal interventions vigilance harmonization in low- and middle-income countries: stakeholder meeting report; Amsterdam, May 1-2, 2018. Vaccine 2019;37(20):2643–50. https://doi.org/10.1016/j.vaccine.2019.03.060. 30955981. PMCPMC6546955.
[64] Krishnaswamy S, Lambach P, Giles ML. Key considerations for successful implementation of maternal immunization programs in low and middle income countries. Hum Vaccin Immunother 2019;15(4):942–50. https://doi.org/10.1080/21645515.2018.1564433. 30676250.
[65] Walsh EE, Mariani TJ, Chu C, Grier A, Gill SR, Qiu X, et al. Aims, study design, and enrollment results from the assessing predictors of infant respiratory syncytial virus effects and severity study. JMIR Res Protoc 2019;8(6):e12907. https://doi.org/10.2196/12907. 31199303.
[66] Advancing RSV maternal immunization: A gap analysis report. Seattle, WA: PATH; 2018.

CHAPTER

11

Group B *Streptococcus*

Gaurav Kwatra[a,b,c], *Shabir A. Madhi*[a,b]

[a]Medical Research Council: Respiratory and Meningeal Pathogens Research Unit, Faculty of Health Sciences, University of the Witwatersrand, Johannesburg, South Africa, [b]Department of Science and Technology/ National Research Foundation: Vaccine Preventable Diseases Unit, Faculty of Health Sciences, University of the Witwatersrand, Johannesburg, South Africa, [c]Department of Clinical Microbiology, Christian Medical College, Vellore, India

Group B *Streptococcus*

Streptococcus agalactiae, commonly referred to as Group B *Streptococcus* (GBS), is a facultative Gram-positive organism. Rebecca Lancefield first described vaginal colonization by GBS in 1935, and the first report of GBS as a human pathogen and cause of invasive disease was described in 1964 [1, 2]. Subsequently, GBS was recognized as a leading cause of sepsis during early infancy (<3 months of age) and particularly in neonates (<28 days) in high-income countries [3–5].

A key virulence factor of GBS is the capsular polysaccharide (CPS) which enables host evasion by protecting against phagocytosis. Despite being composed of only four sugars: glucose, galactose, *N*-acetylglucosamine, and sialic acid, there is structural and antigenic heterogenicity between the CPS of different serotypes [6]. Based on variations in the CPS composition, GBS is immunologically classified into 10 serotypes (Ia, Ib, II, III, IV, V, VI, VII, VIII, IX) [7]. Although the majority of these serotypes have been associated with maternal recto-vaginal colonization, and been described to cause invasive GBS disease, they vary in prevalence of colonization and invasive disease potential. The serotypes with the highest invasive disease potential are III and Ia [8]. Serotype III accounts for 61.5% of GBS invasive disease in young infants, followed by serotype Ia (19.1%). Globally, five serotypes (Ia, Ib, II, III and V) account for 97% of invasive GBS disease in

https://doi.org/10.1016/B978-0-12-814582-1.00012-7

young infants [9]. Furthermore, although there might be temporal fluctuation in the relative contribution of one serotype compared to another in causing invasive GBS disease, the dominant serotypes changes are mainly due to variations in the relative contribution of serotype Ia and III [10].

Molecular based methods such as multi-locus sequence typing (MLST) have been used to identify genetically relatedness among GBS isolates on the basis of their similarities or differences in seven conserved housekeeping genes: alcohol dehydrogenase (adhP), phenylalanyl tRNA synthetase (PheS), amino acid transporter (atr), glutamine synthetase (glnA), serine dehydratase (sdhA), glucose kinase (glcK) and transkelotase (tkt) [11]. Genetically similar GBS isolates are arranged into distinct populations as sequence types (ST) which are representative of alleles assigned to different sequences. Sequence types are further distinguished into clonal complexes (CC), which represents a subset of genetically related sequence types. Currently there are five major clonal complexes (CC1, CC10, CC17, CC19, and CC23) that describe the GBS population structure globally [12]. Clonal complexes CC1, CC12, CC17, CC23, and CC19 have been observed in 95–100%, 81–88% and 92–97% of invasive GBS disease in young infants in Canada, Norway and Italy, respectively [13–15]. Invasive GBS disease occurring between 7 and 90 days age (late onset disease; LOD) is mainly due to CC17 and ST17 strains that express serotype-III capsule [13, 14, 16, 17]. Recently, an increase in invasive GBS disease due to serotype IV has been observed in some high-income settings, possibly related to virulent serotype-III strains having undergone a capsular switch under immunological pressure [18].

Risk factors for Group B *Streptococcus* disease

The categorization of invasive GBS cases as early-onset disease (within 7 days of birth; EOD) and LOD, is partly premised on current understanding of the pathogenesis of invasive GBS disease in young infants. Recto-vaginal GBS colonization of pregnant women is putatively a pre-requisite and strongest identified risk factor for EOD [19]. Acquisition of GBS by the newborn likely occurs mainly through ascending in utero infection of the fetus, or following vaginal passage at the time of birth in newborns born to women with recto-vaginal colonization. Prolonged rupture of the amniotic sac membranes (>18 h) prior to delivery of the newborn, is an additional risk factor for invasive GBS disease. In utero infection of the fetus can also occur in the presence of a macroscopically intact amniotic sac following translocation of GBS from the vagina to amniotic fluid through micro-tears of the amniotic membrane [20]. In the absence of intrapartum antibiotic prophylaxis (IAP) targeted at GBS colonized women during labor, approximately 50–60% of newborns born to recto-vaginally colonized women would be colonized by GBS at birth. Although the majority of newborns colonized with GBS (>98%) at birth are asymptomatic, approximately 1.1% develop EOD [21]. Other risk factors associated with EOD include maternal bacteriuria (possibly a proxy of high density GBS recto-vaginal colonization), maternal fever (possibly due to chorioamnionitis), increased susceptibility of the second birth in twin pregnancies; and preterm labor [19]. Infants born to women living with human immunodeficiency virus (HIV) compared to those without HIV have a 4.4-fold increased risk for LOD [22].

Despite the association of maternal recto-vaginal GBS colonization as a risk factor for EOD, the dynamic nature of maternal GBS recto-vaginal colonization that could affect this risk has not been studied in-depth. A longitudinal cohort study of pregnant women between 20 and 37+ weeks of gestational age in South Africa, highlighted the dynamic nature of maternal GBS colonization during the second half of pregnancy. This included approximately 50% of women been colonized on at least one occasion when sampled at monthly-intervals; including at least 25% of women becoming colonized by a new serotypes in the second half of pregnancy. Also, the mean duration of colonization varied between 6 and 8.5 weeks for the different serotypes [23]. Although not investigated in this study, acquisition of new colonizing serotypes could theoretically be associated with higher density of GBS colonization in these women; which itself could compound the risk of their newborns being colonized and developing EOD [24]. Understanding the differences in maternal GBS colonization; including prevalence of colonization, strain and serotype distribution, density and duration of colonization, immunological mediators of protection against colonization and interaction with other organisms in this ecological niche, could assist in informing the discordant data on the burden of EOD across different parts of the world.

Maternal recto-vaginal GBS colonization is also a risk factor for LOD, albeit less so than for EOD, suggesting other possible sources of infection of the young infant. This could include neonatal mucosal colonization and subsequent risk for invasive disease, bacterial acquisition from the environment or horizontal transfer from their mothers (including possibly through breastmilk) [25].

Epidemiology of Group B *Streptococcus* in young infants

The overall incidence (per 1000 live births) of GBS invasive disease globally is estimated at 0.49 (95% CI: 0.43–0.56) with highest incidence in Southern Africa (2.00; 95% CI: 0.74–3.26) and lowest in Southern Asia (0.22; 95%CI: 0.02–0.41) and Southeast Asia (0.21; 95% CI: 0.09–0.32) [26]. Using a compartmental model, there were an estimated 320,000 cases of invasive GBS disease in infants, which resulted in approximately 90,000 deaths in 2015 [26]. Furthermore, approximately 4% (57,000 annually) of all stillbirths occurring in 2015 were possibly caused by invasive GBS infection in the fetus [26]. GBS also causes disease in adults, including an estimated 33,000 cases in pregnant women in 2015 [27]; and is being increasingly recognized as an important cause of invasive disease in adults with underlying co-morbidities such as obesity and diabetes [28, 29]. Pneumonia, urosepsis, skin, soft-tissue, and osteoarticular infections are common presentations of GBS disease in adults [30]. This chapter relates to GBS infections and its prevention strategies in infants.

A number of challenges exist, especially in resource-limited settings, in directly establishing the role of GBS as a cause of invasive disease and more so for EOD. This includes, laboratory and health care resource limitations for investigating for invasive GBS cases, resulting in possibly the majority of EOD either not been investigated or dying before any investigations can be undertaken. This is particularly pertinent in settings where births occur outside of health facilities, or where neonates are empirically treated with antibiotics prior to having biological samples sent for investigation. Notably, despite only a two-fold difference in prevalence of GBS recto-vaginal colonization (a major risk factor for EOD) in pregnant women from Southern Africa (29.5%) compared to South Asia (12.5%) or Southeast Asia (14.4%), [31] the rate of invasive GBS disease is more than a 10-fold greater in Southern Africa [9]. This, despite the estimated risk of developing EOD being 1.1% (95% CI, 0.6%–1.5%) for newborns' born to women with GBS recto-vaginal colonization in the absence of IAP [21].

Illustrating the challenge in quantifying the contribution of GBS as a cause of newborn and early-infant sepsis, are two recent studies from South Asia and South Africa [32, 33]. In the study conducted in South Asia (ANISA study), GBS was attributed as the cause of neonatal sepsis in only 1.12% of all suspected sepsis cases (8% of all bacterial causes of sepsis). Although this study attempted to undertake community-based surveillance, 1284 newborns demised prior to study staff being able to enroll the child into the surveillance program due to death on the day or within few days of birth. The non-inclusion of such cases when investigating for the causes of neonatal sepsis, could intrinsically result in under-ascertainment of GBS EOD incidence, considering that approximately 90% of EOD occurs at <24h of age and in the absence of IAP [19, 34, 35]. Furthermore, no pathogen was identified in 54% of 1684 neonatal deaths with suspected sepsis in the ANISA study. In contrast, a methodologically similar study undertaken in South Africa, where the majority of births occur in hospital-based facilities, GBS was the leading cause of early-onset neonatal sepsis (4.8% overall and 35.4% of all microbiological confirmed bacterial sepsis) [33]. The challenges of undertaking surveillance on invasive GBS disease as characterized even by the highly resourced ANISA study, requires alternate approaches for estimating the burden of invasive GBS in infants, especially in low-middle income settings.

Maternal prevention strategies against Group B *Streptococcus* infection

The current effective preventive strategy, in some high income countries, against EOD is based on screening pregnant women for GBS colonization between 35 and 37 weeks gestational age and administering IAP to colonized women at least 2–4h before delivery [36]. The implementation of IAP in low resource countries is constrained due to logistical issues and costs related to screening for maternal GBS colonization. Also, the use of IAP as a strategy for prevention of EOD is controversial in high income settings such as the United Kingdom [37]. IAP would entail administering intravenous antibiotics to approximately one-third of all pregnant women during labor. There is emerging evidence that antibiotic therapy during pregnancy and labor could affect the gastro-intestinal microbiome of the newborn, which could predispose to subsequent risk for allergies and obesity [38, 39]. Additionally, IAP does not impact on late-onset GBS disease, stillbirths or GBS-related prematurity [36].

Maternal Group B *Streptococcus* vaccination

Maternal serotype-specific serum immunoglobulin G (IgG) targeted at the CPS, has been associated with reduced risk for invasive GBS disease in young infants. Baker and Kasper first described an inverse association between maternal serotype-specific capsular IgG in serum and invasive GBS disease in the neonates in the 1970s [40]. Since then several other sero-epidemiological studies have also demonstrated an inverse association between maternal serotype-specific antibodies and EOD in infants [41]. Furthermore, some studies have proposed serotype-specific capsular antibody thresholds to be associated with reduced risk for invasive GBS disease in the infants. Differences in the study designs, including with respect to assay methodologies, study settings, different standard of care and duration of follow ups, however, limit head to head comparisons between these studies [41, 42].

Based on sero-epidemiological studies and risk for invasive GBS disease, experimental vaccines formulated using the CPS have been under development since the 1970s [43, 44]. Although these vaccines were initially designed as polysaccharide-only vaccines, the vaccine candidates have now been formulated as polysaccharide-protein conjugate vaccines aimed at enhancing its immunogenicity. It is expected that vaccination of pregnant women with a GBS vaccine would confer passive immunity through increasing transplacental transfer of protective antibody to the fetuses. The protection of the infant against invasive GBS disease, derived from the transplacental transfer of adequate concentration of GBS anti-capsular antibody, could possibly persist for about 3 months of age when almost all GBS invasive disease episodes occur. Studies have also demonstrated an inverse association between serotype-specific blood antibody levels and risk of homotypic serotype recto-vaginal new acquisition during pregnancy [45]. In addition, natural sero-epidemiology studies have also described an inverse association between maternal serum IgG to GBS common proteins (Alpha C and Rib-A) and invasive disease in their young infants in a single study from Sweden [46]. The development of a protein-only based GBS vaccine that includes common proteins, such as Alpc and Rib-A is also currently under development [47].

Group B *Streptococcus* vaccine development

Capsular polysaccharide vaccines: The first generation of GBS polysaccharide-based experimental vaccines consisted of serotype-specific purified polysaccharides, which were evaluated in 1978 by Baker et al. [43]. Although the vaccine was well tolerated, immunogenicity was unpredictable in adults. Nevertheless, the GBS polysaccharide vaccine studies suggested the potential of vaccination as an approach to prevent invasive GBS disease, and subsequently led to the development of multivalent serotype-specific polysaccharide-protein conjugate vaccines which were more immunogenic (Table 1).

The GBS polysaccharide-protein conjugate vaccines consist of purified serotype-specific CPS covalently coupled to tetanus toxoid (TT) or Carrier Related Molecule-197 (CRM 197), as a carrier protein. Coupling of polysaccharide molecules to protein carriers, elicits a T lymphocyte dependent immune response that includes inducing memory B cell development and long-lived T lymphocyte memory cells [72]. Murine models demonstrated that the transplacental transfer of antibodies induced by the tetravalent CPS conjugate vaccine for serotype Ia, Ib, II and III, protect the pups of the vaccinated pregnant mice from lethal challenge by the homotypic serotypes [73].

Since 1994, polysaccharide-protein conjugate experimental vaccines including each of the clinically significant GBS serotypes have been developed, and more recently an experimental trivalent GBS protein-polysaccharide conjugate vaccine containing serotypes Ia, Ib and III was evaluated in phase I-II clinical trials in pregnant women [69] (Table 1). The multi-valent GBS vaccine has now been expanded to a hexavalent vaccine (including serotypes Ia, Ib, II, III, IV and V), which is currently being evaluated in phase Ib/IIa studies in pregnant women [47] (ClinicalTrials.Gov: NCT03765073).

TABLE 1 Studies reporting on safety and immunogenicity of Group B *Streptococcus* vaccines.

Publication year, author	GBS vaccine serotype/ protein	Type of vaccine
1978, Baker et al. [43]	III	Unconjugated CPS
1979, Edwards et al. [48]	III	Unconjugated CPS
1983, Eisentein et al. [49]	II, III	Unconjugated CPS
1983, De Cueninck et al. [50]	III	Unconjugated CPS
1984, Edwards et al. [51]	III	Unconjugated CPS
1988, Baker et al. [52]	III	Unconjugated CPS (in pregnant women)
1996, Kasper et al. [53]	III	Conjugated CPS vaccine
1996, Kotloff et al. [54]	Ia, Ib, II, III	Unconjugated CPS
1999, Baker et al. [55]	Ia, Ib	CPS-TT Conjugate vaccine
2000, Baker et al. [56]	II	CPS-TT Conjugate vaccine
2001, Paolletti et al. [57]	III	CPS-TT Conjugate vaccine
2002, Brigtsen et al. [58]	Ia, Ib	CPS-TT Conjugate vaccine
2002, Guttormsen et al. [59]	III	CPS-TT Conjugate vaccine
2003, Baker et al. [60]	II, III	CPS-TT Conjugate vaccine
2004, Palazzi et al. [61]	V	CPS-TT Conjugate vaccine
2004, Baker et al. [62]	V	CPS-TT Conjugate vaccine
2007, Baker et al. [63]	V	CPS-TT Conjugate vaccine
2009, Pannaraj et al. [64]	Ia, Ib, II, III, V	CPS-TT Conjugate vaccine (monovalent vaccines)
2012, Edwards et al. [65]	Ia, III, V	CPS-TT Conjugate vaccine (monovalent vaccines)
2016, Donders et al. [66]	Ia, Ib, III	CPS-CRM 197 Conjugate vaccine
2016, Leroux-Rouls et al. [67]	Ia, Ib, III	CPS-CRM 197 Conjugate vaccine
2016, Heyderman et al. [68]	Ia, Ib, III	CPS-CRM 197 Conjugate vaccine
2016, Madhi et al. [69]	Ia, Ib, III	CPS-CRM 197 Conjugate vaccine
2017, Minervax [47]	Alp Protein	Protein vaccine
2019, Hillier et al. [70]	III	CPS-TT Conjugate vaccine
2019, Pfizer (Ongoing) [71] (ClinicalTrials.Gov: NCT03765073)	Ia, Ib, II, III, IV, V	CPS-CRM 197 Conjugate vaccine

The initial study on the trivalent GBS conjugate vaccine in non-pregnant women of childbearing age, demonstrated similar immunogenicity between formulations containing 5 μg compared to 20 μg of serotype-specific polysaccharide, as well as no added benefit to include either alum or MF59 adjuvants [67]. Further dosing schedule studies were undertaken in pregnant women, with the non-adjuvanted 5, 2.5 and 0.5 μg formulations; which led to the selection of the 5 μg formulation as the preferred dosing schedule for this vaccine. Notably, however, immune response in the pregnant women was higher for the 2.5 μg than 5 μg formulation for serotypes Ib and III, albeit not significant [69]. The geometric mean antibody concentrations (GMC) in the vaccinated women (all three dosing formulations) remained persistently high from the time of delivery through to one-year post-partum, whilst no increase was observed during the same period in women who received placebo [69]. This suggested there might have been continuous boosting of the antibody in the vaccinated women possibly from natural GBS exposure from recto-vaginal colonization, due to the memory immune responses induced by the conjugate vaccine.

Furthermore, the trivalent GBS vaccine (5 μg formulation) elicited higher GMC one-month following vaccination in South African compared to Belgian/Canadian women for all three serotypes, albeit only significant for serotype Ia. The higher GMC was also evident at the time of delivery in the South African women than Belgian/Canadian women, for serotypes Ia and a similar trend was observed for serotype III [66, 69]. These population differences in immune responses, with higher antibody concentrations elicited in South African women, was also evident when compared to HIV-uninfected counterparts in Malawi for serotypes Ia and III [68]. These results indicate differences in the immune responses to vaccination, that could span across similar and more heterogeneous populations.

Although not fully interrogated, the population differences in immune responses could be due to differences in underlying humoral immunity prior to vaccination. Notably, in the Belgian/Canadian study, the change in GMC from pre-vaccination to one-month post-vaccination in the vaccinated women was 7- to 8-fold lower in those women without detectable serotype-specific serum IgG pre-vaccination compared to those who had detectable IgG levels [66]. This observation, was similar to an earlier study by Baker et al., who also described poorer immune responses in women without detectable serotypes-specific serum IgG prior to immunization [52]. Women in Belgium/Canada were more likely than South African women to have serotype-specific IgG concentrations below the limit of detection prior to vaccination for serotypes Ia (56% vs 32%), Ib (64% vs 7.5%) and III (76% vs 7.5%). This could have contributed to the lower antibody levels observed in the Belgian/Canadian women post-vaccination, however, immune responses were still generally higher in South African compared to Belgian/Canadian women, even when limited to analyzing women with serotype-specific detectable IgG concentrations at the time of vaccination. In contrast, immune responses were similarly suppressed for serotypes Ib and III in women without detectable IgG prior to vaccination in both studies, albeit remaining higher for serotype Ia in South African women [66, 69].

Among pregnant women with or without detectable serotype-specific IgG prior to vaccination who didn't achieve high antibody levels, neither a second dose (with or without adjuvant) or higher dosing formulations of the trivalent GBS vaccine were effective in inducing further increase in GMC compared to after the first vaccine dose [67]. The implications of this observation for possible immune hypo-responsiveness is unclear, but could represent a particular risk group of women whose infants might be more susceptible to develop invasive GBS disease due to failure of their mothers to mount an adequate immune response to GBS polysaccharide epitopes.

Investigating the immunogenicity of GBS vaccines in women living with HIV is of particular relevance to many sub-Saharan African countries. A study of the trivalent GBS vaccine in Southern Africa, demonstrated poorer immune responses in women living with HIV, irrespective of CD4+ lymphocyte category, compared to those without HIV. Even in women without serotype-specific detectable IgG prior to vaccination, immune responses were higher in women living without compared to those with HIV [68]. Further studies are warranted to determine whether a two-dose GBS vaccine schedule might induce better immunogenicity in women living with HIV.

Group B *Streptococcus* vaccination of pregnant women is primarily geared toward protecting their infants against invasive GBS disease, hence, the efficiency of transplacental transfer of antibody to the fetus/newborn is of great importance. GBS vaccination is mainly targeted from 28 weeks of gestational age onward as the maturation of transplacental antibody transfer occurs mainly from 34 weeks gestational age. However, the key issue would be the relative efficiency of transplacental vaccine-induced antibody transfer to the fetus if born at <34 weeks gestational age. The optimal timing of maternal immunization that would maximize protection against young infants requires further investigation.

The newborn to maternal ratio of serotype-specific capsular antibody following maternal immunization with a 5 μg dose formulation of the trivalent GBS vaccine ranged between 0.58 and 0.79 in South Africa and 0.66 to 0.79 in the Belgian/Canadian study [66, 69]. The slightly lower ratio of transplacental antibody transfer in South Africa, however, could be offset by the heightened immunogenicity of the vaccine in the mothers. The newborn to maternal transfer in the South African was similar between mother-newborn dyads living with and without HIV (range 0.58–0.70) [68]. Although not specifically evaluated for in the studies on the trivalent

GBS conjugate vaccine in South Africa and Belgium/Canada, there are other factors that could also have contributed to the lower ratio of transplacental antibody transfer in South African women, including total IgG levels (inverse association with hyper-gammaglobulinemia), gestational age at vaccination and time to delivery. The clinical relevance of the lower efficiency of transplacental antibody transfer to GBS polysaccharide epitopes will, however, only be clarified once a correlate of protection has been established for invasive GBS disease using a standardized assay.

In addition to preventing invasive GBS disease in the infants through increasing transplacental transfer of anti-capsular IgG by vaccinating pregnant women, it is possible that a conjugate vaccine could reduce recto-vaginal GBS colonization in the women. This reduced exposure to GBS, could also inherently reduce invasive GBS disease in the infant. Results from a phase II randomized controlled trial in non-pregnant women of childbearing age demonstrated GBS serotype III conjugate vaccine significantly delayed acquisition of vaginal and rectal GBS serotype III colonization over 18 months [70]. These data were further corroborated by a longitudinal study of pregnant South African women, in which naturally-acquired functional capsular antibody was inversely associated with risk of homotypic serotype recto-vaginal colonization during pregnancy [45].

Vaccination of pregnant women with the trivalent GBS conjugate vaccine in Belgium, Canada, South Africa and Malawi demonstrated vaccination to be safe and not associated with adverse fetal outcomes such as premature birth, stillbirth or low birth weight [66–69] . Furthermore, vaccination of women from 28 weeks of gestational age onward, was not associated with an increase in severe pyrexia post-vaccination [68]. Also, there was no difference in neuro-development of the infant through to 12 months of age, and maternal GBS vaccination did not interfere with immune responses to either diphtheria toxoid or 13-valent pneumococcal polysaccharide-CRM 197 conjugate vaccines in the infants [74].

Common protein vaccines: Another approach to GBS vaccine development underway is the use of common GBS protein antigens as vaccine epitopes. Most GBS strains express antigenic surface proteins that are present in the majority of invasive disease strains regardless of capsular serotype [75]. Use of GBS common proteins as vaccine epitopes could theoretically provide serotype-independent protection. These vaccines could potentially be developed either as stand-alone protein vaccines or could be included as a conjugating protein in polysaccharide-based vaccines. Some of the common proteins being investigated as vaccine candidates include the surface immunogenic protein (Sip), Group B protective surface protein (BPS), C5a peptidase, Rib and Alpha C [75–78]. Recently, pilus like structures in the human GBS pathogen have been discovered, each encoded by a distinct pathogenicity island [79]. The pilus sequences in each island are conserved and all strains carry at least one of the three pilus islands. An experimental vaccine containing components from the three pilus islands conferred protection against all tested GBS challenge strains in a mice model [80]. A phase I clinical study of a GBS protein vaccine based on a fusion of the N-terminal of two surface proteins Alpha C and Rib is currently underway [47].

Pathways for licensure of maternal Group B *Streptococcus* vaccine

The gold standard for the licensure of a new vaccine is through randomized controlled trials demonstrating efficacy against clinical endpoints such as invasive GBS disease. A major obstacle to the clinical development and licensure of a GBS vaccine, however, is that licensure based on demonstrating vaccine efficacy against invasive GBS disease using a traditional vaccine efficacy trial approach would require 62,000 to 180,000 pregnant women to be enrolled to demonstrate 60–80% efficacy against invasive GBS disease in settings where incidence is 1.0 per 1000 live births [81]. There are limited number of settings where clinical trial capacity and capabilities exists to undertake such an efficacy trial [81, 82]. An alternate pathway to licensure of a maternal GBS vaccine could entail demonstrating safety in pregnant women and their infants, coupled with benchmarking vaccine-induced immunogenicity against an established serological correlate of protection (CoP) for invasive GBS disease in infants. Post-licensure studies could then evaluate vaccine effectiveness against invasive GBS disease and other possible benefits of vaccination [81]. Such a pathway to licensure of vaccines, i.e. based initially on CoP derived from sero-epidemiological studies, has been the hallmark for regulatory approval of *Neisseria meningitides* polysaccharide-based vaccines [83, 84]. Similarly, licensure of new formulations of *Haemophilus influenzae* type b and multi-valent *Streptococcus pneumoniae* polysaccharide-protein conjugate vaccines are also benchmarked on immunogenicity in relation to putative CoP thresholds [85–88].

For GBS polysaccharide conjugate vaccines, evidence from different sero-epidemiological studies have consistently shown serotype-specific anti-capsular IgG being associated with risk reduction of invasive GBS disease due to the homotypic serotype (I and III) [42]. Furthermore, the experimental trivalent GBS conjugate vaccine induced functional antibody responses in pregnant women, which was transferred to the fetus in utero [89]. Of the four adequately designed sero-epidemiology studies that systematically evaluated for a CoP against invasive GBS disease for two major serotypes (Ia and III), three only analyzed for a CoP based on maternal blood [90–92], whereas one cohort study from the United States of America analyzed cord blood and maternal blood obtained at the time of birth

[93, 94]. All four studies identified an inverse association between serotype Ia and III specific IgG, and risk reduction of infant invasive GBS disease. The studies, however, report different thresholds as CoP based on analysis of the maternal blood, although a major caveat to any such head-to-head comparison is the different assays used across the studies. Also, the cohort study by Lin et al. demonstrated a 1.2- and 1.6-fold higher threshold as the CoP in maternal compared to cord blood for serotype Ia and III, respectively [93, 94]. Considering the variability and modest efficiency of transplacental transfer of IgG targeted against polysaccharide epitopes, a CoP based on infant rather than maternal blood might be more robust in its predicative value. This would mitigate needing to account for differences between populations that could affect the efficiency of transplacental transfer of IgG.

Other issues which remain to be resolved in relation to use a CoP as a benchmark to evaluate the likely efficacy of a GBS vaccine against invasive GBS disease include the need for studies across different epidemiological settings, for the findings to be generalizable. Furthermore, it remains to be decided whether a single CoP should apply for EOD and LOD, as well as how to impute the CoP for less common invasive disease-causing serotypes for which studies are not adequately powered to establish a CoP. One approach, as used for multi-valent pneumococcal conjugate vaccine [87], would be to establish a CoP on the composite of invasive GBS disease irrespective of serotype. Also, current and future studies investigating for a CoP should ideally use a standardized laboratory assay for IgG estimation, and one which will also be used when evaluating for immunogenicity in the vaccine studies, as is currently being developed [42].

In vitro bacterial killing assays such as opsonophagocytosis killing assay could be an adjunct measure to CoP based on IgG concentrations. Good correlation has been observed for serotype-specific IgG concentration and functional activity using opsonophagocytic killing activity assay when serotype-specific IgG concentrations are >1 μg/mL. It has been proposed that a sero-correlate of protection against invasive GBS disease might be associated with opsonophagocytic killing activity assay titers of 1/64–1/128 for serotypes Ia and III, respectively [42].

Undertaking studies of experimental vaccines in pregnant women is also compounded by concerns about potential liability issues regarding vaccinating pregnant women, where any adverse maternal or fetal outcome might be attributed to vaccination. Therefore, in addition to evaluating the immunogenicity of a GBS vaccine (benchmarked against a CoP on infant blood), licensure of the vaccine would also require robust evidence on the safety of the vaccine in pregnant women and the effect on the fetus and subsequent well-being of the offspring. In 2015, the World Health Organization in collaboration with the Brighton group published harmonized definitions of key terms related to pregnancy to support monitoring of vaccine safety in pregnant women and their child. The Global Alignment of Immunization Safety Assessment (GAIA) in pregnancy has developed a set of case definitions according to the Brighton Collaboration process and format, which should be used as a benchmark when evaluating the safety of the vaccine [95, 96].

Conclusion

In conclusion, GBS remains a leading cause of neonatal sepsis in high and low-middle income countries, contributing to an estimated 90,000 deaths in early-infancy and at least 57,000 stillbirths [26]. Sero-epidemiological studies in high-income and low-middle income settings have consistently showed an inverse association between maternal-derived serotype-specific anti-capsular IgG to serotypes Ia and III and invasive disease by the homotypic serotype in their infants. A major challenge to the licensure of a GBS vaccine, is the large sample size of pregnant women (up to 180,000) that would be required to demonstrate vaccine efficacy against invasive GBS disease in their infants. As an alternative, currently under consideration is to license the vaccine based on assessing immunogenicity benchmarked against a CoP for invasive disease in the young infant and demonstrating the safety of the vaccine in the mother-newborn dyads. This could then be followed by phase IV studies in which vaccine effectiveness can be evaluated against clinical endpoints such as invasive GBS disease, GBS associated stillbirths and other possible GBS related complications such as preterm birth.

References

[1] Lancefield RC, Hare R. The serological differentiation of pathogenic and non-pathogenic strains of hemolytic streptococci from parturient women. J Exp Med 1935;61(3):335–49.
[2] Eickhoff TC, Klein JO, Daly AK, Ingall D, Finland M. Neonatal sepsis and other infections due to group B beta-hemolytic streptococci. N Engl J Med 1964;271:1221–8.
[3] Lloyd DJ, Reid TM. Group B streptococcal infection in the newborn. Criteria for early detection and treatment. Acta Paediatr Scand 1976;65(5):585–91.
[4] Bergqvist G. Neonatal infections caused by Group B streptococci. 3. Incidence in Sweden 1970-71. Scand J Infect Dis 1974;6(1):29–31.
[5] McCracken Jr. GH. Group B streptococci: the new challenge in neonatal infections. J Pediatr 1973;82(4):703–6.
[6] Kogan G, Uhrin D, Brisson JR, Paoletti LC, Blodgett AE, Kasper DL, et al. Structural and immunochemical characterization of the type VIII Group B Streptococcus capsular polysaccharide. J Biol Chem 1996;271(15):8786–90.
[7] Slotved HC, Kong F, Lambertsen L, Sauer S, Gilbert GL. Serotype IX, a proposed new *Streptococcus agalactiae* serotype. J Clin Microbiol 2007;45(9):2929–36.
[8] Madzivhandila M, Adrian PV, Cutland CL, Kuwanda L, Schrag SJ, Madhi SA. Serotype distribution and invasive potential of Group B streptococcus isolates causing disease in infants and colonizing maternal-newborn dyads. PLoS One 2011;6(3):e17861.
[9] Madrid L, Seale AC, Kohli-Lynch M, Edmond KM, Lawn JE, Heath PT, et al. Infant Group B streptococcal disease incidence and serotypes worldwide: systematic review and meta-analyses. Clin Infect Dis 2017;65(Suppl. 2):S160–72.

[10] Dangor Z, Cutland CL, Izu A, Kwatra G, Trenor S, Lala SG, et al. Temporal changes in invasive Group B Streptococcus serotypes: implications for vaccine development. PLoS One 2016;11(12):e0169101.
[11] Jones N, Bohnsack JF, Takahashi S, Oliver KA, Chan MS, Kunst F, et al. Multilocus sequence typing system for Group B streptococcus. J Clin Microbiol 2003;41(6):2530–6.
[12] Honsa E, Fricke T, Stephens AJ, Ko D, Kong F, Gilbert GL, et al. Assignment of *Streptococcus agalactiae* isolates to clonal complexes using a small set of single nucleotide polymorphisms. BMC Microbiol 2008;8:140.
[13] Imperi M, Gherardi G, Berardi A, Baldassarri L, Pataracchia M, Dicuonzo G, et al. Invasive neonatal GBS infections from an area-based surveillance study in Italy. Clin Microbiol Infect 2011;17(12):1834–9.
[14] Manning SD, Springman AC, Lehotzky E, Lewis MA, Whittam TS, Davies HD. Multilocus sequence types associated with neonatal Group B streptococcal sepsis and meningitis in Canada. J Clin Microbiol 2009;47(4):1143–8.
[15] Bergseng H, Afset JE, Radtke A, Loeseth K, Lyng RV, Rygg M, et al. Molecular and phenotypic characterization of invasive Group B streptococcus strains from infants in Norway 2006-2007. Clin Microbiol Infect 2009;15(12):1182–5.
[16] Fluegge K, Wons J, Spellerberg B, Swoboda S, Siedler A, Hufnagel M, et al. Genetic differences between invasive and noninvasive neonatal Group B streptococcal isolates. Pediatr Infect Dis J 2011;30(12):1027–31.
[17] Kao Y, Tsai MH, Lai MY, Chu SM, Huang HR, Chiang MC, et al. Emerging serotype III sequence type 17 Group B streptococcus invasive infection in infants: the clinical characteristics and impacts on outcomes. BMC Infect Dis 2019;19(1):538.
[18] Shabayek S, Spellerberg B. Group B streptococcal colonization, molecular characteristics, and epidemiology. Front Microbiol 2018;9:437.
[19] Heath PT, Balfour G, Weisner AM, Efstratiou A, Lamagni TL, Tighe H, et al. Group B streptococcal disease in UK and Irish infants younger than 90 days. Lancet 2004;363(9405):292–4.
[20] Nishihara Y, Dangor Z, French N, Madhi S, Heyderman R. Challenges in reducing Group B Streptococcus disease in African settings. Arch Dis Child 2017;102(1):72–7.
[21] Russell NJ, Seale AC, O'Sullivan C, Le Doare K, Heath PT, Lawn JE, et al. Risk of early-onset neonatal Group B streptococcal disease with maternal colonization worldwide: systematic review and meta-analyses. Clin Infect Dis 2017;65(Suppl. 2):S152–9.
[22] Cools P, van de Wijgert J, Jespers V, Crucitti T, Sanders EJ, Verstraelen H, et al. Role of HIV exposure and infection in relation to neonatal GBS disease and rectovaginal GBS carriage: a systematic review and meta-analysis. Sci Rep 2017;7(1):13820.
[23] Kwatra G, Adrian PV, Shiri T, Buchmann EJ, Cutland CL, Madhi SA. Serotype-specific acquisition and loss of Group B streptococcus recto-vaginal colonization in late pregnancy. PLoS One 2014;9(6):e98778.
[24] Seedat F, Brown CS, Stinton C, Patterson J, Geppert J, Freeman K, et al. Bacterial load and molecular markers associated with early-onset Group B Streptococcus: a systematic review and meta-analysis. Pediatr Infect Dis J 2018;37(12):e306–14.
[25] Le Doare K, Faal A, Jaiteh M, Sarfo F, Taylor S, Warburton F, et al. Association between functional antibody against Group B Streptococcus and maternal and infant colonization in a Gambian cohort. Vaccine 2017;35(22):2970–8.
[26] Seale AC, Bianchi-Jassir F, Russell NJ, Kohli-Lynch M, Tann CJ, Hall J, et al. Estimates of the burden of Group B streptococcal disease worldwide for pregnant women, stillbirths, and children. Clin Infect Dis 2017;65(Suppl. 2):S200–19.
[27] Seale AC, Blencowe H, Bianchi-Jassir F, Embleton N, Bassat Q, Ordi J, et al. Stillbirth with Group B Streptococcus disease worldwide: systematic review and meta-analyses. Clin Infect Dis 2017;65(Suppl. 2):S125–32.
[28] Pitts SI, Maruthur NM, Langley GE, Pondo T, Shutt KA, Hollick R, et al. Obesity, diabetes, and the risk of invasive Group B streptococcal disease in nonpregnant adults in the United States. Open Forum Infect Dis 2018;5(6):ofy030.

[29] Francois Watkins LK, McGee L, Schrag SJ, Beall B, Jain JH, Pondo T, et al. Epidemiology of invasive Group B streptococcal infections among nonpregnant adults in the United States, 2008-2016. JAMA Intern Med 2019;179(4):479–88.
[30] Farley MM. Group B streptococcal disease in nonpregnant adults. Clin Infect Dis 2001;33(4):556–61.
[31] Russell NJ, Seale AC, O'Driscoll M, O'Sullivan C, Bianchi-Jassir F, Gonzalez-Guarin J, et al. Maternal colonization with Group B Streptococcus and serotype distribution worldwide: systematic review and meta-analyses. Clin Infect Dis 2017;65(Suppl. 2): S100–11.
[32] Saha SK, Schrag SJ, El Arifeen S, Mullany LC, Shahidul Islam M, Shang N, et al. Causes and incidence of community-acquired serious infections among young children in south Asia (ANISA): an observational cohort study. Lancet 2018;392(10142):145–59.
[33] Velaphi SC, Westercamp M, Moleleki M, Pondo T, Dangor Z, Wolter N, et al. Surveillance for incidence and etiology of early-onset neonatal sepsis in Soweto, South Africa. PLoS One 2019;14(4):e0214077.
[34] Schuchat A. Epidemiology of Group B streptococcal disease in the United States: shifting paradigms. Clin Microbiol Rev 1998;11(3):497–513.
[35] Dangor Z, Lala SG, Cutland CL, Koen A, Jose L, Nakwa F, et al. Burden of invasive Group B Streptococcus disease and early neurological sequelae in South African infants. PLoS One 2015;10(4):e0123014.
[36] Schrag SJ, Verani JR. Intrapartum antibiotic prophylaxis for the prevention of perinatal Group B streptococcal disease: experience in the United States and implications for a potential Group B streptococcal vaccine. Vaccine 2013;31(Suppl. 4):D20–6.
[37] Seedat F, Geppert J, Stinton C, Patterson J, Freeman K, Johnson SA, et al. Universal antenatal screening for Group B streptococcus may cause more harm than good. BMJ 2019;364:l463.
[38] Mazzola G, Murphy K, Ross RP, Di Gioia D, Biavati B, Corvaglia LT, et al. Early gut microbiota perturbations following intrapartum antibiotic prophylaxis to prevent Group B streptococcal disease. PLoS One 2016;11(6):e0157527.
[39] Fujimura KE, Sitarik AR, Havstad S, Lin DL, Levan S, Fadrosh D, et al. Neonatal gut microbiota associates with childhood multisensitized atopy and T cell differentiation. Nat Med 2016;22(10):1187–91.
[40] Baker CJ, Kasper DL. Immunological investigation of infants with septicemia or meningitis due to Group B Streptococcus. J Infect Dis 1977;136(Suppl. 1):S98–104.
[41] Dangor Z, Kwatra G, Izu A, Lala SG, Madhi SA. Review on the association of Group B Streptococcus capsular antibody and protection against invasive disease in infants. Expert Rev Vaccines 2015;14(1):135–49.
[42] Le Doare K, Kampmann B, Vekemans J, Heath PT, Goldblatt D, Nahm MH, et al. Serocorrelates of protection against infant Group B streptococcus disease. Lancet Infect Dis 2019;19(5):e162–71.
[43] Baker CJ, Edwards MS, Kasper DL. Immunogenicity of polysaccharides from type III, Group B Streptococcus. J Clin Invest 1978;61(4):1107–10.
[44] Dzanibe S, Madhi SA. Systematic review of the clinical development of Group B Streptococcus serotype-specific capsular polysaccharide-based vaccines. Expert Rev Vaccines 2018;17(7):635–51.
[45] Kwatra G, Adrian PV, Shiri T, Buchmann EJ, Cutland CL, Madhi SA. Natural acquired humoral immunity against serotype-specific Group B Streptococcus rectovaginal colonization acquisition in pregnant women. Clin Microbiol Infect 2015;21(6):568. e13–21.
[46] Larsson C, Lindroth M, Nordin P, Stalhammar-Carlemalm M, Lindahl G, Krantz I. Association between low concentrations of antibodies to protein alpha and rib and invasive neonatal Group B streptococcal infection. Arch Dis Child Fetal Neonatal Ed 2006;91(6):F403–8.
[47] Lin SM, Zhi Y, Ahn KB, Lim S, Seo HS. Status of Group B streptococcal vaccine development. Clin Exp Vaccine Res 2018;7(1):76–81.

[48] Edwards MS, Baker CJ, Kasper DL. Opsonic specificity of human antibody to the type III polysaccharide of Group B Streptococcus. J Infect Dis 1979;140(6):1004–8.

[49] Eisenstein TK, De Cueninck BJ, Resavy D, Shockman GD, Carey RB, Swenson RM. Quantitative determination in human sera of vaccine-induced antibody to type-specific polysaccharides of Group B streptococci using an enzyme-linked immunosorbent assay. J Infect Dis 1983;147(5):847–56.

[50] De Cueninck BJ, Eisenstein TK, McIntosh TS, Shockman GD, Swenson RM. Quantitation of in vitro opsonic activity of human antibody induced by a vaccine consisting of the type III-specific polysaccharide of Group B streptococcus. Infect Immun 1983;39(3):1155–60.

[51] Edwards MS, Fuselier PA, Rench MA, Kasper DL, Baker CJ. Class specificity of naturally acquired and vaccine-induced antibody to type III Group B streptococcal capsular polysaccharide: determination with a radioimmunoprecipitin assay. Infect Immun 1984;44(2):257–61.

[52] Baker CJ, Rench MA, Edwards MS, Carpenter RJ, Hays BM, Kasper DL. Immunization of pregnant women with a polysaccharide vaccine of Group B streptococcus. N Engl J Med 1988;319(18):1180–5.

[53] Kasper DL, Paoletti LC, Wessels MR, Guttormsen HK, Carey VJ, Jennings HJ, et al. Immune response to type III Group B streptococcal polysaccharide-tetanus toxoid conjugate vaccine. J Clin Invest 1996;98(10):2308–14.

[54] Kotloff KL, Fattom A, Basham L, Hawwari A, Harkonen S, Edelman R. Safety and immunogenicity of a tetravalent Group B streptococcal polysaccharide vaccine in healthy adults. Vaccine 1996;14(5):446–50.

[55] Baker CJ, Paoletti LC, Wessels MR, Guttormsen HK, Rench MA, Hickman ME, et al. Safety and immunogenicity of capsular polysaccharide-tetanus toxoid conjugate vaccines for Group B streptococcal types Ia and Ib. J Infect Dis 1999;179(1):142–50.

[56] Baker CJ, Paoletti LC, Rench MA, Guttormsen HK, Carey VJ, Hickman ME, et al. Use of capsular polysaccharide-tetanus toxoid conjugate vaccine for type II Group B Streptococcus in healthy women. J Infect Dis 2000;182(4):1129–38.

[57] Paoletti LC, Rench MA, Kasper DL, Molrine D, Ambrosino D, Baker CJ. Effects of alum adjuvant or a booster dose on immunogenicity during clinical trials of Group B streptococcal type III conjugate vaccines. Infect Immun 2001;69(11):6696–701.

[58] Brigtsen AK, Kasper DL, Baker CJ, Jennings HJ, Guttormsen HK. Induction of cross-reactive antibodies by immunization of healthy adults with types Ia and Ib Group B streptococcal polysaccharide-tetanus toxoid conjugate vaccines. J Infect Dis 2002;185(9):1277–84.

[59] Guttormsen HK, Baker CJ, Nahm MH, Paoletti LC, Zughaier SM, Edwards MS, et al. Type III Group B streptococcal polysaccharide induces antibodies that cross-react with Streptococcus pneumoniae type 14. Infect Immun 2002;70(4):1724–38.

[60] Baker CJ, Rench MA, Fernandez M, Paoletti LC, Kasper DL, Edwards MS. Safety and immunogenicity of a bivalent Group B streptococcal conjugate vaccine for serotypes II and III. J Infect Dis 2003;188(1):66–73.

[61] Palazzi DL, Rench MA, Edwards MS, Baker CJ. Use of type V Group B streptococcal conjugate vaccine in adults 65-85 years old. J Infect Dis 2004;190(3):558–64.

[62] Baker CJ, Paoletti LC, Rench MA, Guttormsen HK, Edwards MS, Kasper DL. Immune response of healthy women to 2 different Group B streptococcal type V capsular polysaccharide-protein conjugate vaccines. J Infect Dis 2004;189(6):1103–12.

[63] Baker CJ, Rench MA, Paoletti LC, Edwards MS. Dose-response to type V Group B streptococcal polysaccharide-tetanus toxoid conjugate vaccine in healthy adults. Vaccine 2007;25(1):55–63.

[64] Pannaraj PS, Edwards MS, Ewing KT, Lewis AL, Rench MA, Baker CJ. Group B streptococcal conjugate vaccines elicit functional antibodies independent of strain O-acetylation. Vaccine 2009;27(33):4452–6.

[65] Edwards MS, Lane HJ, Hillier SL, Rench MA, Baker CJ. Persistence of functional antibodies to Group B streptococcal capsular polysaccharides following immunization with glycoconjugate vaccines. Vaccine 2012;30(28):4123–6.
[66] Donders GG, Halperin SA, Devlieger R, Baker S, Forte P, Wittke F, et al. Maternal immunization with an investigational trivalent Group B streptococcal vaccine: a randomized controlled trial. Obstet Gynecol 2016;127(2):213–21.
[67] Leroux-Roels G, Maes C, Willekens J, De Boever F, de Rooij R, Martell L, et al. A randomized, observer-blind phase Ib study to identify formulations and vaccine schedules of a trivalent Group B Streptococcus vaccine for use in non-pregnant and pregnant women. Vaccine 2016;34(15):1786–91.
[68] Heyderman RS, Madhi SA, French N, Cutland C, Ngwira B, Kayambo D, et al. Group B streptococcus vaccination in pregnant women with or without HIV in Africa: a non-randomised phase 2, open-label, multicentre trial. Lancet Infect Dis 2016;16(5):546–55.
[69] Madhi SA, Cutland CL, Jose L, Koen A, Govender N, Wittke F, et al. Safety and immunogenicity of an investigational maternal trivalent Group B Streptococcus vaccine in healthy women and their infants: a randomised phase 1b/2 trial. Lancet Infect Dis 2016;16(8):923–34.
[70] Hillier SL, Ferrieri P, Edwards MS, Ewell M, Ferris D, Fine P, et al. A phase 2, randomized, control trial of Group B Streptococcus (GBS) type III capsular polysaccharide-tetanus toxoid (GBS III-TT) vaccine to prevent vaginal colonization with GBS III. Clin Infect Dis 2019;68(12):2079–86.
[71] Trial to evaluate the safety, tolerability, and immunogenicity of a multivalent Group B Streptococcus vaccine in healthy nonpregnant women and pregnant women and their infants, https://clinicaltrials.gov; 2019. ClinicalTrials.gov Identifier: NCT03765073.
[72] Avci FY, Li X, Tsuji M, Kasper DL. A mechanism for glycoconjugate vaccine activation of the adaptive immune system and its implications for vaccine design. Nat Med 2011;17(12):1602–9.
[73] Paoletti LC, Wessels MR, Rodewald AK, Shroff AA, Jennings HJ, Kasper DL. Neonatal mouse protection against infection with multiple Group B streptococcal (GBS) serotypes by maternal immunization with a tetravalent GBS polysaccharide-tetanus toxoid conjugate vaccine. Infect Immun 1994;62(8):3236–43.
[74] Madhi SA, Koen A, Cutland CL, Jose L, Govender N, Wittke F, et al. Antibody kinetics and response to routine vaccinations in infants born to women who received an investigational trivalent Group B Streptococcus polysaccharide CRM197-conjugate vaccine during pregnancy. Clin Infect Dis 2017;65(11):1897–904.
[75] Erdogan S, Fagan PK, Talay SR, Rohde M, Ferrieri P, Flores AE, et al. Molecular analysis of Group B protective surface protein, a new cell surface protective antigen of Group B streptococci. Infect Immun 2002;70(2):803–11.
[76] Brodeur BR, Boyer M, Charlebois I, Hamel J, Couture F, Rioux CR, et al. Identification of Group B streptococcal sip protein, which elicits cross-protective immunity. Infect Immun 2000;68(10):5610–8.
[77] Stalhammar-Carlemalm M, Stenberg L, Lindahl G. Protein rib: a novel Group B streptococcal cell surface protein that confers protective immunity and is expressed by most strains causing invasive infections. J Exp Med 1993;177(6):1593–603.
[78] Cheng Q, Carlson B, Pillai S, Eby R, Edwards L, Olmsted SB, et al. Antibody against surface-bound C5a peptidase is opsonic and initiates macrophage killing of Group B streptococci. Infect Immun 2001;69(4):2302–8.
[79] Lauer P, Rinaudo CD, Soriani M, Margarit I, Maione D, Rosini R, et al. Genome analysis reveals pili in Group B Streptococcus. Science 2005;309(5731):105.
[80] Margarit I, Rinaudo CD, Galeotti CL, Maione D, Ghezzo C, Buttazzoni E, et al. Preventing bacterial infections with pilus-based vaccines: the Group B streptococcus paradigm. J Infect Dis 2009;199(1):108–15.

[81] Vekemans J, Crofts J, Baker CJ, Goldblatt D, Heath PT, Madhi SA, et al. The role of immune correlates of protection on the pathway to licensure, policy decision and use of Group B Streptococcus vaccines for maternal immunization: considerations from World Health Organization consultations. Vaccine 2019;37(24):3190–8.
[82] Madhi SA, Dangor Z, Heath PT, Schrag S, Izu A, Sobanjo-Ter Meulen A, et al. Considerations for a phase-III trial to evaluate a Group B Streptococcus polysaccharide-protein conjugate vaccine in pregnant women for the prevention of early- and late-onset invasive disease in young-infants. Vaccine 2013;31(Suppl. 4):D52–7.
[83] Goldschneider I, Gotschlich EC, Artenstein MS. Human immunity to the meningococcus. I. The role of humoral antibodies. J Exp Med 1969;129(6):1307–26.
[84] Borrow R, Andrews N, Goldblatt D, Miller E. Serological basis for use of meningococcal serogroup C conjugate vaccines in the United Kingdom: reevaluation of correlates of protection. Infect Immun 2001;69(3):1568–73.
[85] Kayhty H, Peltola H, Karanko V, Makela PH. The protective level of serum antibodies to the capsular polysaccharide of Haemophilus influenzae type b. J Infect Dis 1983;147(6):1100.
[86] Kayhty H. Difficulties in establishing a serological correlate of protection after immunization with Haemophilus influenzae conjugate vaccines. Biologicals 1994;22(4):397–402.
[87] Jodar L, Butler J, Carlone G, Dagan R, Goldblatt D, Kayhty H, et al. Serological criteria for evaluation and licensure of new pneumococcal conjugate vaccine formulations for use in infants. Vaccine 2003;21(23):3265–72.
[88] Siber GR, Chang I, Baker S, Fernsten P, O'Brien KL, Santosham M, et al. Estimating the protective concentration of anti-pneumococcal capsular polysaccharide antibodies. Vaccine 2007;25(19):3816–26.
[89] Fabbrini M, Rigat F, Tuscano G, Chiarot E, Donders G, Devlieger R, et al. Functional activity of maternal and cord antibodies elicited by an investigational Group B Streptococcus trivalent glycoconjugate vaccine in pregnant women. J Infect 2018;76(5):449–56.
[90] Baker CJ, Carey VJ, Rench MA, Edwards MS, Hillier SL, Kasper DL, et al. Maternal antibody at delivery protects neonates from early onset Group B streptococcal disease. J Infect Dis 2014;209(5):781–8.
[91] Dangor Z, Kwatra G, Izu A, Adrian P, Cutland CL, Velaphi S, et al. Correlates of protection of serotype-specific capsular antibody and invasive Group B Streptococcus disease in south African infants. Vaccine 2015;33(48):6793–9.
[92] Fabbrini M, Rigat F, Rinaudo CD, Passalaqua I, Khacheh S, Creti R, et al. The protective value of maternal Group B Streptococcus antibodies: quantitative and functional analysis of naturally acquired responses to capsular polysaccharides and pilus proteins in European maternal sera. Clin Infect Dis 2016;63(6):746–53.
[93] Lin FY, Philips 3rd JB, Azimi PH, Weisman LE, Clark P, Rhoads GG, et al. Level of maternal antibody required to protect neonates against early-onset disease caused by Group B Streptococcus type Ia: a multicenter, seroepidemiology study. J Infect Dis 2001;184(8):1022–8.
[94] Lin FY, Weisman LE, Azimi PH, Philips 3rd JB, Clark P, Regan J, et al. Level of maternal IgG anti-Group B streptococcus type III antibody correlated with protection of neonates against early-onset disease caused by this pathogen. J Infect Dis 2004;190(5):928–34.
[95] Bonhoeffer J, Kochhar S, Hirschfeld S, Heath PT, Jones CE, Bauwens J, et al. Global alignment of immunization safety assessment in pregnancy—the GAIA project. Vaccine 2016;34(49):5993–7.
[96] Kochhar S, Bauwens J, Bonhoeffer J. Safety assessment of immunization in pregnancy. Vaccine 2017;35(48 Pt A):6469–71, http://www.gaia-consortium.net.

CHAPTER

12

Cytomegalovirus

Mark R. Schleiss

Division of Pediatric Infectious Diseases and Immunology, Center for Infectious Diseases and Microbiology Translational Research, University of Minnesota Medical School, Minneapolis, MN, United States

Introduction to CMV virology and structure

Cytomegalovirus (CMV) is a member of the *Herpesviridae* family of viruses. These viruses are ubiquitous in nature; in humans, eight distinct members of this family cause disease (reviewed in [1]): Herpes Simplex viruses type 1 and 2, Varicella-Zoster virus, Epstein-Barr virus, CMV (or Human Herpes Virus 5), Human Herpes Viruses 6, 7 and 8. CMV is subcategorized in the *Herpesviridae* as a *betaherpesvirus*, based on its ability to infect leukocytes. CMV, like all members of this family of viruses, has a large, double-stranded DNA genome. Like other *Herpesviridae*, CMV establishes lifelong, latent infection, with periodic reactivations occurring during the lifetime of the infected individual (particularly in the context of immune compromise). The coding sequence of CMV encompasses >230 kilobase pairs making CMV the largest human pathogen with respect to total genetic content [2, 3]. The CMV genome is segmented into a unique long (UL) and a unique short (US) segment, with each segment bracketed by terminal repeats; hence, four different isomeric configurations of these two segments are possible in any given virus particle. CMV gene products are, by convention, designated by whether they are encoded by the UL or US segment, and numbered sequentially from "left-to-right" in the viral genome; for example, the glycoprotein B (gB) gene product (described in more detail below) is designated as UL55, since it is the 55th open reading frame annotated in the UL region of the genome. Although ~200 gene products of >100 amino acids were originally predicted by DNA sequence analysis of the viral genome [4], more recent transcriptome studies have revealed a larger number and

https://doi.org/10.1016/B978-0-12-814582-1.00013-9

far greater potential diversity of CMV gene products, generated by alternative splicing [5,6]. Many CMV genes appear to have evolved to promote evasion of the host immune response [7,8], which may explain why reactivation and re-infection are so common over the life course. As discussed below, this aspect of the biology of CMV—the fact that naturally acquired immunity does not confer protection against re-infection—creates major challenges for vaccine development.

CMV replication is complex, and requires the coordinated, sequential expression of three families of viral transcripts. Viral gene expression, including establishment of and reactivation from latency, is regulated through interactions between both virally-encoded proteins and host cellular and transcription factors [9,10]. The replication cycle of CMV is divided temporally into three regulated phases: the immediate early phase, the early phase, and the late phase. Immediate early (IE) mRNA is transcribed within the first few hours after infection of the host cell and the encoded IE proteins, which include multiple isoforms due to extensive splicing, modulate both host and viral gene expression. An isoform of the IE protein family, IE1 (also known as IE72 because of its molecular mass of 72 kDa), is also a predominate target of the host T cell response to CMV infection (discussed below). Early gene products include DNA replication proteins and other structural proteins. Late gene products are typically transcribed at ~72 h post-infection and include structural proteins and other proteins involved in virion assembly, structure, and egress.

The three distinct regions of the CMV viral particle include: an icosahedral capsid, the tegument layer, and the outer lipid bilayer envelope. The CMV virion is schematically represented in Fig. 1. The capsid consists of 162 capsomere subunits arranged in an icosahedral symmetry; it surrounds and encloses the viral dsDNA genome (forming a nucleocapsid) and has a highly electron-dense appearance when imaged by electron microscopy [11]. In the mature virus particle, the nucleocapsid is surrounded by the tegument, a protein-rich layer containing several proteins that are targets of the host T lymphocyte response to infection, and hence are relevant to vaccine development. One leading candidate protein for inclusion in a subunit CMV vaccine is a 65-kilodalton (kDa) phosphoprotein referred to as pp65 (also known as ppUL83, or UL83) [12,13]. Finally, surrounding the tegument is the envelope, which contains several virally-encoded glycoproteins (g): the gB complex (which is expressed as a trimer in its membrane-associated conformation), the gM/gN complex, the gH/gL/gO "trimeric" complex, and the so-called "pentameric complex" (PC) or pentamer consisting of proteins gH/gL/UL128, 130, and 131. CMV-seropositive individuals engender an immune response characterized by neutralizing antibodies targeting these glycoproteins [14–20]; hence, recombinant forms of some of these proteins have emerged as potential candidates for subunit vaccine development. The virally-encoded proteins

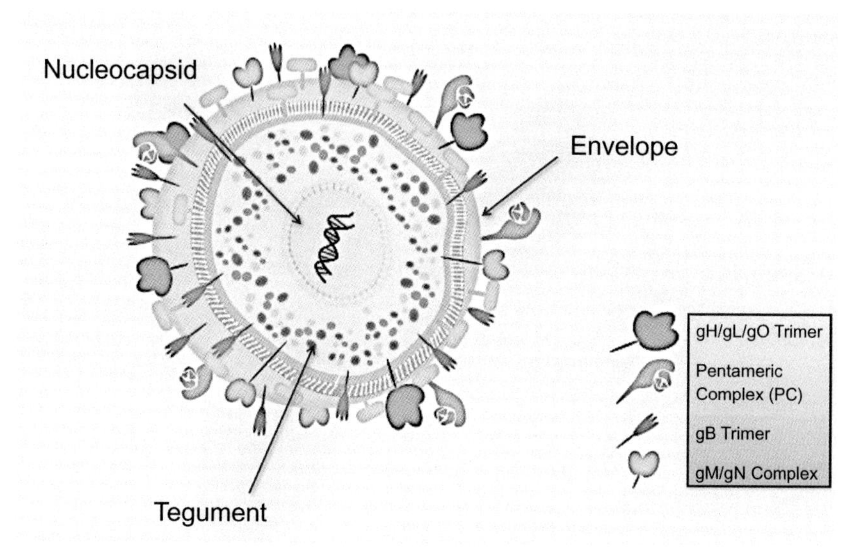

FIG. 1 Schematic representation of CMV virion. Cut-away view of protypical virus particle demonstrating envelope, tegument, and nucleocapsid. The viral *envelope* consists of a phospholipid bilayer derived from the host cell, containing multiple virally-encoded glycoprotein species. Inset provides key for some of the more important glycoproteins being evaluated as potential CMV vaccine targets, including gB complex (trimer), gH/gL/gO complex (trimer), gM/gN complex (dimer), and gH/gL/UL128/130/131 complex (pentamer). Viral *tegument* is the amorphous region between viral envelope and nucleocapsid and containing many viral phosphoproteins, including ppUL83 (pp65), a major target of CD8+ T cell responses in the setting of CMV infection. Inner *nucleocapsid* contains proteins and viral genome.

that have emerged as potential vaccine candidates (CMV-encoded immunogens) are summarized in Table 1.

The mechanism(s) by which anti-glycoprotein antibodies neutralize viral infectivity depends on how the specific CMV glycoproteins interact with cellular receptors during the binding and entry stages of infection [21]. CMV employs at least two distinct cell-type specific mechanisms of cell entry [22]. Entry of CMV into endothelial and epithelial cells is mediated by endocytosis in a pH-dependent fashion; in contrast, entry into fibroblasts is non-endocytic, and pH-independent [22]. Fibroblast entry is initiated by binding to a cellular receptor in a process that requires a trimeric gH, gL, and gO complex [23]. In contrast, endocytic entry in endothelial and epithelial cells requires interaction between the CMV PC and cellular receptors [24]. Neuropilin-2 and platelet-derived growth factor receptor alpha (PDGFRα) have been identified as receptors, respectively, for these trimeric and pentameric complexes that govern cell tropism; additionally, the ligand CD90 (THY-1) plays a role in endocytic entry, and the cellular receptor CD147 interacts with the CMV PC to facilitate virus entry [25].

TABLE 1 Cytomegalovirus-encoded immunogens serving as candidate targets for inclusion in a CMV vaccine.

CMV gene product	Rationale for inclusion
Envelope glycoproteins	
gB	Major target of neutralizing antibodies; target of cytotoxic T lymphocytes (CTLs); cornerstone of CMV vaccine development programs; demonstrated efficacy as subunit vaccine in clinical trials
gH/gL	Important target of neutralizing antibodies; target of CTLs
gH, gL, UL128-131 (Pentamer)	Pentameric complex (PC) of gH/gL/UL128/UL130/UL131 on viral envelope; target of neutralizing antibodies; antibodies neutralize CMV infection at epithelial and endothelial cell surfaces; current focus of clinical trials (in combination with gB)
gH, gL, gO (Trimer)	Target of neutralizing antibody; antibodies to this complex can block cell entry/fusion
gM/gN	Target of antibody neutralizing antibody responses; has not been tested as CMV subunit vaccine; hypervariability of gN in clinical isolates suggests this protein is under immune selective pressure
Structural proteins	
UL83 (pp65)	Major target of CTLs; target of non-neutralizing antibody responses; has been evaluated (in combination with gB) in clinical trials
pp150, pp28	Target of CTLs and antibody responses; no data available from clinical trials
pp50	Target of CTLs; no data available from clinical trials
pp71, pp52	Targets of antibody responses; no data available from clinical trials
Nonstructural proteins	
IE1	Important target of CTLs; target of non-neutralizing antibody responses; has been evaluated (alone, or in combination with gB and UL83 [pp65] in clinical trials)

A general mechanistic model has emerged in which gB functions as the viral protein that mediates the membrane fusion event between the viral and host cell membranes, with either the gH/gL/gO trimer or the PC providing the trigger to gB [25–27]. However, there is evidence that gB also binds cellular ligands necessary to promote virus entry into or signaling of cells. In addition to these CMV entry pathways being mediated at the cellular surface, CMV also employs cell-to-cell fusion as a mechanism to promote an intracellular pathway of spread of infection in vivo; this may, by avoiding extracellular release of virus (which could otherwise be neutralized by circulating antibody), serve as an immune evasion strategy [25,26]. The PC plays a key role in the cell-associated spread pathway, a pathway that is resistant to antibody-mediated neutralization; in contrast, the gH/gL/gO trimer, which is required for the infectivity of cell-free virions [24], may be a pathway more amenable to immune clearance. Trimer- and PC-specific antibodies in sera from CMV-seropositive individuals act synergistically to prevent cell-to-cell spread of virus, but the protection conferred by neutralizing antibodies targeting these complexes may be insufficient to prevent congenital transmission of virus [28]. These glycoproteins are nonetheless highly relevant to both active and passive immunization strategies that are aimed at preventing maternal-to-fetal transmission of CMV (discussed below).

Clinical burden of CMV disease

Most individuals are asymptomatically infected with CMV, often early in life. Transmission of CMV occurs chiefly via contact with infected secretions. Early acquisition of infection commonly occurs through breastfeeding, and CMV may also be transmitted perinatally, by aspiration of cervicovaginal secretions in the birth canal [29]. In infancy and toddlerhood, virus may be readily transmitted to susceptible children via saliva, urine, and fomites [30–32]. Conditions such as crowding, lower socioeconomic status, and attendance at group day-care centers [33–35], all contribute to spread of infection in young children. CMV infections acquired by toddlers are often in turn transmitted to their parents in the household setting [36–38]; if such infections are transmitted from toddler-to-mother in the setting of pregnancy, this can set the stage for congenital CMV transmission in that pregnancy. This has implications for how clinical trials might be designed and how candidate CMV vaccines might be used in clinical practice, once they are licensed (discussed below). In addition to acquisition of CMV from their toddlers, young adults can be infected through sexual activity, via behaviors ranging from kissing [39] to sexual intercourse [40,41]. Blood transfusion, once a common source of CMV infection, is rarely associated with transmission today, since leukofiltration has dramatically reduced the risk of transfusion-associated CMV disease [42,43].

CMV is typically an asymptomatic infection in the healthy immunocompetent host. Studies in identical twins discordant for CMV antibodies suggest that early-life acquisition of CMV may provide benefits to the chronically asymptomatically infected individual, chiefly mediated by induction of pro-inflammatory cytokines that may aid in protection against other infections [44] and in augmenting vaccine responsiveness [45]. In settings where CMV causes disease, a protean range of clinical manifestations can be observed and has been described [46], but usually a primary infection goes unnoticed by the infected individual. Newborns who acquire CMV infection from breastfeeding are typically asymptomatic, unless they are very-low birth weight premature infants; in this setting, post-natally acquired CMV infections have been associated with a sepsis-like syndrome, enterocolitis, and chronic lung disease [47–49]. Infants and toddlers who acquire CMV infections in early childhood may suffer a self-limiting febrile illness [50]. There is little evidence to indicate that primary CMV infections in toddlers and young children have any significant health implications, although one report suggests there may be an increased risk of symptomatic respiratory virus infections in CMV-infected toddlers in group day-care compared to their CMV-negative classmates [51]. In older children and adolescents, a "heterophile-negative" mononucleosis syndrome can be observed with a primary CMV infection [52–54]. The hallmark symptoms of CMV mononucleosis are fever and severe malaise, with atypical lymphocytosis and mild elevation of liver enzymes. It may be difficult to clinically differentiate CMV mononucleosis from the more typical Epstein Barr virus-induced "heterophile-positive" mononucleosis. CMV mononucleosis is, however, associated with less adenopathy and hepatosplenomegaly, and a reduced magnitude of atypical lymphocytosis, than Epstein Barr virus mononucleosis [55]. As with Epstein Barr virus mononucleosis, the use of ß-lactam antibiotics for CMV mononucleosis may precipitate a generalized morbilliform rash [56]. It remains unclear what host or viral factors predict whether a primary CMV infection in an immunocompetent individual will be asymptomatic, or symptomatic (i.e., mononucleosis).

There have been some attempts to characterize primary CMV infection in pregnant patients. As in non-pregnant individuals, most primary (as well as non-primary) infections are asymptomatic in nature. Primary CMV infections during pregnancy have been emphasized for several reasons. First, it is reasonably straightforward to define that a woman is undergoing a primary CMV infection, through a combination of positive immunoglobulin (Ig) M serology; seroconversion from IgG negative to IgG positive status; demonstration of CMV in blood, saliva, urine and other body fluids (including amniotic fluid); and evolution from low-avidity to high-avidity anti-CMV antibody titer [57–60]. In contrast, it is very challenging to define whether a CMV-seropositive woman undergoes a non-primary infection (most likely due to re-infection [61–64]) during pregnancy; demonstration of re-infection would require molecular evidence of infection with distinct, new viral strains, something that is currently available only in the context of a research laboratory. Second, primary maternal infections during pregnancy are more likely to lead to *transmission* of infection to the fetus. It is estimated that primary infection occurs in approximately 1–4% of pregnancies [65], and leads to transmission in probably 30–40% of such cases [66], although ranges of 24–75% have been published [67].

As noted, signs and symptoms of maternal infection during gestation are believed to be rare, although precise estimates are largely unknown, since women are typically not followed longitudinally during pregnancy for serological or virologic evidence of primary infection. It has been reported that <5% of pregnant women with primary infection are symptomatic, and an even smaller percentage recall having a mononucleosis syndrome. Among women who are symptomatic, the most frequent reported symptoms include malaise, persistent fever, myalgia, cervical lymphadenopathy, and, less commonly, pneumonia and hepatitis [67]. Laboratory tests may sometimes reveal atypical lymphocytosis or elevated transaminase levels. Symptomatic women, particularly those at high risk (i.e., mothers of toddlers attending group day-care), should have virological studies, including polymerase chain reaction (PCR) of blood, as well as serological studies (IgM, IgG, and avidity index) if the possibility of CMV infection is being considered by the clinician. It should be noted that the absence of maternal DNAemia does not necessarily provide reassurance, since the test has a suboptimal predictive value in heralding fetal infection. Although ultrasound is generally insensitive in identifying maternal-fetal CMV transmission, some ultrasonographic findings, in particular echogenic fetal bowel, are highly suggestive, and should prompt further investigation for CMV [68]. Fetal magnetic resonance imaging (MRI) may also be considered, particularly when ultrasound identifies abnormalities in the developing fetal brain. The diagnosis of fetal CMV infection is most reliably determined by amniocentesis with PCR testing of amniotic fluid. It is noted that this procedure should optimally be performed >7 weeks after a presumed onset of maternal infection, and after 21 weeks of gestation, since this appears to be the minimal time interval for CMV to be shed in the amniotic fluid [67]. Amniocentesis performed too soon after serological evidence of primary maternal infection therefore runs the risk of producing false-negative results. Of course, if ultrasonographic or MRI findings are strongly suggestive of fetal disease (i.e., echogenic bowel or brain anomalies) then amniocentesis can be considered irrespective of the evolving maternal serological profile.

In contrast to the ubiquitous and largely asymptomatic profile of CMV infection in heathy individuals, the clinical burden of CMV disease is substantial in several special populations. These include individuals living with human immunodeficiency virus (HIV) [69,70], solid organ and hematopoietic stem cell transplant patients [71,72], and congenitally infected infants [73]. Of these, the greatest impact—both with respect to the numbers of affected individuals as well as the economic impact on society [74–77]—is the impact elicited by congenital CMV infection. Although the majority of congenital CMV infections are asymptomatic, at least 10% of the 20,000–30,000 CMV-infected infants born in the United States annually will have long-term adverse neurodevelopmental sequelae, including cerebral palsy, sensorineural hearing loss (SNHL), seizure disorders, microcephaly, and learning disabilities [78]. Clinical manifestations in symptomatic infants include growth restriction, petechiae, hepatosplenomegaly, microcephaly, jaundice, seizures, rash, and periventricular calcifications [73]. The risk of neurological sequelae is greatest when the fetus is infected in the first trimester of gestation [79], although second trimester infections can be associated with neuroimaging abnormalities [80] and sequelae such as SNHL [81]. Notably, those infants that are symptomatic at birth are at greater risk for adverse neurodevelopmental sequelae and, among these sequelae, SNHL is the most common. Indeed, it is estimated that between 22% and 65% of children with symptomatic CMV disease at birth, and 6–23% of children with asymptomatic congenital CMV infections, will have or develop SNHL [82–84]. CMV-induced SNHL may be present at birth and may be demonstrable as a failure on the newborn hearing screening assessment, or hearing loss may become clinically manifest later in childhood, where it may be stable or progressive in nature [85,86]. It is estimated that congenital CMV is the etiology responsible for over 20% of all cases of pediatric SNHL observed at birth [87]. Thus, the argument can be made that the most substantial impact that CMV has on health is its impact on hearing loss in congenitally infected infants.

Epidemiology of CMV

As noted above, CMV infections are ubiquitous and thus, infection is rarely avoided over the life-course. However, worldwide the prevalence of CMV antibodies varies considerably among populations. An overall global seroprevalence of CMV of approximately 80% has been estimated in a meta-analysis (Fig. 2; [88]). This study identified higher rates of seropositivity in residents of the developing world, particularly from the Eastern Mediterranean region, and lower rates (~65%) in Europe. In the United States, an analysis of serum collected by the National Health and Nutrition Survey (NHANES) indicated an overall age-adjusted seroprevalence of CMV of 50.4% (ages, 6–49 years), with higher rates in non-Hispanic black and Mexican-American children, compared with non-Hispanic white children [89]. In this study the likelihood of CMV seropositivity was associated with older age, being female, foreign birthplace, low household income, higher rates of household crowding, and low household education level.

An enigmatic but critical concept in CMV epidemiology relevant to pregnancy and childbirth is the fact that the prevalence of congenital CMV infection is *directly*, and not *inversely*, proportional to the overall prevalence of CMV antibodies in the population being studied. Sometimes described in the literature as a "paradox" or an "enigma" [90–93], this phenomenon reflects the fact that preconception seropositivity does not protect against maternal re-infection during pregnancy. Hence, the biggest burden engendered by congenital CMV infection appears to be in infants born to women with preconception immunity. This is likely due to maternal re-infections during pregnancy with new strains of virus [61–63,

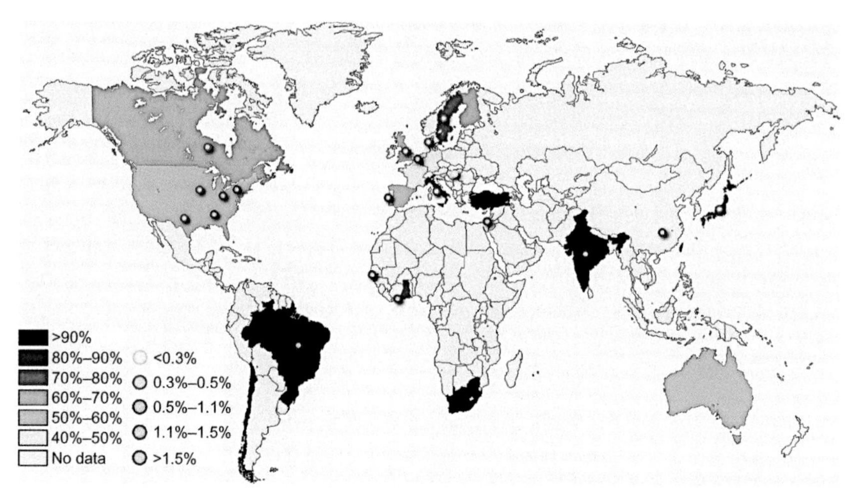

FIG. 2 Global epidemiology of maternal and neonatal CMV infection. Figure provides an overview of available data regarding worldwide CMV seroprevalence rates among women of reproductive age (age 12–49 years) and the birth prevalence of congenital CMV. Birth prevalence studies from published summaries used PCR or culture on saliva or urine, except for Portugal and the Netherlands for which data on newborn dried blood spots was utilized. The countries for which data on maternal seroprevalence and/or birth prevalence of congenital CMV infection were available include Brazil, Canada, England, India, Israel, Italy, Japan, Sweden, and the United States. Reproduced with permission from Manicklal S, Emery VC, Lazzarotto T, Boppana SB, Gupta RK. The "silent" global burden of congenital cytomegalovirus. Clin Microbiol Rev 2013;26(1):86–102.

94, 95], although re-activation of latent maternal viral infection with subsequent transmission of CMV in sequential pregnancies has been suggested by molecular virological analyses [96,97]. While congenital transmission in mothers with pre-existing immunity occurs at a lower rate than that associated with primary maternal infection in pregnancy, because of the high rate of seropositivity in lower and middle-income countries, CMV transmission to the fetuses of seropositive mothers is believed globally to be the most common scenario leading to congenital infection [98]. Since most congenital infections occur in the context of non-primary (recurrent) maternal infection [99,100], it has been estimated that approximately 75% of congenital CMV infections occur during recurrent maternal infection during pregnancy [101].

Recognizing the variables that impact on the birth prevalence of congenital CMV in various populations, efforts have been made to provide informed estimates from meta-analyses and, in some cases, prospective surveillance studies. An estimate of prevalence gleaned from a meta-analysis indicated that approximately 0.65% of all newborns in the United States have congenital CMV infection [102]. A more recent, prospective multi-center study of ~100,000 newborns in six geographic sites in the United States, the CMV Hearing Multicenter Screening (CHIMES) Study, identified a birth prevalence of congenital CMV of ~0.5% [103]. The CHIMES study demonstrated significant racial and ethnic differences in the prevalence of congenital CMV, with black infants noted to be significantly more likely to have infection compared with non-Hispanic white infants [103,104]. As noted above, although the majority of these infections are asymptomatic, it is estimated that 13.5% of the 20,000–30,000 CMV-infected infants born in the United States annually will have long-term neurodevelopmental sequelae, particularly SNHL [78]. Hence, efforts to prevent or modify the course of maternal-fetal CMV transmission, through vaccination or immune-based therapies, are highly desirable.

Vaccines in development

Passive immunization: Anti-CMV IgG

One strategy to prevent transplacental transmission of CMV involves the administration of high-titer CMV immune globulin (CMV-IG) during pregnancy. A major challenge to this approach is the fact that, since women are not routinely monitored for CMV seroconversion or evidence of primary or re-infection, there would not currently be a readily implementable strategy for such prophylactic/therapeutic approaches. Antibodies to virally-encoded envelope glycoproteins have also been shown to be protective against congenital guinea pig CMV infection [105-107]. The use of human CMV-IG is commonplace in solid organ transplant recipients, for which good evidence for protection against CMV disease has been shown [108-111]. A protective effect of maternally-acquired CMV IgG was also demonstrated in past studies of transfusion-acquired CMV infections in premature infants [112]. Major targets of protective IgG responses appear to be the CMV-encoded envelope glycoproteins, particularly the PC proteins [113]. Unfortunately, clinical trials examining this strategy in an attempt to modify CMV transmission in pregnant subjects have yielded mixed results [114-116]. In an uncontrolled study in pregnant women with evidence of primary CMV infection, hyperimmune CMV-IG therapy was associated with both a significantly lower risk of congenital CMV transmission and reduction in CMV disease in the already-infected fetus [117-119]. In contrast, in a controlled (phase IIB, randomized, double blind) study, known as the "CHIP" study, CMV hyperimmune globulin, when compared to a saline placebo, did not demonstrate a statistically significant impact on the rate of transplacental CMV transmission in pregnant women with primary CMV infection [120,121]; moreover, the number of obstetrical adverse events was higher in the CMV-IG group than in placebo recipients (13% vs. 2%). Histological analyses of placentas did not demonstrate any differences between the treatment and placebo groups [122]. An NIH-supported multi-center trial in the United States has recently been conducted to study the unresolved question of the impact of passive CMV-IG therapy on congenital CMV transmission in pregnant women with primary CMV infections [123]; although data are in analysis, this study does not appear to be recruiting subjects at this time (https://clinicaltrials.gov/ct2/show/NCT01376778?term=Nct01376778&draw=2&rank=1). CMV-IG is a pooled, polyclonal preparation of IG obtained from multiple donors with high titers of anti-CMV antibody. The CMV hyperimmune globulin used in the uncontrolled studies conducted in Italy (and published in 2005) was produced by Cytotect Biotest [117] at a prophylaxis dose of up to 100 units/kg monthly. In the CHIP study, the hyperimmune globulin preparation was also purchased from Cytotect (Biotest Pharma) and administered at a similar dosage and dose interval [120]. In the NIH sponsored trial, Cytogam (CSL Behring) was utilized, at an intravenously administered dose of 100 units/kg body weight. Although it is not clear what the impact of dose, dose interval, lot-to-lot variability, or manufacturer source of CMV hyperimmune globulin might have on its putative therapeutic benefit, at this time there exists no information from any controlled clinical trial that supports the use of this intervention to prevent maternal-to-fetal transmission of CMV or to prevent CMV-associated disease in the newborn infant.

To avoid the problems associated with using a pooled human blood product as an antiviral therapy, some efforts have been to commercialize potent CMV-neutralizing monoclonal antibodies targeting key envelope glycoprotein epitopes. Although the humoral correlate(s) of protection against congenital transmission remain unknown, a number of candidate therapies are in development. Of particular interest are monoclonal antibodies targeting the AD2 domain of the gB glycoprotein [124], with one agent, monoclonal TCN-202, progressing to phase II studies [125]. The manufacturer of another anti-AD2 monoclonal, TRL345, currently appears poised to commence clinical trials in the near future [126].

Active immunization: Candidate CMV vaccines

Knowledge of the maternal immunological responses that confer protection against congenital transmission of CMV infection remains elusive. An added layer of complexity comes from the recognition that an effective vaccine against congenital transmission will likely not only require induction of responses that protect against primary infection, but also responses that will protect against re-infection with novel strains in seropositive women. These challenges notwithstanding, there has been progress in recent years in defining some of the putative correlates of protection against congenital transmission. Protection appears to require both innate (natural killer cell) and adaptive (CD4+ and CD8+ T cell, virus-neutralizing antibody) responses [127]. Vaccines evaluated in clinical trials to date have generally focused on eliciting humoral responses and T cell-mediated responses against several key immune targets encoded by the virus (Table 1). In broad terms, this has been approached through two general methods: *live, attenuated* (or replication-disabled) vaccines; or *subunit vaccines* (based on combinations of specific CMV recombinant proteins and/or peptides, or vectored vaccines expressing these immunogens of interest). These are both highlighted below and summarized in Table 2.

TABLE 2 CMV vaccines that have been evaluated in clinical trials.

Subunit vaccines and vectored vaccines based on key cmv immunogens	
Subunit glycoprotein B (CHO cell expression) MF5 or AS01 adjuvant	• Favorable safety profile • High-titer neutralizing antibody and strong cell-mediated immune responses; augments humoral immunity in seropositives • The gB/MF59 demonstrated efficacy in phase II study in young women against primary infection; limited efficacy in adolescents; protection against CMV disease in solid organ transplant patients; gB/AS01 demonstrated immunogenicity in phase I but no efficacy data • Phase I studies are planned combining gB with soluble pentamer (PC)
Enveloped virus-like particles vaccine (gB)	• Currently in phase I studies • gB polypeptide engineered as fusion to vesicular stomatitis virus (VSV) glycoprotein to improve immunogenicity • Vaccine forms enveloped virus-like particles • gB molecule may form authentic trimeric configuration enhancing authenticity of antibody response
Recombinant Lymphocytic Choriomeningitis Virus (LCMV) vectored vaccine	• Currently in phase I studies • Expresses gB and UL83 (pp65) • Robust antibody and cell-mediated immune responses • Capable of boosting seropositives and past recipients of LCMV vaccines
PADRE-pp65-CMV fusion peptide vaccines +/− CpG DNA adjuvant	• Lipidated fusion peptides constructed from UL83 (pp65) CTL epitopes • Linked to either a synthetically derived pan-DR epitope peptide • Phase I studies ongoing
Glycoprotein B/ Canarypox Vector	• Favorable safety profile in human volunteers • Suboptimal immunogenicity • "Prime-Boost" effect on combined administration with Towne • Not currently in clinical development
UL83 (pp65)/ Canarypox Vector	• Favorable safety profile in human volunteers • Strong antibody and cell-mediated immune responses • Not currently in clinical development

TABLE 2 CMV vaccines that have been evaluated in clinical trials—cont'd

gB/pp65/IE1 Trivalent DNA Vaccine; gB/pp65 Bivalent DNA Vaccine	• DNA vaccine adjuvanted with poloxamer adjuvant and benzalkonium chloride • Phase III study of bivalent vaccine failed to protect against CMV disease in transplant patients • Trivalent vaccine has been evaluated with Towne in prime-boost vaccination • Future development plan for this vaccine is uncertain in light of failure in phase III trial
mRNA Vaccine; gB and PC (mRNA-1647); UL83 (pp65; mRNA-1443)	• mRNA encapsulated in lipid nanoparticles • Excellent immunogenicity in preclinical studies (mice and non-human primates) • Phase I study in volunteers in progress
gB/pp65/IE1 Alphavirus Replicon Vectored Trivalent Vaccine	• Engineered using replication-deficient alphavirus technology • Generation of virus-like replicon particles • Phase I clinical trial demonstrated safety and immunogenicity (virus-neutralizing antibody and cell-mediated immune responses) • Vaccine is not currently in clinical development
Modified Vaccinia Virus Ankara (MVA) Vaccines	• "Triplex" vaccine based on gB, UL83 (pp65), IE1/2 • Currently in phase II studies in transplant patients • Pathway for development for women of childbearing age unclear
Live attenuated and disabled virus vaccines	
Towne Vaccine (+/− rhIL12)	• Elicits humoral and cellular immune responses • Favorable safety profile; no evidence for latency or viral shedding in recipients • Reduced CMV disease but not infection in renal transplant recipients • Augmented immunogenicity with recombinant IL-12 in phase I studies • Vaccine is not currently in clinical development
Towne/Toledo Chimera Vaccines	• Recombinant live virus vaccines restoring genes deleted from Towne virus during tissue culture attenuation to enhance immunogenicity • Favorable safety profile; no evidence for latency or viral shedding in CMV seropositive subjects • Attenuated compared to Toledo strain of HCMV • Phase I studies in seropositives (safety) and seronegatives (immunogenicity) have been completed • Vaccine is not currently in clinical development
V160-001 Replication-Defective Vaccine	• AD169 backbone with restoration of UL129/130/131 PC components • Rendered replication-incompetent by inclusion of ddFKBP/Shld1 • Administered with alum-based adjuvant • Phase I studies completed; safe, immunogenic • Phase II studies in progress

Live, attenuated or disabled, single cycle (DISC) vaccines

The first approach to CMV vaccine development utilized live, attenuated strains of virus passaged in tissue culture. The first of such vaccines was based on the strain "AD169," a highly passaged clinical isolate. Limited studies in volunteers were undertaken in the early 1970s with AD169 vaccine [128], but this vaccine was not further developed. The attenuated CMV strain that advanced the furthest in vaccine development, the Towne strain, was originally isolated from a congenitally infected infant and passaged extensively in human embryo fibroblasts until the 125th passage. Notably, both AD169 and Towne strains acquired mutations upon serial tissue culture passage in PC protein coding sequences: for Towne, a 2-base pair insertion (TT) occurred, leading to a frameshift mutation in UL130, and for AD169, a 1-base pair insertion (A) generated a frameshift mutation in UL131 [129,130]. In spite of this mutation (which abrogated the ability of Towne virions to express a PC), studies of intramuscular or subcutaneous administration of Towne vaccine demonstrated seroconversion in nearly 100% of volunteers. Since no vaccine virus was recovered from saliva, urine or blood, the Towne vaccine was assumed to be highly attenuated and hence suitable for human use, although questions were raised about suboptimal immunogenicity (years before the PC had even been discovered and the UL128, 130 and 131A open reading frames characterized). One strategy to improve the immunogenicity of Towne vaccine was to co-administer it with recombinant human interleukin 12 (rhIL-12 [131]). Another approach aimed at improving the immunogenicity of Towne was to prime the vaccinee for memory responses by first immunizing with plasmid VCL-CT02, containing the three CMV genes pp65, IE1, and gB (Table 1). This approach conferred a more rapid induction of antigen-specific responses, with VCL-CT02 priming for an anamnestic response following subsequent administration of Towne [132].

Another modification of Towne vaccine was inspired by discoveries by scientists at MedImmune (formerly Aviron) laboratories, who noted that Towne variants had a substantial deletion (~13.5 kb) in a region of the viral genome designated as the ULb′ region. This region, present in low-passage clinical isolates (in both human and highly related primate CMVs), contains at least 17 genes not found in Towne, some of which are postulated to be important targets for immune responses [133-136]. In light of these observed differences, it was hypothesized that insertion of the ULb′ sequence from a less attenuated CMV strain (the "Toledo" strain) in the background of Towne virus might provide one or more genes coding for proteins that would increase the immunogenicity of Towne, resulting in a more efficacious vaccine. To address this possibility, four recombinant chimeric variants of Towne and the low-passage virulent Toledo strain were generated, using cosmid technology, in which different segments of Towne were replaced by the analogous sequences of

Toledo. Subsequent to the initiation of these trials, it was discovered that the Toledo genome carries a disrupted copy of the UL128 coding sequence (another constituent of the CMV PC), which also abrogates generation of PC [137]. Immunogenicity and safety studies are described below.

A variant of the live, attenuated vaccine strategy is the *Di*sabled, *S*ingle *C*ycle (DISC) vaccine approach, in which the full viral genome is present, but the vaccine virus is rendered incapable of replication, hence ensuring safety. Recently, Merck laboratories have initiated studies of a candidate CMV vaccine, V160 [138,139]. In this vaccine, wild-type UL128/UL130/131 PC protein sequences are restored in the AD169 genome, allowing for bona fide generation of PC. To render the vaccine replication-incompetent, the UL51 and IE1/2 proteins were modified, by linking them through in-frame fusions with the destabilization domain of the FK506 binding protein, ddFKBP [140-142]. This modification renders protein stability dependent on the presence of the small molecule Shield 1 (Shld 1), supplied in *trans*. Hence, virus can only form in cell culture in an environment when Shld 1 is provided; otherwise, the linkage of the essential UL51 and IE1/2 proteins to the FK506 destabilization protein generates a "dead end" in viral replication and no progency viruses can be produced. If the Shld 1 protein is provided, viral replication is rescued and progeny virus generated, but the destabilization domains are encoded in the viral genome, so progeny viruses are replication defective. Hence, vaccine can be generated in the laboratory, but since Shld 1 is a purely synthetic ligand that does not exist in nature, the vaccine virus can undergo only one cycle of (abortive) replication in the human host following vaccine administration, ensuring safety.

Subunit (recombinant/vectored) vaccines

General categories of CMV subunit vaccines include adjuvanted recombinant protein vaccines based on recombinant forms of gB protein (Fig. 1); vaccines based on gB alone, or in combination with the tegument protein, UL83 (also known as pp65) and the major immediate early protein 1 (IE1) generated using viral or nucleic acid-based vectors; vaccines based on generation of virus-like particles (VLP) expressing combinations of these CMV gene products; and peptide-based vaccines based on T cell epitopes (usually from UL83) important in protection against CMV disease. This section focuses only on those vaccines that have been evaluated in human volunteers (Table 2); updated recent information is available about other novel platforms evaluated in animal models or preclinical studies, such as dense body vaccines.

Purified recombinant gB subunit vaccines

Subunit approaches utilizing adjuvanted, recombinant formulations of gB have arguably advanced the furthest in clinical trials. Several phase I and phase II clinical trials utilizing a recombinant CMV gB in microfluidized adjuvant 59 (MF59), a proprietary oil-in-water emulsion first used in influenza vaccines, have been completed [143-146]. Most studies have utilized a three-dose vaccination series, with doses administered at 0, 1 and 6 months. The gB was expressed as a truncated, secreted polypeptide purified by chromatography from Chinese hamster ovary (CHO) cells.

Another gB vaccine with a slightly different configuration and sequence (but also based on a truncated, secreted form of the gB polypeptide), GSK1492903A, was similarly purified from CHO cells and admixed with Adjuvant System (ASO1). This vaccine has been evaluated in CMV-seronegative volunteers [147] and has demonstrated robust immunogenicity. Efficacy evaluations of this vaccine are considered in the following section. At least two major pharmaceutical manufacturers are planning to evaluate "next-generation" gB subunit vaccines, likely to be co-administered with soluble PC, in the near future (Table 2).

Enveloped virus-like particles vaccines

Enveloped virus-like particles (eVLPs) mimic wild-type virus particles but do not have a viral genome, hence ensuring safety. An eVLP vaccine, manufactured by VBI laboratories and expressing CMV gB, is currently in clinical trials in CMV-seronegative subjects. The gB is expressed as an in-frame fusion protein with the transmembrane and cytoplasmic domains of the vesicular stomatitis virus (VSV) G protein, a strategy that improves immune responses [148]. The vaccine, designated VBI-1501A, (https://clinicaltrials.gov/ct2/show/NCT02826798?term=NCT02826798&draw=2&rank=1) is undergoing safety and immunogenicity comparisons of four different dose formulations, with and without alum as an adjuvant. Preliminary data suggest excellent immunogenicity, including induction of neutralizing antibodies, in volunteers.

Vectored CMV vaccines

There are several so-called "vectored" CMV vaccines that have been evaluated in clinical studies. In this strategy, the viral immunogens of interest are expressed, following vaccination, via their incorporation into heterologous "vectors," usually viruses, that carry the gene of interest. One strategy is based on canarypox, an avian pox virus that replicates abortively in mammalian cells. Two canarypox vaccines have delivered gB (ALVAC-CMV[vCP139]) or pp65 (ALVAC-CMV[vCP260]) subunit via this platform to volunteers [149,150]. Another vectored vaccine approach utilized for CMV was based on the alphavirus, Venezuelan equine encephalitis (VEE) virus. In this strategy, VEE genes encoding structural proteins were replaced with genes expressing the extracellular domain of the gB glycoprotein and a pp65-IE1 fusion protein in a double-promoter replicon [151,152]. Progeny virus-like particles generated using this strategy were replication-deficient, but found to undergo extensive transcription, and hence were highly immunogenic, both in pre-clinical evaluation in animal models [153] and in phase I studies in human volunteers [154]. The platform was originally developed by AlphaVax vaccines, which was acquired by Novartis Corporation in 2008, and is now held by GlaxoSmithKline following their purchase of Novartis vaccines. Plans for future studies of these vaccines are unclear at this time.

Another vectored CMV subunit vaccine design uses the attenuated recombinant lymphocytic choriomeningitis virus (LCMV) platform [155-157]. This vector employs a producer cell line constitutively expressing the LCMV viral glycoprotein (GP), therefore making it possible to replace the gene encoding LCMV GP in the viral genome with vaccine antigens of interest. Production of resulting recombinant LCMV in this cell line yields vaccines that are replication-defective, since they do not carry the LCMV GP gene required for full replication of the vector. These virus vectors are, however, able to undergo a single cycle of replication, inducing immune responses. LCMV vectored vaccines are able to elicit potent cytotoxic T lymphocyte (CTL) and CD4+ T cell responses, as well as high levels of neutralizing antibody; therefore, this is a potentially very attractive vaccine approach. Hookipa Biotech AG is currently evaluating a replication-deficient LCMV-vectored CMV vaccine, designated HB-101, which is a bivalent vaccine containing two vectors expressing the pp65 and gB proteins, respectively; with the recent demonstration of efficacy in a "proof-of-concept" study for prevention of CMV transmission in a pre-clinical guinea pig model of congenital infection, these vaccines are now moving forward in phase I study in volunteers [157].

Modified vaccinia virus Ankara (MVA) has been used to express a variety of CMV antigens, including pp65, gB, IE1, IE2, and the PC proteins [158-160]. A vectored MVA vaccine, designated "Triplex", was constructed to encode three of the key, immunodominant T cell antigens associated with protective cell-mediated immunity to CMV: UL83 (pp65), UL123 (IE1-exon4), and UL122 (IE2-exon5). The triplex MVA vectored vaccine is currently in a phase II trial, employing a placebo-controlled, two-dose study design of vaccine delivered by intramuscular route (https://clinicaltrials.gov/ct2/show/NCT02506933?term=nct02506933&draw=2&rank=1). Since the primary aim of this work is to induce protection against CMV disease in hematopoietic stem cell transplantation (HSCT) recipients, the relevance of these vaccines to protection of women of child-bearing age against congenital CMV transmission is less clear.

Nucleic acid-based CMV vaccines

Nucleic acid-based CMV vaccines have been developed and tested in human volunteers using two approaches: expression of key immunogens as DNA molecules (typically, recombinant plasmids with an immunostimulatory adjuvant) or as RNA molecules. Most experience to date has been accrued with DNA vaccines. Most studies of DNA CMV vaccines have focused on the goal of ameliorating or preventing CMV disease in the

HSCT and solid organ transplant patient populations. A vaccine designated ASP0113 (previously known as VCL-CB01 and TransVax) is a DNA vaccine using gB and pp65 encoded by two plasmids, VCL-6368 and VCL-6365, and formulated with poloxamer CRL1005 and a cationic surfactant, benzalkonium chloride [161-164]. Another vaccine candidate is a non-adjuvanted, trivalent DNA vaccine (VCL-CT02), which includes exons 2 and 4 from the T cell target IE1, in addition to gB and pp65, and has also been examined in phase I clinical trials (https://clinicaltrials.gov/ct2/show/NCT00370006?term=NCT00370006&draw=2&rank=1 and https://clinicaltrials.gov/ct2/show/NCT00373412?term=NCT00373412&draw=2&rank=1). This vaccine was developed with a stated development goal for utilization as a vaccine against congenital CMV infection. This vaccine was tested in human volunteers in a "prime-boost" format in which subjects who had been inoculated (intradermally or intramuscularly) with VCL-CT02 plasmid were then "boosted" with attenuated Towne vaccine [132,165]. ASP0113 was recently evaluated in a double-blind, placebo-controlled study in CMV-seronegative kidney transplant recipients receiving a kidney from a CMV-seropositive donor. Transplant recipients were randomized to receive 5 doses of ASP0113 (5 mg, $n=75$) or placebo ($n=74$) on days 30/60/90/120/180 post-transplant, with a primary study endpoint of CMV viremia (≥1000 IU/mL from day 100 through to 1 year after the first dose). Unfortunately, there was no statistically significant difference in the primary endpoint between the ASP0113 and placebo groups [166]. This result has made the future of the VCL-CT02 congenital CMV vaccine program uncertain.

Another nucleic acid-based CMV vaccine strategy uses RNA, and not DNA, as the gene expression platform. This is based on a "self-amplifying" mRNA vaccine platform, developed by Moderna Therapeutics. The platform has been used to express of CMV immunogens, and has been designated by the manufacturer as vaccine candidate mRNA-1647. The vaccine contains six mRNA species—five that express the PC, and one that expresses gB. The mRNA-1647 vaccine is currently being evaluated in a phase I clinical trial. Preclinical studies of this vaccine, in which mice and non-human primates were immunized with lipid nanoparticles encapsulating these RNA molecules encoding the vaccine genes of interest, demonstrated excellent immunogenicity [167]. The current phase I trials are randomized, placebo-controlled, dose-ranging studies that are being conducted in CMV-seronegative women, with the stated goal of evaluating the safety and immunogenicity of mRNA-1647. A second vaccine designated as mRNA-1443, which targets the UL83 protein (https://clinicaltrials.gov/ct2/show/NCT03382405?term=NCT03382405&draw=2&rank=1), has also been developed, although the clinical trial pathway for moving this vaccine forward is not clear at this time.

Peptide vaccines

Other CMV vaccines currently in clinical trials are mostly focused on the protection of HSCT recipients and have unclear relevance to protection against congenital CMV. It is known that pp65-specific CTL responses can protect HSCT patients from CMV disease and this observation has helped to drive development of peptide-based vaccines. CMV pp65 epitopes fused to either a synthetic pan-DR epitope (PADRE) or to a natural tetanus (Tet) sequence have been evaluated in phase I trials (https://clinicaltrials.gov/ct2/show/NCT00722839?term=NCT00722839&draw=2&rank=1), with and without the synthetic CpG TLR9 agonist adjuvant 7909 [168,169]. In a phase I study, this adjuvant, when administered with PADRE and Tet pp65, was found to enhance immune responses in volunteers [169]. Notably, the particular UL83 epitope used, pp65 495-503, was estimated to cover approximately 35% of the United States population, based on the overall population frequency of the human leukocyte antigen (HLA) A*2010 allele [169]. Thus, it is conceivable that a polyepitope vaccine, if it was capable of spanning enough T cell epitopes, could be a viable option if utilized as an HLA-diverse, population-based vaccine against congenital CMV. Phase II studies of the Tet-pp65 vaccine, designated as CMVpp65-A*0201 (CMVPepVax [169]), are now in progress (https://clinicaltrials.gov/ct2/show/NCT02396134?term=NCT02396134&draw=2&rank=1). Enrollment is targeting HLA-A*0201-positive, CMV-seropositive HSCT recipients; as noted above, the applicability of applying this approach to an HLA-diverse population of women of reproductive age in the context of a vaccine against congenital CMV infection is not clear.

Available evidence of effectiveness/efficacy of CMV vaccines

To date, only one vaccine has been shown to provide efficacy against the acquisition of CMV infection in clinical trials: the MF59 adjuvanted purified recombinant gB vaccine. In a phase II study in post-partum women, gB/MF59 vaccine had a 50% efficacy against acquisition of primary CMV infection in seronegative women (vaccinated within 1 year of giving birth) compared to placebo recipients [143]. This result was ground-breaking in nature, since no prior vaccine study had ever demonstrated evidence of efficacy against acquisition of CMV infection. Notably, women who enrolled in this study that were found to be CMV-immune were also vaccinated with either the gB/MF59 vaccine or a placebo, in a spin-off study aiming to evaluate the potential for boosting of the immune response in seropositives [146]. Intriguingly, gB specific responses (including CD4+ T cell response to gB and levels of IFN-γ producing T cells) were higher for vaccinees compared to controls. Since as noted above most congenital CMV infections occur in the setting of re-infections occurring in the setting of preconception immunity to the virus, these data could support a program of vaccination of CMV-immune women during their child-bearing years, toward the goal of preventing such re-infections during pregnancy.

Another gB/MF59 clinical trial targeting young women was recently reported in healthy, CMV-seronegative adolescents [170]. The incidence of

CMV infection after three vaccinations was reduced in the vaccine group compared to placebo, with a calculated vaccine efficacy of 44%. Although the efficacy did not reach conventional levels of significance, the results were consistent with the previous study in adult women that used the same formulation [145]. CMV viral load in vaccinees and placebo recipients was monitored by PCR of urine and blood samples, and if viremia was detected the subject was placed into a sub-study (https://clinicaltrials.gov/ct2/show/NCT00815165?term=NCT00815165&draw=2&rank=1). Currently, this sub-study, which retained blinding, is active, and is examining CMV-specific cell mediated responses in greater detail. Once this data becomes available, it will be of great interest to determine whether additional correlates of vaccine-mediated protection can be identified. In studies of recombinant gB vaccine performed to date, there has been emphasis on evaluation of ELISA and neutralizing antibody responses as the presumed correlate(s) of protection, but other possible effectors of protection need to be explored in future studies.

There have been other studies of gB/MF59 vaccine, including evaluation of vaccine efficacy in preventing CMV disease in solid organ transplant recipients. A phase II double-blinded study (https://clinicaltrials.gov/ct2/show/NCT00299260?term=NCT00299260&draw=2&rank=1) examined immunogenicity and CMV disease, as assessed by viremia (DNAemia), in kidney or liver transplant patients [171]. Seronegative organ recipients who received gB/MF59 and had seropositive organ donors demonstrated reduction in viremia and days of ganciclovir treatment compared to those who received placebo. Additionally, duration of viremia post-transplantation was inversely correlated to the magnitude of the gB antibody response. The study authors hypothesized that antibodies induced by gB/MF59 vaccination may bind to and neutralize CMV virions released by the donated organs to prevent transmission into the new host [171].

Evidence of safety of CMV vaccines

All CMV vaccines studied in clinical trials, to date, have shown excellent safety profiles. Typical local reactions, including erythema, and transient discomfort, have been described in phase I and phase II studies, but none of these have been of major clinical significance, nor do they stand as impediments to further clinical development of these platforms. Two more significant safety concerns that have been raised in the context of CMV vaccine development, however, merit additional consideration.

The first vaccine concern that has been raised on several occasions, particularly by regulatory bodies, is the concern whether any live, attenuated or DISC vaccine for CMV could establish latent infection in the immunized host. This is a relevant concern for the highly passaged Towne strain vaccine, the genetically engineered Towne/Toledo "chimeric" vaccines, and the V160 DISC vaccine (described above). Vaccination of a seronegative individual with a vaccine that has the potential to cause latent/persistent infection could, in theory, be deleterious to the host, given the hypothesized (but unproven) potential for CMV to be associated with malignancy (such as *glioblastoma multiforme*), atherosclerosis, autoimmune disease, and immunosenescence [172-176]. Indeed, if a live, attenuated CMV vaccine capable of conferring bona fide infection (with replication competent virus) upon the vaccinated host was inadvertently administered to a pregnant woman, the very worrisome possibility of transplacental transmission of a vaccine strain of virus would exist – analogous to concerns that were raised in the early days of rubella vaccination [177]. Fortunately, there is no evidence to support the premise that any of the live, attenuated or DISC virus platforms are capable of generating replication-competent virus in the immunized individual. Neither Towne vaccine (reviewed in [178]) nor the Towne/Toledo "chimera" vaccines [179,180] ever demonstrated any evidence of viral shedding in any vaccine recipient in a clinical trial. Although experience with V160 vaccine in phase I studies is limited, the vaccine can only be propagated, as noted above, in the presence of Shld1, a purely synthetic ligand that does not exist in nature [139]. This fact makes the prospect of generation of replication competent virus biologically untenable.

A second safety issue that has been raised in the context of CMV subunit vaccines relates to safety questions about the glycoprotein B. Specifically, in a study in autoimmune-prone mice (strain MRL/mpj), it was shown that immunization with recombinant CMV gB induced an autoimmune response, including antibodies to the U1-70kDa small nuclear riboprotein component of the spliceosome complex [181]. This study raised the concern that vaccination with CMV gB of individuals genetically predisposed to autoimmunity might be able to induce autoimmune disease. However, a reassuring study evaluated pre and post vaccine sera from a clinical gB vaccine trial for induction of autoantibodies targeting the Smith antigen, ribonucleoprotein complex (RNP), and the U1-70kDa component of the RNP complex. This study found no evidence of auto-antibody induction [182], demonstrating that if CMV infection is responsible for induction of autoimmune disease in some patients [173], it is not via immune response to the gB molecule. This provides further reassurance for the safety of gB-based vaccines, and for the continued evaluation of such vaccines in clinical trials.

Future prospects

There are several unresolved issues and major challenges in the development of CMV vaccines that must be resolved in future studies. These are summarized in Table 3. Perhaps the key unmet need in CMV vaccines is the essential need to improve our understanding of the *correlate(s) of*

TABLE 3 Unresolved questions requiring clarification to move a CMV vaccine program forward.

Unresolved issue	Implications for CMV vaccines

1. Identification of key correlate(s) of protection
• Antibody?
  ◦ Neutralizing vs. non-neutralizing
  ◦ Mucosal vs. systemic response
• Cell-mediated immunity?
  ◦ Cytokine response
  ◦ CD4+ vs. CD8+ cells
• Innate immunity
  ◦ Optimal strategy to induce innate responses
	Inclusion of critical immunogens can inform and direct selection of target antigens for subunit vaccines as well as optimal vaccine platform(s)

2. Define study endpoints required for licensure • Protection against maternal infection? • Protection against congenital transmission? • Protection against disease and disability in newborns, even if perinatal transmission occurs?	Clarification of requirements of regulatory bodies for licensure can help resolve design of pivotal phase III efficacy trials
3. Resolving the "enigma of maternal reinfection" • Immune correlates that predict reinfection risk? • Defining whether pre-conception immunity protects against CMV disease even if reinfection occurs? • Importance of viral strain variation in reinfection? • Identification of whether immune responses in women that are already "immune" to CMV can be boosted by a vaccine, and whether this boost provides any level of additional protection against reinfection	Clarification of this issue will help resolve the question of whether CMV vaccine would be universally indicated in women that may become pregnant or if it should selectively target seronegative women

4. Identifying what age group and sex should be targeted for CMV vaccine
• Toddlers (to avoid day-care, and child-to-mother CMV transmission)?
• Adolescents, as part of routine pre-teen schedule?
• Women anticipating or attempting pregnancy?
  ◦ Vaccine decision driven by drawing screening CMV serology?
• Women (girls) only, or men (boys) and women (girls)?
  ◦ Herd immunity and force of infection
  ◦ Reduce sexual transmission of CMV
  ◦ Efficacy level required to reduce magnitude of CMV disease in newborns?
	Key question that will drive vaccine implementation strategy and use in primary care

protective immunity for the mother, placenta and fetus [127]. A major emphasis to date in the pharmaceutical industry in development of CMV vaccines has been driven by the perceived imperative to design a vaccine that induces high levels of virus-neutralizing antibody. This readout has been emphasized in pre-clinical testing and in clinical trials and is typically interrogated by virus-neutralization assays performed using extracellular workpools of "free virus" titrated in the presence or absence of diluted sera on cells in tissue culture, including fibroblasts, epithelial, and endothelial cells. Although useful, such assays overlook a key aspect of the fundamental biology of CMV: namely, that virus dissemination in vivo is almost always mediated by cell-to-cell spread [183]. Thus, vaccine immune read-outs from both preclinical studies and clinical trials must explore alternative mechanisms of viral neutralization, including intracellular neutralization. Moreover, there have been recent insights into *non-neutralizing* functions of antibody response to the gB/MF59 vaccine that suggest that there are novel mechanisms of protection at play, in particular antibody-dependent cellular phagocytosis, which may, in fact, represent the major effector of protection (and not neutralizing responses) in previously reported phase II studies [184-186]. In addition to exploring the roles of non-neutralizing functions of antibody response to CMV vaccines, future studies should examine in greater depth the extent to which cell-mediated immunity, particularly that conferred by CD4+ T cells, provides protection for the developing placenta and fetus [187,188]. Subunit vaccines based on envelope glycoproteins must also more thoroughly examine the ability of IgG responses to cross-neutralize clinical isolates [189], which may differ in protein sequence in key epitopes compared to the vaccine strain sequence (and therefore be less able to provide protection against genetically diverse variants of CMV).

Another question that needs to be addressed in discussions among regulatory groups relates to the optimal timing for administration of a CMV vaccine. The Institute of Medicine, when it modeled the implementation of a candidate CMV vaccine in its report published in 2000, "Vaccines for the 21st Century" [190], identified a hypothetical CMV vaccine that would be administered to 12 year olds for the prevention of perinatal disease as a "Level 1" (most favorable) priority. The potential implementation of a CMV vaccine into the current pre-adolescent vaccine schedule, which currently includes vaccines against *Neisseria meningitidis* infection and human papillomavirus infection, is appealing. Attack rates of CMV infection in adolescence are high [191], and a vaccine capable of preventing infection in this age group would reduce the burden of congenital CMV. One drawback of immunizing adolescents may be the issue of waning immunity. If protection conferred by a CMV vaccine is not long-lived, then repeated booster immunizations may be required throughout a woman's childbearing years. Another alternative would be to "target" CMV vaccination for woman planning a pregnancy, toward the goal of completing the vaccination series just prior to establishment of pregnancy. However, the rate of unintended pregnancy in the United States, although in decline, is approximately 45% [192]—with a disproportionately high rate in women in poverty, which in turn is a risk factor for congenital CMV infection [193].

Thus, a strategy of targeted vaccination (even if a highly effective vaccine was available) of women just prior to the onset of their planned pregnancies would most likely fail to prevent the majority of cases of maternal-fetal CMV transmission in the United States.

An alternative to vaccination of women of reproductive age would be to immunize toddlers. Some models have suggested that immunization of toddlers and young children may be preferable to vaccination of adolescents, since this strategy could reduce circulation of CMV in group daycare centers, and result in protection of young women from the high-risk exposures associated with providing care to their toddlers [194,195]. Indeed, the gB/MF59 demonstrates robust immunogenicity in toddlers [196], making this a very intriguing group to target for universal immunization. Boys and girls could be vaccinated, analogous to how rubella vaccine is used, with an overarching goal of conferring herd immunity and protecting against circulation of the virus. CMV is not a highly infectious virus, and as noted above requires close, even intimate contact with body fluids (saliva, breastmilk, urine, blood) for transmission. Accordingly, the so-called "force of infection" is low compared to other common viral infections, suggesting that even a modestly effective vaccine would, if implemented as a universal immunization in early childhood, have a major impact on disease control [197,198].

Lastly, and perhaps most importantly, efforts must be made to increase knowledge and awareness of congenital CMV—among (but by no means limited to) young women, primary care physicians, obstetricians, ancillary healthcare providers, parents, daycare providers, policy-makers, elected officials, and public health investigators. Many state legislatures in the United States have passed bills directing state health departments to provide education about the risks of CMV [199]. Increased knowledge of the magnitude of the problem and the risk to maternal and neonatal health is essential; this in turn will drive demand for a vaccine that can help resolve this unmet need.

Acknowledgments

Support from the March of Dimes Birth Defects Foundation (FY17-849) and the National Institutes of Health (HD079918 and HD098866) is acknowledged.

References

[1] Schleiss MR. Persistent and recurring viral infections: the human herpesviruses. Curr Probl Pediatr Adolesc Health Care 2009;39(1):7–23.

[2] Martí-Carreras J, Maes P. Human cytomegalovirus genomics and transcriptomics through the lens of next-generation sequencing: revision and future challenges. Virus Genes 2019. https://doi.org/10.1007/s11262-018-1627-3. 30604286.

[3] Wilkinson GW, Davison AJ, Tomasec P, Fielding CA, Aicheler R, Murrell I, Seirafian S, Wang EC, Weekes M, Lehner PJ, Wilkie GS, Stanton RJ. Human cytomegalovirus: taking the strain. Med Microbiol Immunol 2015;204(3):273–84.

[4] Rigoutsos I, Novotny J, Huynh T, Chin-Bow ST, Parida L, Platt D, Coleman D, Shenk T. In silico pattern-based analysis of the human cytomegalovirus genome. J Virol 2003;77(7):4326–44.

[5] Gatherer D, Seirafian S, Cunningham C, Holton M, Dargan DJ, Baluchova K, Hector RD, Galbraith J, Herzyk P, Wilkinson GW, Davison AJ. High-resolution human cytomegalovirus transcriptome. Proc Natl Acad Sci U S A 2011;108(49):19755–60.

[6] Ma Y, Wang N, Li M, Gao S, Wang L, Zheng B, Qi Y, Ruan Q. Human CMV transcripts: an overview. Future Microbiol 2012;7(5):577–93.

[7] Schleiss MR. Congenital cytomegalovirus infection: molecular mechanisms mediating viral pathogenesis. Infect Disord Drug Targets 2011;11(5):449–65.

[8] Boeckh M, Geballe AP. Cytomegalovirus: pathogen, paradigm, and puzzle. J Clin Invest 2011;121(5):1673–80.

[9] Cheng S, Caviness K, Buehler J, Smithey M, Nikolich-Žugich J, Goodrum F. Transcriptome-wide characterization of human cytomegalovirus in natural infection and experimental latency. Proc Natl Acad Sci U S A 2017;114(49):E10586–95.

[10] Tirosh O, Cohen Y, Shitrit A, Shani O, Le-Trilling VT, Trilling M, Friedlander G, Tanenbaum M, Stern-Ginossar N. The transcription and translation landscapes during human cytomegalovirus infection reveal novel host-pathogen interactions. PLoS Pathog 2015;11(11):e1005288.

[11] Chen DH, Jiang H, Lee M, Liu F, Zhou ZH. Three-dimensional visualization of tegument/capsid interactions in the intact human cytomegalovirus. Virology 1999;260(1):10–6.

[12] Kern F, Bunde T, Faulhaber N, Kiecker F, Khatamzas E, Rudawski IM, Pruss A, Gratama JW, Volkmer-Engert R, Ewert R, Reinke P, Volk HD, Picker LJ. Cytomegalovirus (CMV) phosphoprotein 65 makes a large contribution to shaping the T cell repertoire in CMV-exposed individuals. J Infect Dis 2002;185(12):1709–16.

[13] Gibson L, Piccinini G, Lilleri D, Revello MG, Wang Z, Markel S, Diamond DJ, Luzuriaga K. Human cytomegalovirus proteins pp65 and immediate early protein 1 are common targets for CD8+ T cell responses in children with congenital or postnatal human cytomegalovirus infection. J Immunol 2004;172(4):2256–64.

[14] Landini MP, La Placa M. Humoral immune response to human cytomegalovirus proteins: a brief review. Comp Immunol Microbiol Infect Dis 1991;14(2):97–105.

[15] Liu YN, Kari B, Gehrz RC. Human immune responses to major human cytomegalovirus glycoprotein complexes. J Virol 1988;62(3):1066–70.

[16] Rasmussen L, Matkin C, Spaete R, Pachl C, Merigan TC. Antibody response to human cytomegalovirus glycoproteins gB and gH after natural infection in humans. J Infect Dis 1991;164(5):835–42.

[17] Shimamura M, Mach M, Britt WJ. Human cytomegalovirus infection elicits a glycoprotein M (gM)/gN-specific virus-neutralizing antibody response. J Virol 2006;80(9):4591–600.

[18] Burkhardt C, Himmelein S, Britt W, Winkler T, Mach M. Glycoprotein N subtypes of human cytomegalovirus induce a strain-specific antibody response during natural infection. J Gen Virol 2009;90(Pt 8):1951–61.

[19] Ha S, Li F, Troutman MC, Freed DC, Tang A, Loughney JW, Wang D, Wang IM, Vlasak J, Nickle DC, Rustandi RR, Hamm M, DePhillips PA, Zhang N, McLellan JS, Zhu H, Adler SP, McVoy MA, An Z, Fu TM. Neutralization of diverse human cytomegalovirus strains conferred by antibodies targeting viral gH/gL/pUL128-131 pentameric complex. J Virol 2017;91(7). https://doi.org/10.1128/JVI.02033-16. pii: e02033-16, 28077654.

[20] Ourahmane A, Cui X, He L, Catron M, Dittmer DP, Al Qaffasaa A, Schleiss MR, Hertel L, McVoy MA. Inclusion of antibodies to cell culture media preserves the integrity of genes encoding RL13 and the pentameric complex components during fibroblast passage of human cytomegalovirus. Viruses 2019;11(3). https://doi.org/10.3390/v11030221. pii: E221, 30841507.

[21] Liu J, Jardetzky TS, Chin AL, Johnson DC, Vanarsdall AL. The human cytomegalovirus trimer and pentamer promote sequential steps in entry into epithelial and endothelial cells at cell surfaces and endosomes. J Virol 2018;92(21). https://doi.org/10.1128/JVI.01336-18. pii: e01336-18 30111564.
[22] Wang D, Shenk T. Human cytomegalovirus virion protein complex required for epithelial and endothelial cell tropism. Proc Natl Acad Sci U S A 2005;102:18153–8.
[23] Scrivano L, Sinzger C, Nitschko H, Koszinowski UH, Adler B. HCMV spread and cell tropism are determined by distinct virus populations. PLoS Pathog 2011;7(1):e1001256.
[24] Nguyen CC, Kamil JP. Pathogen at the gates: human cytomegalovirus entry and cell tropism. Viruses 2018;10(12). https://doi.org/10.3390/v10120704. pii: E704, 30544948.
[25] Kinzler ER, Compton T. Characterization of human cytomegalovirus glycoprotein induced cell-cell fusion. J Virol 2005;79(12):7827–37.
[26] Zhou M, Lanchy JM, Ryckman BJ. Human cytomegalovirus gH/gL/gO promotes the fusion step of entry into all cell types, whereas gH/gL/UL128-131 broadens virus tropism through a distinct mechanism. J Virol 2015;89(17):8999–9009.
[27] Malito E, Chandramouli S, Carfi A. From recognition to execution-the HCMV Pentamer from receptor binding to fusion triggering. Curr Opin Virol 2018;31:43–51.
[28] Vanarsdall AL, Chin AL, Liu J, Jardetzky TS, Mudd JO, Orloff SL, Streblow D, Mussi-Pinhata MM, Yamamoto AY, Duarte G, Britt WJ, Johnson DC. HCMV trimer- and pentamer-specific antibodies synergize for virus neutralization but do not correlate with congenital transmission. Proc Natl Acad Sci U S A 2019;116(9):3728–33.
[29] Schleiss MR. Acquisition of human cytomegalovirus infection in infants via breast milk: natural immunization or cause for concern? Rev Med Virol 2006;16(2):73–82.
[30] Amin MM, Bialek SR, Dollard SC, Wang C. Urinary cytomegalovirus shedding in the United States: The National Health and Nutrition Examination Surveys, 1999-2004. Clin Infect Dis 2018;67(4):587–92.
[31] Mayer BT, Matrajt L, Casper C, Krantz EM, Corey L, Wald A, Gantt S, Schiffer JT. Dynamics of persistent oral cytomegalovirus shedding during primary infection in Ugandan infants. J Infect Dis 2016;214(11):1735–43.
[32] Amin MM, Stowell JD, Hendley W, Garcia P, Schmid DS, Cannon MJ, Dollard SC. CMV on surfaces in homes with young children: results of PCR and viral culture testing. BMC Infect Dis 2018;18(1):391.
[33] Marshall BC, Adler SP. The frequency of pregnancy and exposure to cytomegalovirus infections among women with a young child in day care. Am J Obstet Gynecol 2009;200(2):163.e1-5.
[34] Noyola DE, Valdez-López BH, Hernández-Salinas AE, Santos-Díaz MA, Noyola-Frías MA, Reyes-Macías JF, Martínez-Martínez LG. Cytomegalovirus excretion in children attending day-care centers. Arch Med Res 2005;36(5):590–3.
[35] Pass RF, Hutto SC, Reynolds DW, Polhill RB. Increased frequency of cytomegalovirus infection in children in group day care. Pediatrics 1984;74(1):121–6.
[36] Pass RF, Hutto C, Ricks R, Cloud GA. Increased rate of cytomegalovirus infection among parents of children attending day-care centers. N Engl J Med 1986;314(22):1414–8.
[37] Yarrish RL, Wormser GP, Bittker SJ, Aron-Hott L, Cabello F, Huang ES. The febrile father with a cytomegalovirus infection. A family affair. Postgrad Med 1989;85(1):251–4.
[38] Adler SP. Molecular epidemiology of cytomegalovirus: viral transmission among children attending a day care center, their parents, and caretakers. J Pediatr 1988;112(3):366–72.
[39] Hendrie CA, Brewer G. Kissing as an evolutionary adaptation to protect against human cytomegalovirus-like teratogenesis. Med Hypotheses 2010;74(2):222–4.
[40] Fowler KB, Pass RF. Risk factors for congenital cytomegalovirus infection in the offspring of young women: exposure to young children and recent onset of sexual activity. Pediatrics 2006;118:e286–92.

[41] Staras SA, Flanders WD, Dollard SC, Pass RF, McGowan Jr JE, Cannon MJ. Influence of sexual activity on cytomegalovirus seroprevalence in the United States, 1988-1994. Sex Transm Dis 2008;35:472–9.

[42] Furui Y, Yamagishi N, Morioka I, Taira R, Nishida K, Ohyama S, Matsumoto H, Nakamachi Y, Hasegawa T, Matsubayashi K, Nagai T, Satake M. Sequence analyses of variable cytomegalovirus genes for distinction between breast milk- and transfusion transmitted infections in very-low-birth-weight infants. Transfusion 2018;58(12):2894–902.

[43] Ziemann M, Thiele T. Transfusion-transmitted CMV infection—current knowledge and future perspectives. Transfus Med 2017;27(4):238–48.

[44] Brodin P, Jojic V, Gao T, Bhattacharya S, Angel CJ, Furman D, Shen-Orr S, Dekker CL, Swan GE, Butte AJ, Maecker HT, Davis MM. Variation in the human immune system is largely driven by non-heritable influences. Cell 2015;160(1–2):37–47.

[45] Furman D, Jojic V, Sharma S, Shen-Orr SS, Angel CJ, Onengut-Gumuscu S, Kidd BA, Maecker HT, Concannon P, Dekker CL, Thomas PG, Davis MM. Cytomegalovirus infection enhances the immune response to influenza. Sci Transl Med 2015;7(281):281ra43.

[46] Weller TH. The cytomegaloviruses: ubiquitous agents with protean clinical manifestations. I.II. N Engl J Med 1971;285(4):203–14. 285(5):267–74.

[47] Bardanzellu F, Fanos V, Reali A. Human breast milk-acquired cytomegalovirus infection: certainties, doubts and perspectives. Curr Pediatr Rev 2018. https://doi.org/10.2174/1573396315666181126105812. 30474531.

[48] Tengsupakul S, Birge ND, Bendel CM, Reed RC, Bloom BA, Hernandez N, Schleiss MR. Asymptomatic DNAemia heralds CMV-associated NEC: case report, review, and rationale for preemption. Pediatrics 2013;132(5):e1428–34.

[49] Hamprecht K, Goelz R. Postnatal cytomegalovirus infection through human milk in preterm infants: transmission, clinical presentation, and prevention. Clin Perinatol 2017;44(1):121–30.

[50] Adler SP, Marshall B. Cytomegalovirus infections. Pediatr Rev 2007;28(3):92–100.

[51] Chomel JJ, Allard JP, Floret D, Honneger D, David L, Lina B, Aymard M. Role of cytomegalovirus infection in the incidence of viral acute respiratory infections in children attending day-care centers. Eur J Clin Microbiol Infect Dis 2001;20(3):167–72.

[52] Cohen JI, Corey GR. Cytomegalovirus infection in the normal host. Medicine (Baltimore) 1985;64:100–14.

[53] Nolan N, Halai UA, Regunath H, Smith L, Rojas-Moreno C, Salzer W. Primary cytomegalovirus infection in immunocompetent adults in the United States—a case series. IDCases 2017;10:123–6.

[54] Orasch C, Conen A. Severe primary cytomegalovirus infection in the immunocompetent adult patient: a case series. Scand J Infect Dis 2012;44(12):987–91.

[55] Ishii T, Sasaki Y, Maeda T, Komatsu F, Suzuki T, Urita Y. Clinical differentiation of infectious mononucleosis that is caused by Epstein-Barr virus or cytomegalovirus: a single-center case-control study in Japan. J Infect Chemother 2019. https://doi.org/10.1016/j.jiac.2019.01.012. 30773381.

[56] Klemola E. Hypersensitivity reactions to ampicillin in cytomegalovirus mononucleosis. Scand J Infect Dis 1970;2(1):29–31.

[57] Yinon Y, Farine D, Yudin MH. Screening, diagnosis, and management of cytomegalovirus infection in pregnancy. Obstet Gynecol Surv 2010;65(11):736–43.

[58] Emery VC, Lazzarotto T. Cytomegalovirus in pregnancy and the neonate. F1000Res 2017;6:138.

[59] Tanimura K, Yamada H. Maternal and neonatal screening methods for congenital cytomegalovirus infection. J Obstet Gynaecol Res 2019;45(3):514–21.

[60] Tanimura K, Tairaku S, Morioka I, Ozaki K, Nagamata S, Morizane M, Deguchi M, Ebina Y, Minematsu T, Yamada H. Universal screening with use of immunoglobulin G avidity for congenital cytomegalovirus infection. Clin Infect Dis 2017;65(10):1652–8.

[61] Boppana SB, Rivera LB, Fowler KB, Mach M, Britt WJ. Intrauterine transmission of cytomegalovirus to infants of women with preconceptional immunity. N Engl J Med 2001;344(18):1366–71.
[62] Yamamoto AY, Mussi-Pinhata MM, Boppana SB, Novak Z, Wagatsuma VM, Oliveira Pde F, Duarte G, Britt WJ. Human cytomegalovirus reinfection is associated with intrauterine transmission in a highly cytomegalovirus-immune maternal population. Am J Obstet Gynecol 2010;202(3):297.e1-8.
[63] Ross SA, Arora N, Novak Z, Fowler KB, Britt WJ, Boppana SB. Cytomegalovirus reinfections in healthy seroimmune women. J Infect Dis 2010;201(3):386–9.
[64] Boucoiran I, Mayer BT, Krantz EM, Marchant A, Pati S, Boppana S, Wald A, Corey L, Casper C, Schiffer JT, Gantt S. Nonprimary maternal cytomegalovirus infection after viral shedding in infants. Pediatr Infect Dis J 2018;37(7):627–31.
[65] Davis NL, King CC, Kourtis AP. Cytomegalovirus infection in pregnancy. Birth Defects Res 2017;109(5):336–46.
[66] Stagno S, Pass RF, Cloud G, Britt WJ, Henderson RE, Walton PD, Veren DA, Page F, Alford CA. Primary cytomegalovirus infection in pregnancy. Incidence, transmission to fetus, and clinical outcome. JAMA 1986;256(14):1904–8.
[67] Bonalumi S, Trapanese A, Santamaria A, D'Emidio L, Mobili L. Cytomegalovirus infection in pregnancy: review of the literature. J Prenat Med 2011;5(1):1–8.
[68] Nyholm JL, Schleiss MR. Prevention of maternal cytomegalovirus infection: current status and future prospects. Int J Womens Health 2010;2:23–35.
[69] Slyker JA. Cytomegalovirus and paediatric HIV infection. J Virus Erad 2016;2(4):208–14.
[70] Falconer O, Newell ML, Jones CE. The effect of human immunodeficiency virus and cytomegalovirus infection on infant responses to vaccines: a review. Front Immunol 2018;9:328.
[71] Razonable RR, Humar A. Cytomegalovirus in solid organ transplant recipients—guidelines of the American Society of Transplantation Infectious Disease Community of Practice. Clin Transplant 2019;e13512. https://doi.org/10.1111/ctr.13512. 30817026.
[72] Danziger-Isakov L, Englund J, Green M, Posfay-Barbe KM, Zerr DM. Cytomegalovirus in pediatric hematopoietic stem cell transplantation: a case-based panel discussion of current challenges. J Pediatric Infect Dis Soc 2018;7(suppl_2):S72–4.
[73] Swanson EC, Schleiss MR. Congenital cytomegalovirus infection: new prospects for prevention and therapy. Pediatr Clin North Am 2013;60(2):335–49.
[74] Korndewal MJ, Weltevrede M, van den Akker-van Marle ME, Oudesluys-Murphy AM, de Melker HE, Vossen ACTM. Healthcare costs attributable to congenital cytomegalovirus infection. Arch Dis Child 2018;103(5):452–7.
[75] Gantt S, Dionne F, Kozak FK, Goshen O, Goldfarb DM, Park AH, Boppana SB, Fowler K. Cost-effectiveness of universal and targeted newborn screening for congenital cytomegalovirus infection. JAMA Pediatr 2016;170(12):1173–80.
[76] Retzler J, Hex N, Bartlett C, Webb A, Wood S, Star C, Griffiths P, Jones CE. Economic cost of congenital CMV in the UK. Arch Dis Child 2018. https://doi.org/10.1136/archdischild-2018-316010. 30472664.
[77] Williams EJ, Gray J, Luck S, Atkinson C, Embleton ND, Kadambari S, Davis A, Griffiths P, Sharland M, Berrington JE, Clark JE. First estimates of the potential cost and cost saving of protecting childhood hearing from damage caused by congenital CMV infection. Arch Dis Child Fetal Neonatal Ed 2015;100(6):F501–6.
[78] Dollard SC, Grosse SD, Ross DS. New estimates of the prevalence of neurological and sensory sequelae and mortality associated with congenital cytomegalovirus infection. Rev Med Virol 2007;17:355–63.
[79] Faure-Bardon V, Magny JF, Parodi M, Couderc S, Garcia P, Maillotte AM, Benard M, Pinquier D, Astruc D, Patural H, Pladys P, Parat S, Guillois B, Garenne A, Bussières L, Guilleminot T, Stirnemann J, Ghout I, Ville Y, Leruez-Ville M. Sequelae of congenital cytomegalovirus (cCMV) following maternal primary infection are limited to those acquired in the first trimester of pregnancy. Clin Infect Dis 2018. https://doi.org/10.1093/cid/ciy1128. 30596974.

[80] Lipitz S, Yinon Y, Malinger G, Yagel S, Levit L, Hoffman C, Rantzer R, Weisz B. Risk of cytomegalovirus-associated sequelae in relation to time of infection and findings on prenatal imaging. Ultrasound Obstet Gynecol 2013;41(5):508–14.
[81] Pass RF, Fowler KB, Boppana SB, Britt WJ, Stagno S. Congenital cytomegalovirus infection following first trimester maternal infection: symptoms at birth and outcome. J Clin Virol 2006;35(2):216–20.
[82] Fowler KB, Boppana SB. Congenital cytomegalovirus (CMV) infection and hearing deficit. J Clin Virol 2006;35:226–31.
[83] Goderis J, De Leenheer E, Smets K, Van Hoecke H, Keymeulen A, Dhooge I. Hearing loss and congenital CMV infection: a systematic review. Pediatrics 2014;134:972–82.
[84] Goderis J, Keymeulen A, Smets K, Van Hoecke H, De Leenheer E, Boudewyns A, Desloovere C, Kuhweide R, Muylle M, Royackers L, Schatteman I, Dhooge I. Hearing in children with congenital cytomegalovirus infection: results of a longitudinal study. J Pediatr 2016;172:110–5.
[85] Dahle AJ, Fowler KB, Wright JD, Boppana SB, Britt WJ, Pass RF. Longitudinal investigation of hearing disorders in children with congenital cytomegalovirus. J Am Acad Audiol 2000;11:283–90.
[86] Fowler KB, McCollister FP, Dahle AJ, Boppana S, Britt WJ, Pass RF. Progressive and fluctuating sensorineural hearing loss in children with asymptomatic congenital cytomegalovirus infection. J Pediatr 1997;130:624–30.
[87] Fowler KB. Congenital cytomegalovirus infection: audiologic outcome. Clin Infect Dis 2013;57:S182–4.
[88] Zuhair M, Smit GSA, Wallis G, Jabbar F, Smith C, Devleesschauwer B, Griffiths P. Estimation of the worldwide seroprevalence of cytomegalovirus: a systematic review and meta-analysis. Rev Med Virol 2019;e2034, 30706584.
[89] Bate SL, Dollard SC, Cannon MJ. Cytomegalovirus seroprevalence in the United States: the national health and nutrition examination surveys, 1988-2004. Clin Infect Dis 2010;50(11):1439–47.
[90] de Vries JJ, van Zwet EW, Dekker FW, Kroes AC, Verkerk PH, Vossen AC. The apparent paradox of maternal seropositivity as a risk factor for congenital cytomegalovirus infection: a population-based prediction model. Rev Med Virol 2013;23(4):241–9.
[91] Britt W. Controversies in the natural history of congenital human cytomegalovirus infection: the paradox of infection and disease in offspring of women with immunity prior to pregnancy. Med Microbiol Immunol 2015;204(3):263–71.
[92] Britt WJ. Maternal immunity and the natural history of congenital human cytomegalovirus infection. Viruses 2018;10(8). https://doi.org/10.3390/v10080405. 30081449.
[93] Britt WJ. Congenital human cytomegalovirus infection and the enigma of maternal immunity. J Virol 2017;91(15). https://doi.org/10.1128/JVI.02392-16. 28490582.
[94] Barbosa NG, Yamamoto AY, Duarte G, Aragon DC, Fowler KB, Boppana S, Britt WJ, Mussi-Pinhata MM. Cytomegalovirus shedding in seropositive pregnant women from a high-seroprevalence population: the Brazilian cytomegalovirus hearing and maternal secondary infection study. Clin Infect Dis 2018;67(5):743–50.
[95] Mussi-Pinhata MM, Yamamoto AY, Aragon DC, Duarte G, Fowler KB, Boppana S, Britt WJ. Seroconversion for cytomegalovirus infection during pregnancy and fetal infection in a highly seropositive population: "The BraCHS Study". J Infect Dis 2018;218(8):1200–4.
[96] Stagno S, Reynolds DW, Lakeman A, Charamella LJ, Alford CA. Congenital cytomegalovirus infection: consecutive occurrence due to viruses with similar antigenic compositions. Pediatrics 1973;52(6):788–94.
[97] Nagamori T, Koyano S, Inoue N, Yamada H, Oshima M, Minematsu T, Fujieda K. Single cytomegalovirus strain associated with fetal loss and then congenital infection of a subsequent child born to the same mother. J Clin Virol 2010;49(2):134–6.
[98] Manicklal S, Emery VC, Lazzarotto T, Boppana SB, Gupta RK. The "silent" global burden of congenital cytomegalovirus. Clin Microbiol Rev 2013;26(1):86–102.

[99] Mussi-Pinhata MM, Yamamoto AY, Moura-Britto RM, de Lima Isaac M, de Carvalho e Oliveira PF, Boppana S, Britt WJ. Birth prevalence and natural history of congenital cytomegalovirus infection in a highly seroimmune population. Clin Infect Dis 2009;49:522–8.
[100] Boppana SB, Fowler KB, Britt WJ, Stagno S, Pass RF. Symptomatic congenital cytomegalovirus infection in infants born to mothers with preexisting immunity to cytomegalovirus. Symptomatic congenital cytomegalovirus infection in infants born to mothers with preexisting immunity to cytomegalovirus. Pediatrics 1999;104(1 Pt 1):55–60.
[101] Wang C, Zhang X, Bialek S, Cannon MJ. Attribution of congenital cytomegalovirus infection to primary versus non-primary maternal infection. Clin Infect Dis 2011;52:e11–3.
[102] Kenneson A, Cannon MJ. Review and meta-analysis of the epidemiology of congenital cytomegalovirus (CMV) infection. Rev Med Virol 2007;17:253–76.
[103] Boppana SB, Ross SA, Shimamura M, Palmer AL, Ahmed A, Michaels MG, Sánchez PJ, Bernstein DI, Tolan Jr RW, Novak Z, Chowdhury N, Britt WJ, Fowler KB. Saliva polymerase-chain-reaction assay for cytomegalovirus screening in newborns. N Engl J Med 2011;364:2111–8.
[104] Fowler KB, Ross SA, Shimamura M, Ahmed A, Palmer AL, Michaels MG, Bernstein DI, Sánchez PJ, Feja KN, Stewart A, Boppana S. Racial and ethnic differences in the prevalence of congenital cytomegalovirus infection. J Pediatr 2018;200:196–201.
[105] Bratcher DF, Bourne N, Bravo FJ, Schleiss MR, Slaoui M, Myers MG, Bernstein DI. Effect of passive antibody on congenital cytomegalovirus infection in Guinea pigs. J Infect Dis 1995;172:944–50.
[106] Chatterjee A, Harrison CJ, Britt WJ, Bewtra C. Modification of maternal and congenital cytomegalovirus infection by anti-glycoprotein b antibody transfer in guinea pigs. J Infect Dis 2001;183:1547–53.
[107] Auerbach MR, Yan D, Vij R, Hongo JA, Nakamura G, Vernes JM, Meng YG, Lein S, Chan P, Ross J, Carano R, Deng R, Lewin-Koh N, Xu M, Feierbach B. A neutralizing anti-gH/gL monoclonal antibody is protective in the guinea pig model of congenital CMV infection. PLoS Pathog 2014;10(4):e1004060.
[108] Snydman DR. Cytomegalovirus immunoglobulins in the prevention and treatment of cytomegalovirus disease. Rev Infect Dis 1990;12(Suppl 7):S839–48.
[109] Snydman DR. Prevention of cytomegalovirus-associated diseases with immunoglobulin. Transplant Proc 1991;23(Suppl 3):131–5.
[110] Snydman DR, Werner BG, Dougherty NN, Griffith J, Rubin RH, Dienstag JL, Rohrer RH, Freeman R, Jenkins R, Lewis WD, Hammer S, O'Rourke E, Grady GF, Fawaz K, Kaplan MM, Hoffman MA, Katz AT, Doran M, Boston Center for Liver Transplantation CMVIG Study Group. Cytomegalovirus immune globulin prophylaxis in liver transplantation. A randomized, double-blind, placebo-controlled trial. Ann Intern Med 1993;119:984–91.
[111] Meyers JD. Critical evaluation of agents used in the treatment and prevention of cytomegalovirus infection in immunocompromised patients. Transplant Proc 1991; 23:139–42. discussion 142–3.
[112] Yeager AS, Grumet FC, Hafleigh EB, et al. Prevention of transfusion-acquired cytomegalovirus infections in newborn infants. J Pediatr 1981;98:281–7.
[113] Fouts AE, Chan P, Stephan JP, Vandlen R, Feierbach B. Antibodies against the gH/gL/UL128/UL130/UL131 complex comprise the majority of the anti-cytomegalovirus (anti-CMV) neutralizing antibody response in CMV hyperimmune globulin. J Virol 2012;86(13):7444–7.
[114] Jückstock J, Rothenburger M, Friese K, Traunmüller F. Passive immunization against congenital cytomegalovirus infection: current state of knowledge. Pharmacology 2015;95:209–17.
[115] Visentin S, Manara R, Milanese L, Da Roit A, Forner G, Salviato E, Citton V, Magno FM, Orzan E, Morando C, Cusinato R, Mengoli C, Palu G, Ermani M, Rinaldi R, Cosmi E,

Gussetti N. Early primary cytomegalovirus infection in pregnancy: maternal hyperimmunoglobulin therapy improves outcomes among infants at 1 year of age. Clin Infect Dis 2012;55(4):497–503.

[116] Buxmann H, Stackelberg OM, Schlößer RL, Enders G, Gonser M, Meyer-Wittkopf M, Hamprecht K, Enders M. Use of cytomegalovirus hyperimmunoglobulin for prevention of congenital cytomegalovirus disease: a retrospective analysis. J Perinat Med 2012;40(4):439–46.

[117] Nigro G, Adler SP, La Torre R, Best AM, Congenital Cytomegalovirus Collaborating Group. Passive immunization during pregnancy for congenital cytomegalovirus infection. N Engl J Med 2005;353(13):1350–62.

[118] Nigro G, Adler SP, Parruti G, Anceschi MM, Coclite E, Pezone I, Di Renzo GC. Immunoglobulin therapy of fetal cytomegalovirus infection occurring in the first half of pregnancy—a case-control study of the outcome in children. J Infect Dis 2012;205(2):215–27.

[119] Maidji E, Nigro G, Tabata T, McDonagh S, Nozawa N, Shiboski S, Muci S, Anceschi MM, Aziz N, Adler SP, Pereira L. Antibody treatment promotes compensation for human cytomegalovirus-induced pathogenesis and a hypoxia-like condition in placentas with congenital infection. Am J Pathol 2010;177(3):1298–310.

[120] Revello MG, Lazzarotto T, Guerra B, Spinillo A, Ferrazzi E, Kustermann A, Guaschino S, Vergani P, Todros T, Frusca T, Arossa A, Furione M, Rognoni V, Rizzo N, Gabrielli L, Klersy C, Gerna G, CHIP Study Group. A randomized trial of hyperimmune globulin to prevent congenital cytomegalovirus. N Engl J Med 2014;370:1316–26.

[121] Revello MG, Fornara C, Arossa A, Zelini P, Lilleri D. Role of human cytomegalovirus(HCMV)-specific antibody in HCMV-infected pregnant women. Early Hum Dev 2014;90(Suppl 1):S32–4.

[122] Gabrielli L, Bonasoni MP, Foschini MP, Silini EM, Spinillo A, Revello MG, Chiereghin A, Piccirilli G, Petrisli E, Turello G, Simonazzi G, Gibertoni D, Lazzarotto T. Histological analysis of term placentas from hyperimmune globulin-treated and untreated mothers with primary cytomegalovirus infection. Fetal Diagn Ther 2019;45(2):111–7.

[123] Johnson J, Anderson B. Screening, prevention, and treatment of congenital cytomegalovirus. Obstet Gynecol Clin North Am 2014;41:593–9.

[124] Kauvar LM, Liu K, Park M, DeChene N, Stephenson R, Tenorio E, Ellsworth SL, Tabata T, Petitt M, Tsuge M, Fang-Hoover J, Adler SP, Cui X, McVoy MA, Pereira L. A high- affinity native human antibody neutralizes human cytomegalovirus infection of diverse cell types. Antimicrob Agents Chemother 2015;59:1558–68.

[125] Theraclone Sciences, Inc. Safety study of human anti-cytomegalovirus monoclonal antibody [Internet]. In: ClinicalTrials.gov [Internet]. Bethesda, MD: National Library of Medicine (US); 2014. Available from http://clinicaltrials.gov/ct2/show/NCT01594437.

[126] McVoy MM, Tenorio E, Kauvar LM. A native human monoclonal antibody targeting HCMV gB (AD-2 site I). Int J Mol Sci 2018;19(12). https://doi.org/10.3390/ijms19123982. 30544903.

[127] Schleiss MR. Congenital cytomegalovirus infection: improved understanding of maternal immune responses that reduce the risk of transplacental transmission. Clin Infect Dis 2017;65(10):1666–9.

[128] Elek SD, Stern H. Development of a vaccine against mental retardation caused by cytomegalovirus infection in utero. Lancet 1974;1(7845):1–5.

[129] Murphy E, Yu D, Grimwood J, Schmutz J, Dickson M, Jarvis MA, Hahn G, Nelson JA, Myers RM, Shenk TE. Coding potential of laboratory and clinical strains of human cytomegalovirus. Proc Natl Acad Sci U S A 2003;100:14976–81.

[130] Murphy E, Shenk T. Human cytomegalovirus genome. Curr Top Microbiol Immunol 2008;325:1–19.

[131] Jacobson MA, Sinclair E, Bredt B, Agrillo L, Black D, Epling CL, Carvidi A, Ho T, Bains R, Girling V, Adler SP. Safety and immunogenicity of Towne cytomegalovirus vaccine with or without adjuvant recombinant interleukin-12. Vaccine 2006;24:5311–9.
[132] Jacobson MA, Adler SP, Sinclair E, Black D, Smith A, Chu A, Moss RB, Wloch MK. A CMV DNA vaccine primes for memory immune responses to live-attenuated CMV (Towne strain). Vaccine 2009;27:1540–8.
[133] Cha TA, Tom E, Kemble GW, Duke GM, Mocarski ES, Spaete RR. Human cytomegalovirus clinical isolates carry at least 19 genes not found in laboratory strains. J Virol 1996;70:78–83.
[134] Bradley AJ, Lurain NS, Ghazal P, Trivedi U, Cunningham C, Baluchova K, Gatherer D, Wilkinson GW, Dargan DJ, Davison AJ. High-throughput sequence analysis of variants of human cytomegalovirus strains Towne and AD169. J Gen Virol 2009;90:2375–80.
[135] Goodrum F, Reeves M, Sinclair J, High K, Shenk T. Human cytomegalovirus sequences expressed in latently infected individuals promote a latent infection in vitro. Blood 2007;110:937–45.
[136] Assaf BT, Mansfield KG, Strelow L, Westmoreland SV, Barry PA, Kaur A. Limited dissemination and shedding of the UL128 complex-intact, UL/b′-defective rhesus cytomegalovirus strain 180.92. J Virol 2014;88:9310–20.
[137] Dolan A, Cunningham C, Hector RD, Hassan-Walker AF, Lee L, Addison C, Dargan DJ, McGeoch DJ, Gatherer D, Emery VC, Griffiths PD, Sinzger C, McSharry BP, Wilkinson GW, Davison AJ. Genetic content of wild-type human cytomegalovirus. J Gen Virol 2004;85:1301–12.
[138] Fu TM, An Z, Wang D. Progress on pursuit of human cytomegalovirus vaccines for prevention of congenital infection and disease. Vaccine 2014;32:2525–33.
[139] Wang D, Freed DC, He X, Li F, Tang A, Cox KS, Dubey SA, Cole S, Medi MB, Liu Y, Xu J, Zhang ZQ, Finnefrock AC, Song L, Espeseth AS, Shiver JW, Casimiro DR, Fu TM. A replication-defective human cytomegalovirus vaccine for prevention of congenital infection. Sci Transl Med 2016;8:362ra145.
[140] Banaszynski LA, Chen LC, Maynard-Smith LA, Ooi AG, Wandless TJ. A rapid, reversible, and tunable method to regulate protein function in living cells using synthetic small molecules. Cell 2006;126:995–1004.
[141] Glass M, Busche A, Wagner K, Messerle M, Borst EM. Conditional and reversible disruption of essential herpesvirus proteins. Nat Methods 2009;6:577–9.
[142] Borst EM, Kleine-Albers J, Gabaev I, Babic M, Wagner K, Binz A, Degenhardt I, Kalesse M, Jonjic S, Bauerfeind R, Messerle M. The human cytomegalovirus UL51 protein is essential for viral genome cleavage-packaging and interacts with the terminase subunits pUL56 and pUL89. J Virol 2013;87:1720–32.
[143] Pass RF, Duliegè AM, Boppana S, Sekulovich R, Percell S, Britt W, Burke RL. A subunit cytomegalovirus vaccine based on recombinant envelope glycoprotein B and a new adjuvant. J Infect Dis 1999;180:970–5.
[144] Frey SE, Harrison C, Pass RF, Yang E, Boken D, Sekulovich RE, Percell S, Izu AE, Hirabayashi S, Burke RL, Duliège AM. Effects of antigen dose and immunization regimens on antibody responses to a cytomegalovirus glycoprotein B subunit vaccine. J Infect Dis 1999;180(5):1700–3.
[145] Pass RF, Zhang C, Evans A, Simpson T, Andrews W, Huang ML, Corey L, Hill J, Davis E, Flanigan C, Cloud G. Vaccine prevention of maternal cytomegalovirus infection. N Engl J Med 2009;360:1191–9.
[146] Sabbaj S, Pass RF, Goepfert PA, Pichon S. Glycoprotein B vaccine is capable of boosting both antibody and CD4 T-cell responses to cytomegalovirus in chronically infected women. J Infect Dis 2011;203:1534–41.
[147] Anderholm KM, Bierle CJ, Schleiss MR. Cytomegalovirus vaccines: current status and future prospects. Drugs 2016;76(17):1625–45.

[148] Kirchmeier M, Fluckiger AC, Soare C, Bozic J, Ontsouka B, Ahmed T, Diress A, Pereira L, Schodel F, Plotkin S, Dalba C, Klatzmann D, Anderson DE. Enveloped virus-like particle expression of human cytomegalovirus glycoprotein B antigen induces antibodies with potent and broad neutralizing activity. Clin Vaccine Immunol 2014;21:174–80.
[149] Berencsi K, Gyulai Z, Gönczöl E, Pincus S, Cox WI, Michelson S, Kari L, Meric C, Cadoz M, Zahradnik J, Starr S, Plotkin S. A canarypox vector-expressing cytomegalovirus (CMV) phosphoprotein 65 induces long-lasting cytotoxic T cell responses in human CMV-seronegative subjects. J Infect Dis 2001;183:1171–9.
[150] Bernstein DI, Schleiss MR, Berencsi K, Gonczol E, Dickey M, Khoury P, Cadoz M, Meric C, Zahradnik J, Dullege AM, Plotkin S. Effect of previous or simultaneous immunization with canarypox expressing cytomegalovirus (CMV) glycoprotein B (gB) on response to subunit gB vaccine plus MF59 in healthy CMV-seronegative adults. J Infect Dis 2002;185:686–90.
[151] Reap EA, Morris J, Dryga SA, Maughan M, Talarico T, Esch RE, Negri S, Burnett B, Graham A, Olmsted RA, Chulay JD. Development and preclinical evaluation of an alphavirus replicon particle vaccine for cytomegalovirus. Vaccine 2007;25:7441–9.
[152] Reap EA, Dryga SA, Morris J, Rivers B, Norberg PK, Olmsted RA, Chulay JD. Cellular and humoral immune responses to alphavirus replicon vaccines expressing cytomegalovirus pp65, IE1, and gB proteins. Clin Vaccine Immunol 2007;14:748–55.
[153] Schleiss MR, Lacayo JC, Belkaid Y, McGregor A, Stroup G, Rayner J, Alterson K, Chulay JD, Smith JF. Preconceptual administration of an alphavirus replicon UL83 (pp65 homolog) vaccine induces humoral and cellular immunity and improves pregnancy outcome in the guinea pig model of congenital cytomegalovirus infection. J Infect Dis 2007;195(6):789–98.
[154] Bernstein DI, Reap EA, Katen K, Watson A, Smith K, Norberg P, Olmsted RA, Hoeper A, Morris J, Negri S, Maughan MF, Chulay JD. Randomized, double-blind, phase 1 trial of an alphavirus replicon vaccine for cytomegalovirus in CMV seronegative adult volunteers. Vaccine 2009;28:484–93.
[155] Flatz L, Hegazy AN, Bergthaler A, Verschoor A, Claus C, Fernandez M, Gattinoni L, Johnson S, Kreppel F, Kochanek S, Broek M, Radbruch A, Lévy F, Lambert PH, Siegrist CA, Restifo NP, Lohning M, Ochsenbein AF, Nabel GJ, Pinschewer DD. Development of replication-defective lymphocytic choriomeningitis virus vectors for the induction of potent CD8+ T cell immunity. Nat Med 2010;16:339–45.
[156] Ring S, Flatz L. Generation of lymphocytic choriomeningitis virus based vaccine vectors. Methods Mol Biol 2016;1404:351–64.
[157] Schleiss MR, Berka U, Watson E, Aistleithner M, Kiefmann B, Mangeat B, Swanson EC, Gillis PA, Hernandez-Alvarado N, Fernández-Alarcón C, Zabeli JC, Pinschewer DD, Lilja AE, Schwendinger M, Guirakhoo F, Monath TP, Orlinger KK. Additive protection against congenital cytomegalovirus conferred by combined glycoprotein B/pp65 vaccination using a lymphocytic choriomeningitis virus vector. Clin Vaccine Immunol 2017;24(1). pii: e00300-16.
[158] Wussow F, Chiuppesi F, Martinez J, Campo J, Johnson E, Flechsig C, Newell M, Tran E, Ortiz J, La Rosa C, Hermann A, Longmate J, Chakraborty R, Barry P, Diamond D. Human cytomegalovirus vaccine based on the envelope gH/gL pentamer complex. PLoS Pathog 2014;10:e1004524.
[159] Chiuppesi F, Wussow F, Johnson E, Bian C, Zhuo M, Rajakumar A, Barry PA, Britt WJ, Chakraborty R, Diamond DJ. Vaccine-derived neutralizing antibodies to the human cytomegalovirus gH/gL pentamer potently block primary cytotrophoblast infection. J Virol 2015;89:11884–98.
[160] La Rosa C, Longmate J, Martinez J, Zhou Q, Kaltcheva TI, Tsai W, Drake J, Carroll M, Wussow F, Chiuppesi F, Hardwick N, Dadwal S, Aldoss I, Nakamura R, Zaia JA,

Diamond DJ. MVA vaccine encoding CMV antigens safely induces durable expansion of CMV-specific T cells in healthy adults. Blood 2017;160(129):114–25.
[161] Selinsky C, Luke C, Wloch M, Geall A, Hermanson G, Kaslow D, Evans T. A DNA based vaccine for the prevention of human cytomegalovirus-associated diseases. Hum Vaccin 2005;1:16–23.
[162] Wloch MK, Smith LR, Boutsaboualoy S, Reyes L, Han C, Kehler J, Smith HD, Selk L, Nakamura R, Brown JM, Marbury T, Wald A, Rolland A, Kaslow D, Evans T, Boeck M. Safety and immunogenicity of a bivalent cytomegalovirus DNA vaccine in healthy adult subjects. J Infect Dis 2008;197:1634–42.
[163] Kharfan-Dabaja MA, Boeckh M, Wilck MB, Langston AA, Chu AH, Wloch MK, Guterwill DF, Smith LR, Rolland AP, Kenney RT. A novel therapeutic cytomegalovirus DNA vaccine in allogeneic haemopoietic stem-cell transplantation: a randomised, double- blind, placebo controlled, phase 2 trial. Lancet Infect Dis 2012;12:290–9.
[164] Mori T, Kanda Y, Takenaka K, Okamoto S, Kato J, Kanda J, Yoshimoto G, Gondo H, Doi S, Inaba M, Kodera Y. Safety of ASP0113, a cytomegalovirus DNA vaccine, in recipients undergoing allogeneic hematopoietic cell transplantation: an open-label phase 2 trial. Int J Hematol 2017;105:206–12.
[165] Bernstein DI. Vaccines for cytomegalovirus. Infect Disord Drug Targets 2011;11(5): 514–25.
[166] Vincenti F, Budde K, Merville P, Shihab F, Ram Peddi V, Shah M, Wyburn K, Cassuto-Viguier E, Weidemann A, Lee M, Flegel T, Erdman J, Wang X, Lademacher C. A randomized, phase 2 study of ASP0113, a DNA-based vaccine, for the prevention of CMV in CMV-seronegative kidney transplant recipients receiving a kidney from a CMV seropositive donor. Am J Transplant 2018;18(12):2945–54.
[167] John S, Yuzhakov O, Woods A, Deterling J, Hassett K, Shaw CA, Ciaramella G. Multiantigenic human cytomegalovirus mRNA vaccines that elicit potent humoral and cell-mediated immunity. Vaccine 2018;36(12):1689–99.
[168] ANON. CpG 7909: PF 3512676, PF-3512676. Drugs R D 2006;7:312–6.
[169] La Rosa C, Longmate J, Lacey SF, Kaltcheva T, Sharan R, Marsano D, Kwon P, Drake J, Williams B, Denison S, Broyer S, Couture L, Nakamura R, Kelsey MI, Krieg AM, Diamond DJ, Zaia JA. Clinical evaluation of safety and immunogenicity of PADRE cytomegalovirus (CMV) and tetanus-CMV fusion peptide vaccines with or without PF03512676 adjuvant. J Infect Dis 2012;205:1294–304.
[170] Bernstein DI, Munoz FM, Callahan ST, Rupp R, Wootton SH, Edwards KM, Turley CB, Stanberry LR, Patel SM, McNeal MM, Pichon S, Amegashie C, Bellamy AR. Safety and efficacy of a cytomegalovirus glycoprotein B (gB) vaccine in adolescent girls: a randomized clinical trial. Vaccine 2016;34:313–9.
[171] Griffiths PD, Stanton A, McCarrell E, Smith C, Osman M, Harber M, Davenport A, Jones G, Wheeler DC, O'Beirne J, Thorburn D, Patch D, Atkinson CE, Pichon S, Sweny P, Lanzman M, Woodford E, Rothwell E, Old N, Kinyanjui R, Haque T, Atabani S, Luck S, Prideaux S, Milne RS, Emery VC, Burroughs AK. Cytomegalovirus glycoprotein-B vaccine with MF59 adjuvant in transplant recipients: a phase 2 randomised placebo controlled trial. Lancet 2011;377:1256–63.
[172] Effros RB. The silent war of CMV in aging and HIV infection. Mech Ageing Dev 2016;158:46–52.
[173] Söderberg-Nauclér C. Does cytomegalovirus play a causative role in the development of various inflammatory diseases and cancer? J Intern Med 2006;259:219–46.
[174] Frasca D, Blomberg BB. Aging, cytomegalovirus (CMV) and influenza vaccine responses. Hum Vaccin Immunother 2016;12:682–90.
[175] Simanek AM, Dowd JB, Pawelec G, Melzer D, Dutta A, Aiello AE. Seropositivity to cytomegalovirus, inflammation, all-cause and cardiovascular disease-related mortality in the United States. PLoS One 2011;6:e16103.

[176] Gkrania-Klotsas E, Langenberg C, Sharp SJ, Luben R, Khaw KT, Wareham NJ. Seropositivity and higher immunoglobulin G antibody levels against cytomegalovirus are associated with mortality in the population-based European prospective investigation of Cancer Norfolk cohort. Clin Infect Dis 2013;56:1421–7.
[177] Castillo-Solórzano C, Reef SE, Morice A, Vascones N, Chevez AE, Castalia-Soares R, Torres C, Vizzotti C, Ruiz Matus C. Rubella vaccination of unknowingly pregnant women during mass campaigns for rubella and congenital rubella syndrome elimination, the Americas 2001–2008. J Infect Dis 2011;204(Suppl 2):S713–7.
[178] Schleiss MR. Cytomegalovirus vaccines under clinical development. J Virus Erad 2016;2(4):198–207.
[179] Heineman TC, Schleiss M, Bernstein DI, Spaete RR, Yan L, Duke G, Prichard M, Wang Z, Yan Q, Sharp MA, Klein N, Arvin AM, Kemble G. A phase 1 study of 4 live, recombinant human cytomegalovirus Towne/Toledo chimeric vaccines. J Infect Dis 2006;193(10):1350–60.
[180] Adler SP, Manganello AM, Lee R, McVoy MA, Nixon DE, Plotkin S, Mocarski E, Cox JH, Fast PE, Nesterenko PA, Murray SE, Hill AB, Kemble G. A phase 1 study of 4 live, recombinant human cytomegalovirus Towne/Toledo chimera vaccines in cytomegalovirus seronegative men. J Infect Dis 2016;214(9):1341–8.
[181] Curtis HA, Singh T, Newkirk MM. Recombinant cytomegalovirus glycoprotein gB (UL55) induces an autoantibody response to the U1–70 kDa small nuclear ribonucleoprotein. Eur J Immunol 1999;29(11):3643–53.
[182] Schleiss MR, Bernstein DI, Passo M, Parker S, Meric C, Verdier F, Newkirk MM. Lack of induction of autoantibody responses following immunization with cytomegalovirus (CMV) glycoprotein B (gB) in healthy CMV-seronegative subjects. Vaccine 2004;23(5):687–92.
[183] Jackson JW, Sparer T. There is always another way! cytomegalovirus' multifaceted dissemination schemes. Viruses 2018;10(7).
[184] Nelson CS, Huffman T, Jenks JA, Cisneros de la Rosa E, Xie G, Vandergrift N, Pass RF, Pollara J, Permar SR. HCMV glycoprotein B subunit vaccine efficacy mediated by nonneutralizing antibody effector functions. Proc Natl Acad Sci U S A 2018;115(24):6267–72.
[185] Baraniak I, Kropff B, Ambrose L, McIntosh M, McLean GR, Pichon S, Atkinson C, Milne RSB, Mach M, Griffiths PD, Reeves MB. Protection from cytomegalovirus viremia following glycoprotein B vaccination is not dependent on neutralizing antibodies. Proc Natl Acad Sci U S A 2018;115(24):6273–8.
[186] Schleiss MR. Recombinant cytomegalovirus glycoprotein B vaccine: rethinking the immunological basis of protection. Proc Natl Acad Sci U S A 2018;115(24):6110–2.
[187] Bialas KM, Tanaka T, Tran D, Varner V, Cisneros de la Rosa E, Chiuppesi F, Wussow F, Kattenhorn L, Macri S, Kunz EL, Estroff JA, Kirchherr J, Yue Y, Fan Q, Lauck M, O'Connor DH, Hall AH, Xavier A, Diamond DJ, Barry PA, Kaur A, Permar SR. Maternal CD4+ T cells protect against severe congenital cytomegalovirus disease in a novel nonhuman primate model of placental cytomegalovirus transmission. Proc Natl Acad Sci U S A 2015;112(44):13645–50.
[188] Schleiss MR. Preventing congenital cytomegalovirus infection: protection to a 'T'. Trends Microbiol 2016;24(3):170–2.
[189] Nelson CS, Vera Cruz D, Su M, Xie G, Vandergrift N, Pass RF, Forman M, Diener-West M, Koelle K, Arav-Boger R, Permar SR. Intrahost dynamics of human cytomegalovirus variants acquired by seronegative glycoprotein B vaccinees. J Virol 2019;93(5). https://doi.org/10.1128/JVI.01695-18. pii:e01695-18.
[190] Institute of Medicine, Division of Health Promotion and Disease Prevention, Committee to Study Priorities for Vaccine Development, Stratton KR, Durch JS, Lawrence RS. Vaccines for the 21st century: a tool for decision making. The National Academies Press; 2000. https://doi.org/10.17226/5501.

[191] Stanberry LR, Rosenthal SL, Mills L, Succop PA, Biro FM, Morrow RA, Bernstein DI. Longitudinal risk of herpes simplex virus (HSV) type 1, HSV type 2, and cytomegalovirus infections among young adolescent girls. Clin Infect Dis 2004;39(10):1433–8.
[192] Finer LB, Zolna MR. Declines in unintended pregnancy in the United States, 2008–2011. N Engl J Med 2016;374(9):843–52.
[193] Hotez PJ. Neglected infections of poverty in the United States of America. PLoS Negl Trop Dis 2008;2(6):e256.
[194] Lanzieri TM, Bialek SR, Ortega-Sanchez IR, Gambhir M. Modeling the potential impact of vaccination on the epidemiology of congenital cytomegalovirus infection. Vaccine 2014;32:3780–6.
[195] Azevedo RS, Amaku M. Modelling immunization strategies with cytomegalovirus vaccine candidates. Epidemiol Infect 2011;139:1818–26.
[196] Mitchell DK, Holmes SJ, Burke RL, Duliege AM, Adler SP. Immunogenicity of a recombinant human cytomegalovirus gB vaccine in seronegative toddlers. Pediatr Infect Dis J 2002;21(2):133–8.
[197] Hogea C, Dieussaert I, Van Effelterre T, Guignard A, Mols J. A dynamic transmission model with age-dependent infectiousness and reactivation for cytomegalovirus in the United States: potential impact of vaccination strategies on congenital infection. Hum Vaccin Immunother 2015;11(7):1788–802.
[198] Griffiths PD, McLean A, Emery VC. Encouraging prospects for immunisation against primary cytomegalovirus infection. Vaccine 2001;19:1356–62 (11–12).
[199] Schleiss MR. Congenital cytomegalovirus: impact on child health. Contemp Pediatr 2018;35(7):16–24.

CHAPTER

13

Zika virus

Natalie Quanquin[a], Kristina Adachi[b], Karin Nielsen-Saines[b]

[a]Department of Pediatrics, Division of Infectious Diseases, Children's Hospital Los Angeles, Los Angeles, CA, United States, [b]Department of Pediatrics, Division of Infectious Diseases, University of California, Los Angeles David Geffen School of Medicine, Los Angeles, CA, United States

Introduction

Zika virus (ZIKV) was initially discovered seven decades ago in Uganda (Fig. 1), with reported small epidemics in 2007 in Micronesia and later in 2013–2014 in French Polynesia [1–8]. However, following the 2015–2016 Brazilian ZIKV epidemic and the growing recognition of an association of ZIKV with Guillain-Barré syndrome and congenital abnormalities such as microcephaly, ZIKV was declared a "public health emergency" by the World Health Organization (WHO) from February to November 2016 [9–16]. While the number of acute ZIKV and congenital ZIKV infection cases have dramatically declined in the past 2 years, it is clear that ZIKV will remain a continued global threat, particularly for pregnant women and their children, for years to come.

Zika virus virology and structure

Classification and structure

ZIKV is an *Aedes* mosquito-borne arbovirus belonging to the *Flaviviridae* family within the *Flavivirus* genus and *Flavivirus* cluster [17–19]. Together with Spondweni virus, it shares a unique clade (clade X) [20]. Other viruses in the *Flaviviridae* family include yellow fever, Japanese encephalitis, West Nile, dengue, and hepatitis C, and ZIKV shares approximately

https://doi.org/10.1016/B978-0-12-814582-1.00014-0

FIG. 1 **Zika forest in Uganda.** Zika forest in Uganda, where Zika virus was first isolated in 1947 while scientists were studying yellow fever. *Source: Images from the WHO: http://www.portal.pmnch.org/emergencies/zika-virus/history/en/.*

60% of its nucleotides with Japanese encephalitis, West Nile, and dengue virus [20]. ZIKV also has a similar structure to other flaviviruses. It is an icosahedral-shaped, small enveloped single-stranded positive sense RNA virus (+ssRNA) that is approximately 40–45 nm in diameter and composed of 10,794 nucleotides (Fig. 2) [19,21–25]. ZIKV has 2 non-coding regions (5′ and 3′ NCR), and a single long open reading frame encoding a polyprotein. The polyprotein, 5′-C-prM-*E*-NS1-NS2A-NS2B-NS3-NS4A-NS4B-NS5-3′, is cleaved into three smaller structural proteins [capsid (C), precursor of membrane (prM), envelope (E)] as well as seven other non-structural (NS) proteins [NS1, NS2A, NS2B, NS3, NS4A, NS4B, and NS5] that are involved with genome replication, polyprotein and cellular process modulation [26–28]. The 53 kDa envelope protein (also known as E protein), is the major surface protein of ZIKV and a key component in the viral replication cycle that allows for binding and cell membrane fusion [26]. Its nucleocapsid is surrounded by a lipid bilayer derived from the host membrane with E and M envelope proteins. The E protein is believed to be the most common target for neutralizing antibodies [29]. The ZIKV NS1 nonstructural protein functions as a virulence factor that has important roles in viral replication and immune evasion [28,30]. In contrast, NS5 has dual roles in RNA-dependent RNA polymerase (C-terminus) activity and RNA capping [26]. The 3′ non-coding region may also have roles in translation, genome stabilization, RNA packaging, and cellular and viral factor recognition [26].

In general, flaviviruses may be inactivated by high temperatures (>56 °C for at least 30 min), pH ≤6, UV-light, gamma-radiation, and 30 min of contact with disinfectants like 1% sodium hypochlorite, 2% glutaraldehyde,

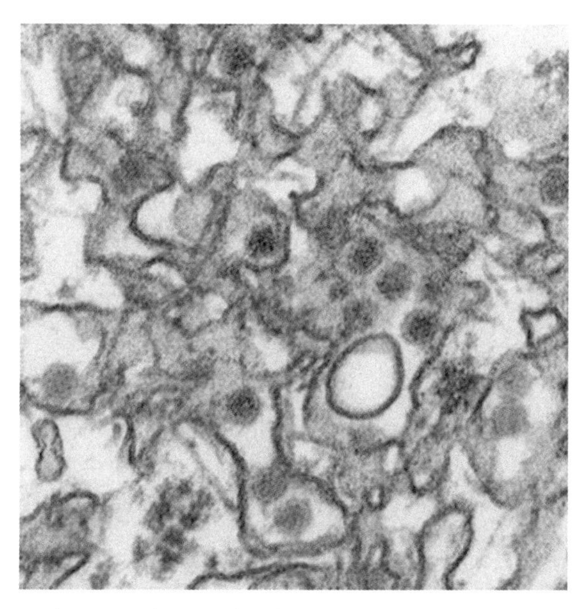

FIG. 2 **ZIKV electron micrograph.** Electron micrograph of ZIKV that has been digitally-colorized, which depicts ZIKV particles in red. The ZIKV particles are 40 nm in diameter with an outer envelope and an inner dense core. *CDC/Cynthia Goldsmith. http://phil.cdc.gov/phil/details.asp.*

70% ethanol, 3–6% hydrogen peroxide, 3–8% formaldehyde, TRIzol and AVL buffer (lysis buffer) [31–35] [36]. Loss of ZIKV infectivity was also observed after exposure to alcohols, 1% hypochlorite (often used to inactivate viruses in BSL-2/3 laboratories), 2% paraformaldehyde, 2% glutaraldehyde, ultraviolet (UV) light radiation, temperatures $\geq$60 °C, and pH $\geq$ 12 or pH $\leq$4 [37]. These studies also observed that dried ZIKV could remain infectious for >3 days, however standard contact barriers including nitrile and latex gloves are protective [37].

Phylogeny and genetics

While the origin of ZIKV can be traced back to Africa, full genome sequencing has identified the emergence of two primary ZIKV lineages: the African lineage (including Nigeria, Senegal, and Uganda strains) and the Asian lineage (including Malaysia, Yap, and Cambodia strains) [17,38]. Sequencing of ZIKV strains circulating in Brazil during the 2015–16 outbreak demonstrated that it had roots in the ZIKV Asian lineage, likely originating from the French Polynesia strain [39–42]. It is believed that the Asian ZIKV strains, which may also cause asymptomatic or mild infection, have been largely responsible for the reported congenital anomalies observed during the 2015–16 global pandemic. Some have speculated that recombination events in the ZIKV genome may have contributed to changes in its pathogenicity [17,26,43,44].

Epidemiology

Although epidemic ZIKV was first observed to circulate in Latin America in 2015, primarily in Brazil, recent phylogenetic analysis suggests that ZIKV may have arrived in Brazil several years prior, possibly as early as 2012–2013 [41,45,46]. From January 2015 to March 2017, there were >754,000 ZIKV cases in the Americas reported to the Pan American Health Organization (PAHO), of which the largest number were from Brazil (>346,000), Colombia (>107,000), and Venezuela (>62,000) [47,48]. ZIKV subsequently spread rapidly throughout the world, particularly in the Americas and the Caribbean [49]. Transmission was reported in over 84 countries, with 100 countries and territories considered to be at risk of ZIKV transmission [41,50–55]. Many countries in Africa, Asia (especially Southeast and South Asia), and the Western Pacific Region also reported active ZIKV transmission since 2016 [56]. The first United States (US) cases of locally transmitted ZIKV were identified in Miami, Florida in July 2016 and in Brownsville, Texas in November 2016 [57–61]. By 2017, ZIKV cases in Brazil and other Latin American countries dropped considerably, with a subsequent decrease in US imported travel-cases in 2017 and 2018. In 2018, there were only 28 travel-associated ZIKV cases in the US, as opposed to >4800 in 2016, and 67 locally-transmitted cases in US Territories (primarily Puerto Rico) in contrast to >36,300 in 2016 [62].

The most recent data from the WHO suggests that there are currently nearly 3700 infants who have laboratory-confirmed congenital Zika syndrome, the most severe form of congenital ZIKV infection, the majority of which are in Brazil [47]. To follow the outcomes of pregnant women living in the US and US Territories, the US Centers for Disease Control and Prevention (CDC) created the US ZIKV Pregnancy Registry in 2016 [63]. As of June 2018, the CDC ZIKV Pregnancy Registry has been following >2400 US pregnant women and >4800 pregnant women in other US Territories (primarily Puerto Rico) [64,65]. Data collected from several studies of ZIKV infection during pregnancy conducted by the CDC have suggested that approximately 5–15% of those fetuses/infants had ZIKV-associated birth defects. The timing of infection during pregnancy appeared to have an impact, as up to 8–15% of fetuses were affected when maternal infection was acquired in the first trimester [63,65,66]. Similar findings of ZIKV-associated neurologic or ocular abnormalities (7%) were also seen in a study published on ZIKV pregnancy outcomes for women living in the French Territories of the Americas (French Guiana, Guadeloupe, Martinique) [67]. Earlier data from a prospective study of PCR-confirmed ZIKV infection in

Brazilian pregnant women found much higher rates of abnormalities in those fetuses/infants (42%), however that study included a broader range of infant functional and structural abnormalities, including seizures and non-specific imaging findings [68]. A study comparing estimated birth defects in the US before (2013–2014) and after the introduction of ZIKV in the Americas also found an increase in the rates of infants born with birth defects, from 2.86 per 1000 live births to 58.8 per 1000 live births, approximately a 20-fold higher rate [69]. Given ongoing ZIKV transmission in many countries, and the risk ZIKV poses, particularly to pregnant women, travel warnings remain in effect in many global regions, particularly for women who are pregnant or planning to conceive.

Clinical aspects of ZIKV infection

Acute infection

Infection with ZIKV is believed to be asymptomatic in the majority of people (80%) [5, 70]. The incubation period of ZIKV is 3–14 days, with about half developing symptoms within 7 days [71]. For those with symptomatic acute ZIKV infection, the symptoms are generally considered to be mild in the majority of patients, and classically characterized by a maculopapular rash, arthralgia, and conjunctivitis (Fig. 3) [5, 72–80]. Interestingly, most patients with acute ZIKV infection do not have fever, which is seen in only one third of patients and is typically brief and low-grade compared to those with chikungunya or dengue infections [42,72,75,79,81,82]. The most common symptoms in ZIKV patients include conjunctivitis and a maculopapular rash, especially one that is pruritic [42,68,75,83–85].

Some studies have tried to assess the sensitivity and specificity of ZIKV diagnosis using only clinical case definitions [PAHO, US CDC, European Centre for Disease Control and Prevention (ECDC), WHO, Brazil Ministry of Health]. A study in Brazil found that a ZIKV clinical definition which included rash, pruritis, conjunctival hyperemia, absence of fever, petechiae, and anorexia was a better predictor of ZIKV infection, with 86.6% sensitivity and 78.3% specificity, than the other definitions, particularly in areas where ZIKV, chikungunya and dengue viruses were endemic [86]. A Singaporean study also reported that many of the currently used official clinical case definitions for ZIKV lacked specificity, but rash was a key clinical characteristic in the identification of symptomatic ZIKV infection [87].

Potential complications following acute ZIKV infection

The majority of people with acute ZIKV infection recover without complications. One review of over 2300 patients with confirmed or probable

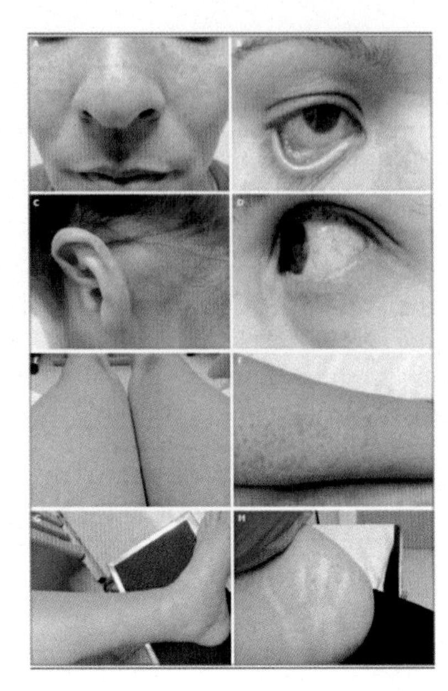

FIG. 3 **Clinical manifestations of Zika virus in a pregnant woman.** Panel A, maculopapular rash on the face; Panel B, conjunctival and palpebral erythema; Panel C, retroauricular lymphadenopathy; Panel D, conjunctival injection with prominence of vasculature; Panel E, a rash on the legs, with a lacy reticular pattern; Panel F, a maculopapular rash on the inner arm; Panel G, edema of the foot, which the patient reported was painful; and Panel H, a blanching macular rash on the gravid abdomen. *Source: Brasil P, Pereira Jr JP, Raja Gabaglia C, et al. Zika virus infection in pregnant women in Rio de Janeiro – preliminary report. N Engl J Med 2016.*

ZIKV found that only 3% required hospitalization [88]. Another study from the CDC also found that there were no severe ZIKV complications such as meningitis, encephalitis, septic shock, Guillain-Barré Syndrome, or death reported in their series [89]. However, during the 2015–16 ZIKV epidemic, there were case reports of unusual potential complications following acute ZIKV infection. These included: eye abnormalities (hypertensive iridocyclitis, uveitis, acute maculopathy, acute chorioretinitis), temporary hearing impairments, dysgeusia (distortion of the sense of taste), hematospermia, prostatitis, myocarditis and mesenteric adenitis [90–102]. In contrast to dengue, hemorrhagic complications such as severe thrombocytopenia with mucosal bleeding and hematomas were rarely reported [99,103–113]. Some cases of ZIKV-associated death have also been reported in the literature from Brazil, Colombia, Puerto Rico, and the US, generally in patients with underlying immunodeficiency [88,108,110,114–117].

In addition, various neurologic complications have also been reported following acute ZIKV infection, including meningoencephalitis, encephalopathy, acute sensory polyneuropathy, seizures, myelitis including Guillain-Barré Syndrome and Acute Disseminated Encephalomyelitis

[9,10,88,117–138]. There was even a report of a patient with a history of depression who suffered neuropsychiatric and cognitive changes for months following acute ZIKV infection [139]. Of these neurologic complications, Guillain-Barré Syndrome, an acute or sub-acute immune-mediated progressive flaccid paralysis, has received the most focus [121,140–142]. In fact, Guillain-Barré Syndrome was first observed as a potential complication after the French Polynesia ZIKV epidemic, where ZIKV was associated with a 34-fold increased risk of Guillain-Barré Syndrome [50,121,123]. Observations in Brazil as well as in other countries in the Americas also documented an increased incidence of Guillain-Barré Syndrome during the 2015–16 ZIKV pandemic [9,10,51,56,88,122,124,130–132,134,143–145]. One review assessing the temporal association between ZIKV disease and the Guillain-Barré Syndrome cases from several countries in Latin America and the Caribbean (Brazil, Colombia, Dominican Republic, El Salvador, Honduras, Suriname, and Venezuela) found a 2- to nearly 10-fold increase in Guillain-Barré Syndrome diagnoses [133].

Congenital ZIKV infection

One of the most alarming aspects of the ZIKV 2015–16 pandemic centered on ZIKV's potential to cause severe teratogenic effects following infection in pregnancy. We now understand that ZIKV infection in pregnancy may cause a spectrum of fetal/infant abnormalities that range from mild to severe, primarily manifesting in the form of central nervous system abnormalities. To describe the most severely affected ZIKV infants, the CDC created the term "congenital Zika syndrome" (CZS) (Fig. 4). Typical CZS infant findings include the following: (1) severe microcephaly (>3 SD below the mean for gestational age and gender); (2) brain abnormalities (subcortical calcifications, ventriculomegaly, cortical thinning, gyral pattern anomalies, hypoplasia of the cerebellum, or corpus callosum anomalies); (3) ocular findings; (4) congenital contractures; (5) neurologic impairment [146]. Currently, it is estimated that there are over 3700 infants with laboratory-confirmed CZS worldwide, the majority of whom reside in Brazil, and this figure is likely an underestimate [147].

Studies of Brazilian infants with ZIKV-associated microcephaly found that it was often accompanied by severe brain abnormalities [148–150]. One such study found that in infants with ZIKV-associated microcephaly who had brain imaging performed, 93% had calcifications, 69% had significant cortical developmental malformations (lissencephaly, pachygyria, agyria), and 66% had ventriculomegaly due to brain atrophy [148]. Congenital contractures such as arthrogryposis have been observed in 5.7–20.7% of CZS cases, which may be so severe that they can even result in congenital hip or knee dislocations [146,149,151]. Clubfoot has also been reported in 3.8–14% of CZS cases [146]. In infants with CZS who

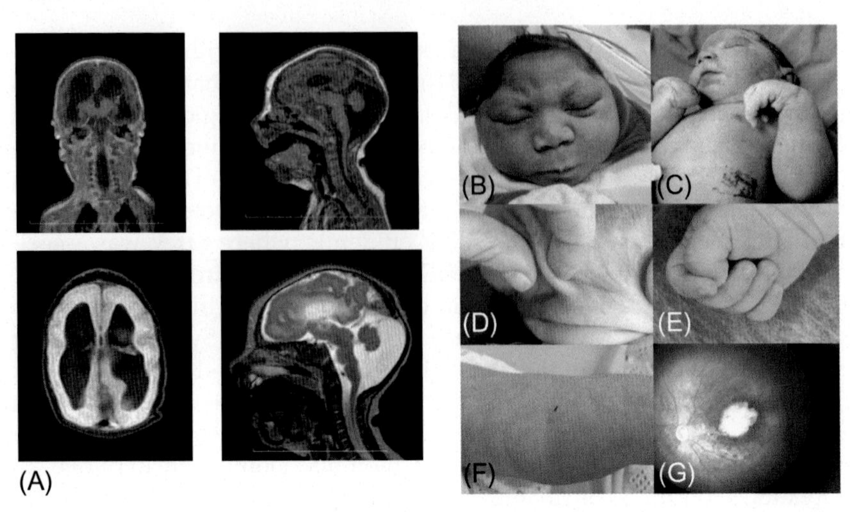

FIG. 4 **Clinical manifestations of congenital ZIKV syndrome.** Panel A, MRI images of cranial facial disproportion with microcephaly. Diffuse reduction in white matter thickness with corpus callosum not identified. Diffuse pachygyria, more evident in the right parietal lobe. Supratentorial ventricular dilatation. Atrophy of the cerebellar vermix with ample communication of the fourth ventricle with the cisterna magna. Widening of the perivascular spaces of the cerebellar hemispheres, with surrounding gliosis. Panel B, disproportionate microcephaly; Panel C, arthrogryposis at birth; Panel D, redundant scalp; Panel E shows a cortical thumb; Panel F, knee fovea; and Panel G shows the left infant retina demonstrating optic disc hypoplasia, peripapillary atrophy; macular chorioretinal atrophy with a colobomatous-like aspect with hyperpigmented halo and pigmentary mottling. *Source: Brasil P, Pereira Jr JP, Moreira ME, et al. Zika virus infection in pregnant women in Rio de Janeiro. N Engl J Med 2016;375(24):2321–34.*

have microcephaly, sensorineural hearing loss has been reported in 6% [152,153], whereas eye abnormalities were more frequently observed, with reported rates ranging from 35% to 46% to as high as 100% of cases [149,154–160]. Ophthalmic findings included focal retinal pigment mottling, chorioretinal atrophy, optic-nerve changes, lens subluxation, loss of foveal reflex, iris colobomas, micro-ophthalmia, cataracts, intraocular calcifications, and myopia [146,154–157,161,162]. Less frequently, eye abnormalities at birth have also been reported in ZIKV exposed infants without other central nervous system abnormalities [163]. Non-life threatening congenital heart defects (10–13.5% with septal defects) in ZIKV-exposed infants have also been reported in at least two studies [164,165] while other case studies have reported findings of unilateral diaphragm paralysis and severe spinal cord injury in some CZS infants [149,166].

Long-term outcomes of infants with CZS or in utero ZIKV exposure are currently under investigation. However, some early studies have provided a preliminary glimpse into some of the chronic problems that may affect infants with CZS [68,146,167,168]. One study of almost fifty Brazilian infants with probable CZS reported common chronic issues including severe irritability (85%), pyramidal/extrapyramidal syndrome (56%) including hypertonia, clonus, hyperreflexia, epileptic seizures (50%), dysphagia (15%), congenital clubfoot (10%), and arthrogryposis (10%) [169]. Another report from the CDC of 13 Brazilian infants with in utero ZIKV exposure without microcephaly but with significant radiographical brain abnormalities in the postnatal course described decreasing rates of head circumference growth, with the majority (85%) developing postnatal or secondary microcephaly [167]. Interestingly, a small study of 37 CZS infants found a decrease in brain calcifications at 1 year of life compared to the time of birth, which is different from other congenital infections [170]. Other potential chronic issues and complications described in the literature include other types of epilepsy including infantile spasms, neurogenic bladder, poor sleep, and even neonatal cerebral infarction [171–173]. Additional work from Brazil has highlighted the notable case-fatality rates for infants with ZIKV-associated microcephaly ranging from 8.3–27.3% [174,175]. A report of longer-term outcomes of CZS infants followed at 19–24 months of age highlights the poor growth observed, with the majority having notable adverse outcomes including seizure disorders, sleeping difficulties, eating or swallowing issues, hearing abnormalities, ophthalmologic abnormalities, and frequent hospitalizations including those for pneumonia and bronchitis [176]. The study also highlighted that the majority of children had severe motor impairment and did not pass the developmental assessment used for children at 6 months of age [176]. In fact, observations of CZS in Brazil have suggested that many infants with CZS at 1 year of age only function at a 2–3 month age developmental level. Based on pre-existing knowledge about other conditions resulting in infant microcephaly, it is probable that the majority of CZS children will have severe to profound developmental delay, with inability to communicate verbally, sit unsupported, or walk, and will demonstrate severe irritability and difficulties regulating sensory input, and will require lifelong care [177].

Particularly in countries in Latin America and the Caribbean, the ZIKV 2015–16 pandemic has created an enormous economic and societal burden, leaving thousands of CZS infants in its wake who will require constant care throughout childhood, adolescence and adulthood [178–180]. Some have estimated that the lifetime cost of caring for one infant with ZIKV-associated microcephaly is approximately $3.8 million but can possibly be as high as $10 million for patients who survive to adulthood [181].

ZIKV vaccine development

While an association between ZIKV infection and Guillain-Barré Syndrome was concerning, it was not until the epidemic became associated with an increase in the incidence of congenital microcephaly that the

WHO was spurred into action, declaring it a "Public Health Emergency of International Concern" [182]. Initial research efforts faced several hurdles, such as the fact that ZIKV had been sparsely studied before the outbreak on Yap Island in 2007 [183], and that animal models for testing ZIKV vaccines had yet to be discovered. After a single year, however, there were already 45 vaccine candidates undergoing *in vitro* or animal testing, with six in phase I human clinical trials and one in a phase II study [184]. At the time of this writing, there are 14 clinical trials reported on the WHO's Vaccine Pipeline Tracker on nine candidate vaccines [185], with one additional candidate and trial from India. This rapid progress was aided by previous experiences with related flavivirus vaccines, such as those for dengue, Japanese encephalitis, West Nile and yellow fever viruses [186]. In addition, this work was stimulated by significant government funding [187] and facilitated by recent advancements in vaccine development technologies.

Challenges to vaccine development

In addition to passing the usual standards for phase I (safety and dosage) and phase II and III trials (efficacy and monitoring for adverse events), ZIKV candidate vaccines will face other rigorous challenges. While Guillain-Barré Syndrome is a concerning sequela of ZIKV infection, the incidence is still considered relatively low at 2.5–3.0 per 10,000 ZIKV-infected cases, affecting mostly those over the age of 60 years [188]. In contrast, congenital microcephaly, the most outwardly visible manifestation of CZS, was reported to have had an incidence as high as 49.9 cases per 10,000 live births in Northeast Brazil during the peak of the ZIKV epidemic (up to 24 times higher than the baseline rate of microcephaly) [189]. This makes pregnant women and fetuses the most at-risk populations for ZIKV-associated disease, and therefore the ones who would benefit most from vaccination. Live vaccines are not recommended for use in pregnancy, because of the potential for replication of virus and therefore the theoretical risk of harm to the mother or infant. The risk of teratogenic effects during critical stages of embryonic development also contraindicate the use of DNA vaccines, which have a theoretical risk of genetic recombination and oncogenesis. However, it is important that women who discover that they are infected with ZIKV during or shortly before pregnancy be given options for quickly suppressing their ZIKV viremia and potentially reducing the risk of CZS (although CZS has also been found to occur even in asymptomatic women with low level viremia) [190]. While drugs or antibody therapies may become an option in the future, developing a vaccine that can be used safely during any stage of pregnancy is still feasible, and can be beneficial if protective antibodies are passed on to the infant. Many ZIKV vaccine candidates are now being tested in pregnant animal models to examine both safety and whether the maternal ZIKV antibodies transferred to offspring can protect them from ZIKV exposure in utero or in the neonatal period [191–194]. However, anticipating that important vaccine modalities might otherwise be abandoned if safety during pregnancy became a requirement for all vaccine candidates, the WHO created guidelines for ZIKV vaccine developers, encouraging an emergency arm for pregnant patients exposed in an outbreak setting (a prioritized target population) and a separate arm for routine immunizations [195,196]. This allows continued testing of promising live attenuated vaccines, which are generally considered an ideal platform due to their stimulation of all branches of the immune system induced by natural infection. This approach improves the likelihood of broad immunity to multiple antigen targets and generates a potential for lifelong protection when given in early childhood. ZIKV vaccines nevertheless should be made available to all age groups, at least initially for catch-up, to ensure coverage of sexually active women who could later become pregnant. In the past, the policy of restricting immunizations to young children and not including women of childbearing age resulted in an entire generation of infants at risk for congenital rubella syndrome even after a vaccine had been discovered [197]. To date, all ZIKV vaccines undergoing clinical trials have yet to take the challenging leap of including pregnant women or young children into their testing cohorts (the youngest age group currently being tested is 15–35 year-old individuals for the VRC DNA vaccine).

This raises the interesting question of the ideal timing for ZIKV vaccination. As mentioned, immunization with live vaccines could provide lasting immunity, however these vaccines are generally contraindicated before the age of 1 year [measles, mumps, and rubella (MMR) vaccine can be given as early as 6 months in cases of anticipated exposure, but the dose is not counted as part of the series due to uncertain efficacy]. In cases where mothers are vaccinated and sufficient effective maternal antibody is transferred, infants may be passively protected for up to 6 months of age, when titers to MMR or Varicella have been noted to diminish [198]. The first 6 months of life also comprises a period of vulnerability to ZIKV infection. In an outbreak setting, it is possible that a dose of a live ZIKV vaccine would be permitted at 6 months of age, similar to the MMR vaccine. Alternatively, nucleic acid or inactivated vaccines are also being developed, and those could be given as early as 6 weeks of age due to their positive safety profile. Ultimately, the most important issue is that children are vaccinated before they reach the age of sexual maturity. Not only is there concern for the effects of ZIKV on the developing fetus, but the virus has also been associated with adverse effects on sperm, with prolonged shedding in semen [199,200]. In cases where mothers require vaccination during pregnancy, it would be important to immunize as early as possible. ZIKV is a neurotropic virus, and critical brain development occurs in the first and early second trimesters [201–203] although complications following ZIKV infection have also been noted in infections occurring later in fetal development [68].

Another difficulty in the development of ZIKV vaccines is the lack of a clear and defined correlate of protection. ZIKV infection is asymptomatic in the majority of human cases, therefore clinical endpoints for vaccine trials would be impractical. Even mouse models must use immunocompromised animals for signs of ZIKV-associated disease to manifest. Although the WHO proposed using viremia as an alternative criteria [196], this presents its own set of challenges, as diagnostic tests have a range of sensitivities based on modality, likelihood of cross-reactivity with other flaviviruses, specimen type (blood, urine, semen) and timing of collection [204]. Some preclinical ZIKV vaccine trials chose to test immunocompetent animals, using a decrease in the level of short-term ZIKV viremia in wild-type mice or non-human primates as the measure of efficacy, while attempting to correlate levels of viremia with the attainment of a particular titer of neutralizing ZIKV antibodies. This approach was used in vaccine studies of other flaviviruses, where a minimum serum titer of 1:10 was defined as protective [205]. Interestingly, preliminary studies in animals infected with ZIKV suggest that the protective titer may need to be as high as 1:100 [206]. This would be in line with observations that ZIKV appears to be more stable than other flaviviruses [207], has a greater number of tissue reservoirs [208], and requires a stronger immune response to suppress infection [206]. The WHO states that if immunological endpoints are indeed used as the basis of vaccine authorization, vaccine effectiveness will still need to be demonstrated as a follow-up commitment [196].

Finally, as clinical trials approach late stages of testing where they require proof of efficacy in locations where ZIKV is endemic, there is the problem of shrinking susceptible and exposed populations. While we have no accurate figures for the incidence of ZIKV infection during the initial years of the outbreak in the Americas and Caribbean (Brazil declared early on that they ceased recording the number of cases due to the sheer magnitude) [182], the infection rate has since dropped dramatically, such that in late 2016 the WHO no longer considered ZIKV a global health emergency [16]. The general belief is that the decreased ZIKV incidence is due to persisting immunity in these populations, which experienced a large rate of infection in 2015–2016.

A distinguishing trait of ZIKV is that unlike other arboviruses, other routes of transmission outside of mosquitoes have been described (vertical, sexual, breastmilk, and transfusion-acquired). High levels of virus are present in body fluids such as saliva and urine, demonstrating the potential for additional alternate routes of transmission. Nevertheless, the primary means of ZIKV spread is the *Aedes* sp. mosquito. Therefore, regions where *Aedes* mosquitoes are prevalent remain ideal locations for testing of ZIKV vaccines. Given that this mosquito is also a carrier for multiple other flaviviruses (such as dengue and yellow fever) or other arboviruses such as chikungunya, there have been considerations toward developing a vaccine platform that can immunize against several of these species at once [207]. Unfortunately, there is also the concern that pre-existing immunity to one of these viruses may result in either a dampened or an enhanced reaction to another related virus. For example, dengue exists in four serotypes, and a prior exposure to one type may aggravate the illness during exposure to another. This is thought to be through a process known as antibody-dependent enhancement, where cross-reactive antibodies bind to the virus but are unable to neutralize it, and their simultaneous binding to the Fc gamma receptor on myeloid cells actually has the unfortunate effect of facilitating their infection [209]. The resulting increased viral burden promotes a proinflammatory environment, coagulation dysfunction, and damage to vasculature endothelial cell linings, resulting in life-threatening hemorrhagic disease [210,211]. There have been in vitro tests showing that immune cells exposed to dengue virus-specific antibodies had an enhanced response when later exposed to ZIKV [212–214]. The significance of these findings in vivo is controversial. While one study showed that transfer of ZIKV-immune plasma enhanced dengue infection in mice [215], antibody-dependent enhancement from prior exposure to heterologous flaviviruses has yet to be shown in humans or non-human primates [216,217]. Nevertheless, the risk for an antibody-dependent enhancement-like response in clinical trial subjects from areas where ZIKV is endemic (and where they also have a greater likelihood of exposure to other flaviviruses) is significant enough that the WHO also recommended testing in these populations [196]. Some vaccine trials have already begun recruiting subjects from regions where other flaviviruses have been circulating, such as Puerto Rico (NCT03008122, NCT03110770 and NCT02887482), or in patients who are primed with licensed flavivirus vaccines (NCT02963909). As a particular domain in the flavivirus E protein was noted to contain immunodominant cross-reactive epitopes associated with antibody-dependent enhancement reactions, another vaccine development strategy has been to remove that particular region [218]. Similarly, some groups specifically examine whether the antibodies generated by their vaccines will subsequently enhance dengue infection in Fc gamma receptor-expressing cells [219,220].

Given the limited number of places where susceptible patients living in ZIKV endemic areas can still be recruited for these studies, one consideration has been whether it is ethical to implement a controlled human infection model. In this circumstance, subjects from non-endemic areas are intentionally infected with the minimal level of virus that is both considered safe and that can generate enough of an infection where protection between vaccinated and unvaccinated groups can be measured and compared. This plan was proposed by researchers at Johns Hopkins University and the University of Vermont in the US, but was rejected by an ethics panel convened by the National Institutes of Health (NIH) and other US institutions, who pointed out that our limited knowledge of ZIKV pathogenesis does not allow us to guarantee a safe dose, especially in light of the risk of Guillain-Barré Syndrome and the possibility of infections in women who are not yet aware that they are pregnant [221,222]. Alternatively, there is the possibility of using the "animal rule," in which after passing phase II trials in humans for safety and immunogenicity, the efficacy of vaccine candidates against ZIKV viral challenge is established in animal models instead of humans [223,224]. As previously discussed, this situation would present its own set of problems, and would still require further testing for long-term adverse effects in humans.

Evidence of vaccine safety and efficacy

Of the 14 ZIKV clinical trials listed by the WHO in 2018 as active or in development [185], one is shown as completed (a recombinant measles vector vaccine), although there have been no results released about its efficacy. Six other studies have published their preliminary findings; all appear to show strong immune responses with only mild to moderate adverse reactions reported. The candidates include four nucleic acid vaccines (three DNA and one messenger RNA [mRNA]), four inactivated whole virus vaccines, a live attenuated recombinant measles vaccine, and a vaccine that targets mosquito salivary antigens and could potentially prevent all mosquito-borne infections (Table 1). Other vaccine platforms, such as subunit proteins and virus-like particles [228,229], live attenuated flavivirus chimeric vaccines [230,231], and adenovirus-vectored vaccines [232,233], have shown promising results in animal models, but have not yet advanced to human trials. Common targets for many ZIKV vaccines are the ZIKV pre-membrane (prM) and envelope (E) proteins, which represent multiple surface epitopes easily accessible to antibodies.

Vaccine candidates that have completed or are currently undergoing clinical trials are as follows:

Nucleic acid vaccines

DNA plasmid-based and mRNA vaccines have the advantage that they can be rapidly designed and cloned into delivery systems for testing. Despite a DNA vaccine's theoretical risk of genetic recombination in the host nucleus leading to oncogenesis (which would not occur in a mRNA vaccine, which remains in the cytoplasm), there has been no evidence of this occurring to date. However, nucleic acid vaccines have the

TABLE 1 Ongoing Zika vaccines in clinical trials.

Developer	Base	Platform	Phase I trial[a]	Phase II trial[a]	Status/published reference	Collaborators
Bharat Biotech	India	PIV (Zikavac)	CTRI/2017/05/008539		Phase I began Jan. 2017	
WRAIR	U.S.	PIV (ZPIV)	NCT02937233		Phase I began Dec. 2016 [225]	BIDMC, NIAID, BARDA
		PIV (ZPIV)	NCT02952833		Phase I began Jun. 2016 [225]	SLU, NIAID, BARDA
		PIV (ZPIV)	NCT02963909		Phase I began Nov. 2016 [225]	NIAID, BARDA
		PIV (ZPIV)	NCT03008122		Phase I began Feb. 2017	Ponce CAIMED, NIAID
US NIAID	U.S.	DNA vaccine encoding prME (VRC5283)	NCT02840487	NCT03110770	Phase I began Aug. 2016 [226] Phase II began Mar. 2017	Various
		DNA vaccine encoding prME from both ZIKV and JEV (VRC5288)	NCT02996461		Phase I began Dec. 2016 [226]	
		Synthetic mosquito salivary proteins (AGS-v)	NCT03055000		Phase I began Feb. 2017	
Inovio/ GeneOne	U.S./ Korea	DNA vaccine (GLS-5700)	NCT02809443		Phase I began Jul. 2016 [227]	Various

Continued

TABLE 1 Ongoing Zika vaccines in clinical trials—cont'd

Developer	Base	Platform	Phase I trial[a]	Phase II trial[a]	Status/published reference	Collaborators
Inovio/ GeneOne	U.S./ Korea	DNA vaccine (GLS-5700)	NCT02887482		Phase I began Aug. 2016	Puerto Rico
Takeda	Japan	PIV (TAK-426)	NCT03343626		Phase I began Nov 2017	Various
Moderna Therapeutics	U.S.	Messenger RNA-based vaccine (mRNA-1325)	NCT03014089		Phase I began Dec. 2016	US DHHS; US BARDA
Themis Bioscience	Austria	Live attenuated measles vaccine virus vector (MV-Zika)	NCT02996890		Phase I completed Apr. 2018	Institut Pasteur
Valneva	France	PIV (VLA 1601)	NCT03425149		Phase I began Feb. 2018	Emergent Biosolutions

WRAIR, Walter Reed Army Institute of Research; *BIDMC*, Beth Israel Deaconess Medical Center; *NIAID*, National Institute of Allergy and Infectious Diseases; *BARDA*, Biomedical Advanced Research and Development Authority; *SLU*, Saint Louis University; *CAIMED*, Centro Ambulatorio de Investigaciones Medicas; *DHHS*, Department of Health and Human Services.

[a]*Listed by identifier number for ClinicalTrials.gov (NCT) or Clinical Trials Registry India (CTRI).*

unfortunate history of showing promising results in animal models without reproducibility in humans [195].

The current ZIKV DNA candidate vaccines include GLS-5700 (consensus sequence of the prM-E genes by GeneOne Life Science, Inc. and Inovio Pharmaceuticals, trials NCT02809443 and NCT02887482) and two vaccines from the NIH Vaccine Research Center [VRC5283 (prME, NCT02840487) and VRC5288 (prM+E from both ZIKV and Japanese encephalitis, NCT02996461)].

An interim report [227] on GLS-5700 (NCT02809443) revealed results of in vitro ZIKV infection assays using serum from 40 vaccinated subjects, showing neutralizing antibodies in 62% of vaccinees and 90% inhibition of ZIKV infection of neural cell in 70% of vaccinees (however 95% of serum samples showed at least 50% inhibition). Injection of human vaccinee serum into susceptible mice also protected 92% of animals from an otherwise lethal ZIKV infection. Adverse reactions were seen in 50% of subjects, but only included injection-site pain, redness, swelling, and itching.

The NIH published the results on their phase I vaccines trials as well, which encoded the prME genes on a previously approved West Nile vaccine backbone [226]. Safety was assessed, and adverse reactions were described as mild to moderate, with pain or tenderness at the injection site being the most frequent local finding (46% in VRC 319 and 80% in VRC 320); malaise or headache were the most frequent systemic symptoms (27% and 22% respectively in VRC 319, and 38% and 33% respectively in VRC 320). Different dosage routes and timing were tested for both vaccines. While VRC5288 showed positive antibody responses in up to 89% of one group of vaccinees (17/19), responses were 100% (14/14) in another group of vaccinees receiving three doses of VRC5283 through a split-dose needle-free delivery system. This vaccine generated ZIKV-neutralizing antibodies. The latter group also produced elevated T cell responses to pooled ZIKV peptides (CD4 $p=0{\cdot}0108$ and CD8 cells $p=0{\cdot}0039$ compared with baseline). The decision was then made to select VRC5283 over VRC5288 for a phase II trial (NCT03110770), which has already opened recruitment at multiple centers in the Americas and Caribbean for 15–35 year-old subjects.

Finally, there is a nucleoside-modified mRNA vaccine expressing ZIKV prME (mRNA-1325) developed by Moderna Therapeutics that entered phase I/II clinical trials in December 2016 (NCT03014089). Results on its efficacy have not been reported, although preclinical results were promising [234–236].

Inactivated whole virus vaccines

To eliminate the risk of infection using viral materials while attempting to preserve the structure of antibody targets, viruses can be inactivated with formalin and purified, leading to the term "purified inactivated vaccine" (PIV). These are considered safe for use during pregnancy, making them ideal for ZIKV immunization. However, given the lack of infectivity, these vaccines usually require multiple doses and adjuvants to make them effective, and protection is seldom lifelong. PIVs have already been approved for Japanese encephalitis and tick-borne encephalitis virus, thus facilitating the development of a ZIKV PIV. All ZIKV PIV candidates currently in clinical trials use an aluminum-based adjuvant (Alum) which has been proven safe and effective in over 80 years of testing [237].

The NIH has entered four phase I clinical trials for a candidate ZIKV PIV derived from a 2015 Puerto Rican strain (PRVABC59). Different testing strategies were divided between collaborators in the US at Beth Israel Deaconess Medical Center (BIDMC), Saint Louis University (SLU), and the Walter Reed Army Institute of Research (WRAIR). The WRAIR trial (NCT02963909) immunized subjects with either the yellow fever or Japanese encephalitis vaccines prior to ZIKV PIV vaccination to test for possible cross-reactive immune effects. The SLU trial (NCT02952833) examined three different ZIKV PIV dose regimens (5·0 µg, 2·5 µg, or 10·0 µg). The BIDMC trial (NCT02937233) assessed three dosing schedules: two doses separated by 4 or 2 weeks, or a single dose.

Results for these three trials were published in December 2017 [225], reflecting initial studies using the 5.0 µg dosing regimen on days 1 and 29. Adverse events were noted to be mild to moderate, with the most frequent local effect being pain (60%) or tenderness (47%) at the injection site, and the most frequent systemic events being fatigue (43%), headache (39%), and malaise (22%). Fourteen days after the second dose of ZIKV PIV, peak anti-ZIKV serum titers exceeding the threshold of protection in animal studies were noted, and by 28 days, 92% of vaccinees had seroconverted (microneutralization titer ≥1:10). There are plans to continue to monitor these subjects for 6 to 18 months, as well as plans to modify the phase I trials to include ZIKV PIV as a boost to its ZIKV DNA vaccine candidate; testing whether a third injection of the vaccine or administration of higher doses can yield more potent and sustained immunogenicity are also approaches under consideration. A new trial investigating the effectiveness of ZIKV PIV in Puerto Rico, a ZIKV-endemic setting, has been initiated (NCT03008122).

Three international companies, Bharat Biotech International based in India, Takeda Vaccines Inc. based in Japan, and Valneva SE based in France, are also working on a ZIKV PIV. Similar to ZIKV PIV, Bharat's candidate Zikavac (derived from ZIKV strain MR8766) is being tested in a 2-dose regimen of 2.5, 5 or 10 µg each (CTRI/2017/05/008539). Phase I trials began in January 2017 on 48 adult subjects. Although results have yet to be published, preclinical data in immunodeficient mice were promising [233]. Takeda's candidate TAK-426 is being tested in a clinical trial in Florida and Puerto Rico (NCT03343626) using the same dosing regimen as Zikavac on subjects aged 18–49 years who are either flavivirus naive or exposed. Recruitment began in November 2017, however preliminary results have not yet been published. Valneva is collaborating with Emergent Biosolutions to test their candidate, VLA 1601, in a clinical trial that also recruits 18–49 year-old subjects in the US (NCT03425149). The regimen includes up to three injections (days 0, 7, 28) of 0.25 mL or 0.5 mL of their vaccine. Although the study began in early 2018 and has achieved primary completion, final results are pending.

Live attenuated recombinant measles vector vaccine

The use of innocuous viruses as delivery systems for genes or antigens from more dangerous pathogens has become a common vaccine strategy. Like nucleic acid vaccines, they lead to the production of immunogenic microbial antigens that can prime the immune system against a threat by the actual microbe. Themis Bioscience took the interesting approach of using the attenuated but immunogenic measles virus, currently used in the MMR vaccine, as a delivery system for select ZIKV antigens. Given that measles is not transmissible by arthropod vectors, this accommodates the requirement that live flavivirus vaccines be proven unable to infect mosquitoes, which otherwise risks spreading the vaccine strain. The resulting vaccine (MV-Zika) was tested as a single high dose or two variable doses, in 18–55 year-old subjects in Vienna, Austria in 2017 (NCT02996890). Although preliminary results are pending, previously promising results were reported using the same model for a chikungunya vaccine currently in phase II trials [238].

Universal mosquito-borne disease vaccine

As multiple flaviviruses, chikungunya, and malaria can be transmitted by female mosquitoes, the NIH developed a vaccine that contains four synthetic mosquito salivary proteins to induce a modified allergic response in humans that will prevent mosquito transmission of infectious agents, and also potentially interfere with the mosquito's feeding or reproduction. The vaccine, labeled AGS-v, was tested alone or with an unspecified adjuvant in two injections 3 weeks apart, and 18–50 year-old subjects will be evaluated for production of antibodies to these antigens, and for any effects their blood may have on feeding *Aedes aegypti* mosquitos. The trial (NCT03055000) began in 2017, but results have not yet been reported.

Conclusion

Despite the fact that 10 ZIKV vaccine candidates are currently being evaluated in clinical trials, and many others are showing promise in animal models and are awaiting further development, licensure of a ZIKV vaccine may still be some distance away. Phase II trials have only begun on a small number of candidates, and none have yet been tested on young children or pregnant women. In addition, given the low incidence of serious potential adverse effects such as Guillain-Barré Syndrome and CZS, large numbers of subjects will need to be vaccinated in stage III or IV trials before safety concerns can be properly evaluated [216]. There is also the possibility that despite promising preliminary findings, some companies may discontinue their ZIKV vaccine development efforts as resources dwindle or the incidence of ZIKV (and the concern for a widespread pandemic) wanes. For instance, in September 2017, Sanofi Pasteur announced that it was withdrawing from further development of ZIKV vaccines [239]. Nevertheless, ZIKV is still circulating at low levels in tropical areas of the Americas, Asia and Africa. As with other flaviviruses, the threat of new outbreaks continues to exist. Therefore, one hopes that other companies and institutions remain committed to finding vaccines against this infection, for which there are no approved therapeutic or prophylactic treatments.

Global warming, globalization and frequent air travel have significantly increased the risk of ZIKV acquisition, due to expansion of the vector population and enhanced risk of human exposure. The fact that ZIKV can be sexually transmitted also poses a risk to pregnant women who may not live or travel to endemic areas, but whose partners may be exposed. Control of mosquito populations, recommendations for protective measures against mosquito exposure, and travel restrictions to endemic areas are the only interventions currently available. The number of ZIKV-infected cases is assumed to be grossly underreported due to the low rate of symptoms, which itself facilitates sexual transmission from asymptomatic individuals [84], which could potentially lead to new infections during pregnancy. Due to reports of ZIKV remaining detectable in semen for up to 188 days after symptom onset [200], it may be unreasonable to expect those living in or traveling to endemic areas to practice consistent condom use or to delay pregnancy for the recommended period of time (currently defined as 6 months for exposed male travelers and 8 weeks for exposed female travelers) [240]. For all of the above reasons, it is urgent that we develop a vaccine that effectively prevents ZIKV infection and disease. Furthermore, future epidemics are likely to appear as herd immunity diminishes with the influx of younger, ZIKV-naive populations [241]. We must hope that by the time this occurs, we will be prepared and have the necessary tools to protect communities from this devastating infection.

References

[1] Posen HJ, Keystone JS, Gubbay JB, Morris SK. Epidemiology of Zika virus, 1947-2007. BMJ Glob Health 2016;1:e000087.

[2] Musso D, Cao-Lormeau VM, Gubler DJ. Zika virus: following the path of dengue and chikungunya? Lancet 2015;386:243–4.

[3] Hennessey M, Fischer M, Staples JE. Zika virus spreads to new areas—region of the Americas, May 2015-January 2016. MMWR Morb Mortal Wkly Rep 2016;65:55–8.
[4] Dick GW, Kitchen SF, Haddow AJ. Zika virus. I. Isolations and serological specificity. Trans R Soc Trop Med Hyg 1952;46:509–20.
[5] Duffy MR, Chen TH, Hancock WT, et al. Zika virus outbreak on Yap Island, Federated States of Micronesia. N Engl J Med 2009;360:2536–43.
[6] Cao-Lormeau VM, Roche C, Teissier A, et al. Zika virus, French Polynesia, South Pacific, 2013. Emerg Infect Dis 2014;20:1085–6.
[7] Musso D, Nilles EJ, Cao-Lormeau VM. Rapid spread of emerging Zika virus in the Pacific area. Clin Microbiol Infect 2014;20:O595–6.
[8] Musso D, Baud D, Gubler DJ. Zika virus: what do we know? Clin Microbiol Infect 2016;22:494–6.
[9] Araujo LM, Ferreira ML, Nascimento OJ. Guillain-Barre syndrome associated with the Zika virus outbreak in Brazil. Arq Neuropsiquiatr 2016;74:253–5.
[10] Brasil P, Sequeira PC, Freitas AD, et al. Guillain-Barre syndrome associated with Zika virus infection. Lancet 2016;387:1482.
[11] Centers for Disease Control and Prevention. Microcephaly in infants, Pernambuco, Brazil, 2015: accumulating evidence of a relationship between microcephaly and Zika virus infection during pregnancy; 2015.
[12] Schuler-Faccini L, Ribeiro EM, Feitosa IM, et al. Possible association between Zika virus infection and microcephaly—Brazil, 2015. MMWR Morb Mortal Wkly Rep 2016;65:59–62.
[13] Brasil P, Nielsen-Saines K. More pieces to the microcephaly—Zika virus puzzle in Brazil. Lancet Infect Dis 2016;16(12):1307–9. https://doi.org/10.1016/S1473-3099(16)30372-3. PMID: 27641776.
[14] Gulland A. Zika virus is a global public health emergency, declares WHO. BMJ 2016;352:i657.
[15] World Health Organization. WHO Director-General summarizes the outcome of the Emergency Committee regarding clusters of microcephaly and Guillain-Barré syndrome. WHO statement on the first meeting of the International Health Regulations (2005). Emergency Committee on Zika virus and observed increase in neurological disorders and neonatal malformations; 2016.
[16] Sun LH. WHO no longer considers Zika a global health emergency. Wash Post 2016. November 18.
[17] Musso D, Gubler DJ. Zika virus. Clin Microbiol Rev 2016;29:487–524.
[18] Gubler DJ, Kuno G, Markoff L. Flaviviruses. 5th ed. Philadelphia, PA: Lippincott Williams and Wilkins; 2007.
[19] Kuno G, Chang GJ, Tsuchiya KR, Karabatsos N, Cropp CB. Phylogeny of the genus Flavivirus. J Virol 1998;72:73–83.
[20] Durbin AP. Dengue antibody and Zika: friend or foe? Trends Immunol 2016;37:635–6.
[21] Cook S, Holmes EC. A multigene analysis of the phylogenetic relationships among the flaviviruses (Family: Flaviviridae) and the evolution of vector transmission. Arch Virol 2006;151:309–25.
[22] Hayes EB. Zika virus outside Africa. Emerg Infect Dis 2009;15:1347–50.
[23] Kuno G, Chang GJ. Full-length sequencing and genomic characterization of Bagaza, Kedougou, and Zika viruses. Arch Virol 2007;152:687–96.
[24] Saiz JC, Vazquez-Calvo A, Blazquez AB, Merino-Ramos T, Escribano-Romero E, Martin-Acebes MA. Zika virus: the latest newcomer. Front Microbiol 2016;7:496.
[25] Dick GW. Zika virus. II. Pathogenicity and physical properties. Trans R Soc Trop Med Hyg 1952;46:521–34.
[26] Faye O, Freire CC, Iamarino A, et al. Molecular evolution of Zika virus during its emergence in the 20(th) century. PLoS Negl Trop Dis 2014;8:e2636.
[27] Pierson TC. Flaviviruses. 6th ed. Philadelphia, PA: Wolter Kluwer; 2013.

[28] Brown WC, Akey DL, Konwerski JR, et al. Extended surface for membrane association in Zika virus NS1 structure. Nat Struct Mol Biol 2016;23:865–7.
[29] Speer SD, Pierson TC. Virology. Diagnostics for Zika virus on the horizon. Science 2016;353:750–1.
[30] Xu X, Song H, Qi J, et al. Contribution of intertwined loop to membrane association revealed by Zika virus full-length NS1 structure. EMBO J 2016;35:2170–8.
[31] Fang Y, Brault AC, Reisen WK. Comparative thermostability of West Nile, St. Louis encephalitis, and western equine encephalomyelitis viruses during heat inactivation for serologic diagnostics. Am J Trop Med Hyg 2009;80:862–3.
[32] Gollins SW, Porterfield JS. The uncoating and infectivity of the flavivirus West Nile on interaction with cells: effects of pH and ammonium chloride. J Gen Virol 1986;67(Pt 9):1941–50.
[33] Aubry M, Richard V, Green J, Broult J, Musso D. Inactivation of Zika virus in plasma with amotosalen and ultraviolet A illumination. Transfusion 2016;56:33–40.
[34] Musso D, Richard V, Broult J, Cao-Lormeau VM. Inactivation of dengue virus in plasma with amotosalen and ultraviolet A illumination. Transfusion 2014;54:2924–30.
[35] Burke DS, Monath TP. Flaviviruses. Philadelphia, PA: Lippincott Williams & Wilkins; 2001.
[36] Blumel J, Musso D, Teitz S, et al. Inactivation and removal of Zika virus during manufacture of plasma-derived medicinal products. Transfusion 2016;57:790–6.
[37] Muller JA, Harms M, Schubert A, et al. Inactivation and environmental stability of Zika virus. Emerg Infect Dis 2016;22:1685–7.
[38] Haddow AD, Schuh AJ, Yasuda CY, et al. Genetic characterization of Zika virus strains: geographic expansion of the Asian lineage. PLoS Negl Trop Dis 2012;6:e1477.
[39] Cunha MS, Esposito DL, Rocco IM, et al. First complete genome sequence of Zika virus (Flaviviridae, Flavivirus) from an autochthonous transmission in Brazil. Genome Announc 2016;4.
[40] Giovanetti M, Faria NR, Nunes MR, et al. Zika virus complete genome from Salvador, Bahia, Brazil. Infect Genet Evol 2016;41:142–5.
[41] Faria NR, Azevedo Rdo S, Kraemer MU, et al. Zika virus in the Americas: early epidemiological and genetic findings. Science 2016;352:345–9.
[42] Brasil P, Calvet GA, Siqueira AM, et al. Zika virus outbreak in Rio de Janeiro, Brazil: clinical characterization, epidemiological and virological aspects. PLoS Negl Trop Dis 2016;10:e0004636.
[43] Palacios R, Poland GA, Kalil J. Another emerging arbovirus, another emerging vaccine: targeting Zika virus. Vaccine 2016;34:2291–3.
[44] Wang L, Valderramos SG, Wu A, et al. From mosquitos to humans: genetic evolution of Zika virus. Cell Host Microbe 2016;19:561–5.
[45] Ayllon T, Campos RM, Brasil P, et al. Early evidence for Zika virus circulation among Aedes aegypti mosquitoes, Rio de Janeiro, Brazil. Emerg Infect Dis 2017;23:1411–2.
[46] Massad E, Burattini MN, Khan K, Struchiner CJ, Coutinho FAB, Wilder-Smith A. On the origin and timing of Zika virus introduction in Brazil. Epidemiol Infect 2017;145:2303–12.
[47] Zika suspected and confirmed cases reported by countries and territories in the Americas Cumulative cases, 2015–2016. PAHO/WHO; 2016. at: https://www.paho.org/hq/dmdocuments/2017/2017-ago-10-phe-ZIKV-casess.pdf; 2016.
[48] Hills SL, Fischer M, Petersen LR. Epidemiology of Zika virus infection. J Infect Dis 2017;216:S868–74.
[49] Ikejezie J, Shapiro CN, Kim J, et al. Zika virus transmission-region of the Americas, May 15, 2015-December 15, 2016. Am J Transplant 2017;17:1681–6.
[50] Petersen LR, Jamieson DJ, Powers AM, Honein MA. Zika virus. N Engl J Med 2016;374:1552–63.
[51] World Health Organization. Zika virus, microcephaly, and Guillain-Barre Syndrome: Zika situation report; June 9, 2016. ed. 2016.

[52] Zika virus and complications. Accessed November 15, 2016, at, http://www.who.int/emergencies/zika-virus/en/; 2016.
[53] Table 1. Countries and territories that have reported mosquito-borne Zika virus transmission. at, http://www.who.int/emergencies/zika-virus/situation-report/15-december-2016/en/; 2016.
[54] Baud D, Gubler DJ, Schaub B, Lanteri MC, Musso D. An update on Zika virus infection. Lancet 2017;390(10107):2099–109. https://doi.org/10.1016/S0140-6736(17)31450-2. PMID: 28647173.
[55] Situation Report. Zika virus, microcephaly, Guillain-Barre syndrome. March 10, 2017. Accessed August 14, 2017, at, http://apps.who.int/iris/bitstream/10665/254714/1/zikasitrep10Mar17-eng.pdf?ua=1; 2017.
[56] The history of Zika virus. Accessed November 15, 2016, at, http://www.who.int/emergencies/zika-virus/history/en/; 2016.
[57] McCarthy M. US officials issue travel alert for Miami area as Zika cases rise to 15. BMJ 2016;354:i4298.
[58] McCarthy M. Four in Florida are infected with Zika from local mosquitoes. BMJ 2016;354:i4235.
[59] Likos A, Griffin I, Bingham AM, et al. Local mosquito-borne transmission of Zika virus—Miami-Dade and Broward counties, Florida, June-August 2016. MMWR Morb Mortal Wkly Rep 2016;65:1032–8.
[60] Centers for Disease Control and Prevention. CDC supporting Texas investigation of possible local Zika transmission; 2016.
[61] Chen L, Hafeez F, Curry CL, Elgart G. Cutaneous eruption in a U.S. woman with locally acquired Zika virus infection. N Engl J Med 2017;376:400–1.
[62] Zika cases in the United States: cumulative Zika virus disease case counts in the United States, 2015–2017. Accessed August 3, 2017, at, https://www.cdc.gov/zika/reporting/case-counts.html; 2017.
[63] Reynolds MR, Jones AM, Petersen EE, et al. Vital signs: update on Zika virus-associated birth defects and evaluation of all U.S. infants with congenital Zika virus exposure—U.S. Zika pregnancy registry, 2016. MMWR Morb Mortal Wkly Rep 2017;66:366–73.
[64] Outcomes of pregnancies with laboratory evidence of possible Zika virus infection in the United States, 2016. Accessed June 26, 2016 and November 16, 2016, at, http://www.cdc.gov/zika/geo/pregnancy-outcomes.html; 2016.
[65] Honein MA, Dawson AL, Petersen EE, et al. Birth defects among fetuses and infants of US women with evidence of possible Zika virus infection during pregnancy. JAMA 2017;317:59–68.
[66] Shapiro-Mendoza CK, Rice ME, Galang RR, et al. Pregnancy outcomes after maternal Zika virus infection during pregnancy—U.S. territories, January 1, 2016-April 25, 2017. MMWR Morb Mortal Wkly Rep 2017;66:615–21.
[67] Hoen B, Schaub B, Funk AL, et al. Pregnancy outcomes after ZIKV infection in French territories in the Americas. N Engl J Med 2018;378:985–94.
[68] Brasil P, Pereira Jr JP, Moreira ME, et al. Zika virus infection in pregnant women in Rio de Janeiro. N Engl J Med 2016;375:2321–34.
[69] Cragan JD, Mai CT, Petersen EE, et al. Baseline prevalence of birth defects associated with congenital Zika virus infection—Massachusetts, North Carolina, and Atlanta, Georgia, 2013-2014. MMWR Morb Mortal Wkly Rep 2017;66:219–22.
[70] Plourde AR, Bloch EM. A literature review of Zika virus. Emerg Infect Dis 2016;22:1185–92.
[71] Krow-Lucal ER, Biggerstaff BJ, Staples JE. Estimated incubation period for Zika virus disease. Emerg Infect Dis 2017;23:841–5.
[72] World Health Organization. Zika virus disease: interim case definitions; 2016.
[73] Zika virus disease and Zika virus, congenital infection 2016 case definition. Accessed July 8, 2016, at, https://wwwn.cdc.gov/nndss/conditions/zika-virus-disease-and-zika-virus-congenital-infection/case-definition/2016/; 2016.

[74] Zika virus disease and Zika virus infection 2016 case definition. Approved June 2016, at, https://wwwn.cdc.gov/nndss/conditions/zika/case-definition/2016/06/; 2016.
[75] Brasil P, Pereira Jr JP, Raja Gabaglia C, et al. Zika virus infection in pregnant women in Rio de Janeiro—preliminary report. N Engl J Med 2016;https://doi.org/10.1056/NEJMoa1602412.
[76] Waggoner JJ, Pinsky BA. Zika virus: diagnostics for an emerging pandemic threat. J Clin Microbiol 2016;54:860–7.
[77] Cabral-Castro MJ, Cavalcanti MG, Peralta RH, Peralta JM. Molecular and serological techniques to detect co-circulation of DENV, ZIKV and CHIKV in suspected dengue-like syndrome patients. J Clin Virol 2016;82:108–11.
[78] Fernandez-Salas I, Diaz-Gonzalez EE, Lopez-Gatell H, Alpuche-Aranda C. Chikugunya and zika virus dissemination in the Americas: different arboviruses reflecting the same spreading routes and poor vector-control policies. Curr Opin Infect Dis 2016;29:467–75.
[79] World Health Organization. Zika virus: fact sheet; 2016.
[80] Guerbois M, Fernandez-Salas I, Azar SR, et al. Outbreak of Zika virus infection, Chiapas State, Mexico, 2015, and first confirmed transmission by Aedes aegypti mosquitoes in the Americas. J Infect Dis 2016;214:1349–56.
[81] Cerbino-Neto J, Mesquita EC, Souza TM, et al. Clinical manifestations of Zika virus infection, Rio de Janeiro, Brazil, 2015. Emerg Infect Dis 2016;22:1318–20.
[82] Thomas DL, Sharp TM, Torres J, et al. Local transmission of Zika virus—Puerto Rico, November 23, 2015-January 28, 2016. MMWR Morb Mortal Wkly Rep 2016;65:154–8.
[83] Farahnik B, Beroukhim K, Blattner CM, Young 3rd J. Cutaneous manifestations of the Zika virus. J Am Acad Dermatol 2016;74:1286–7.
[84] Brooks RB, Carlos MP, Myers RA, et al. Likely sexual transmission of Zika virus from a man with no symptoms of infection—Maryland, 2016. MMWR Morb Mortal Wkly Rep 2016;65:915–6.
[85] Centers for Disease Control and Prevention. CDC responds to ZIKA: Zika virus: information for clinicians; June 13, 2016.
[86] Braga JU, Bressan C, Dalvi APR, et al. Accuracy of Zika virus disease case definition during simultaneous dengue and chikungunya epidemics. PLoS One 2017;12:e0179725.
[87] Chow A, Ho H, Win MK, Leo YS. Assessing sensitivity and specificity of surveillance case definitions for Zika virus disease. Emerg Infect Dis 2017;23:677–9.
[88] Walker WL, Lindsey NP, Lehman JA, et al. Zika virus disease cases—50 states and the District of Columbia, January 1-July 31, 2016. MMWR Morb Mortal Wkly Rep 2016;65:983–6.
[89] Goodman AB, Dziuban EJ, Powell K, et al. Characteristics of children aged <18 years with Zika virus disease acquired postnatally—U.S. states, January 2015-July 2016. MMWR Morb Mortal Wkly Rep 2016;65:1082–5.
[90] Fonseca K, Meatherall B, Zarra D, et al. First case of Zika virus infection in a returning Canadian traveler. Am J Trop Med Hyg 2014;91:1035–8.
[91] Tappe D, Nachtigall S, Kapaun A, Schnitzler P, Gunther S, Schmidt-Chanasit J. Acute Zika virus infection after travel to Malaysian Borneo, September 2014. Emerg Infect Dis 2015;21:911–3.
[92] Foy BD, Kobylinski KC, Chilson Foy JL, et al. Probable non-vector-borne transmission of Zika virus, Colorado, USA. Emerg Infect Dis 2011;17:880–2.
[93] Musso D. Zika virus transmission from French Polynesia to Brazil. Emerg Infect Dis 2015;21:1887.
[94] Weitzel T, Cortes CP. Zika virus infection presenting with postauricular lymphadenopathy. Am J Trop Med Hyg 2016;95(2):255–6. https://doi.org/10.4269/ajtmh.16-0096. PMID: 27694629. PMCID: PMC4973165.
[95] Fontes BM. Zika virus-related hypertensive iridocyclitis. Arq Bras Oftalmol 2016;79:63.
[96] Furtado JM, Esposito DL, Klein TM, Teixeira-Pinto T, da Fonseca BA. Uveitis associated with Zika virus infection. N Engl J Med 2016;375:394–6.

[97] Parke 3rd DW, Almeida DR, Albini TA, Ventura CV, Berrocal AM, Mittra RA. Serologically confirmed Zika-related unilateral acute maculopathy in an adult. Ophthalmology 2016;123:2432–3.
[98] Meltzer E, Leshem E, Lustig Y, Gottesman G, Schwartz E. The clinical spectrum of Zika virus in returning travelers. Am J Med 2016;129:1126–30.
[99] Zea-Vera AF, Parra B. Zika virus (ZIKV) infection related with immune thrombocytopenic purpura (ITP) exacerbation and antinuclear antibody positivity. Lupus 2017;26(8):890–2. https://doi.org/10.1177/0961203316671816. PMID: 27694629.
[100] Aletti M, Lecoules S, Kanczuga V, et al. Transient myocarditis associated with acute Zika virus infection. Clin Infect Dis 2017;64:678–9.
[101] Henry CR, Al-Attar L, Cruz-Chacon AM, Davis JL. Chorioretinal lesions presumed secondary to Zika virus infection in an immunocompromised adult. JAMA Ophthalmol 2017;135:386–9.
[102] Slavov S, Matsuno A, Yamamoto A, et al. Zika virus infection in a pediatric patient with acute gastrointestinal involvement. Pediatr Rep 2017;9:7341.
[103] Karimi O, Goorhuis A, Schinkel J, et al. Thrombocytopenia and subcutaneous bleedings in a patient with Zika virus infection. Lancet 2016;387:939–40.
[104] Sarmiento-Ospina A, Vásquez-Serna H, Jimenez-Canizales CE, Villamil-Gómez WE, Rodriguez-Morales AJ. Zika virus associated deaths in Colombia. Lancet Infect Dis 2016;16(5):523–4. https://doi.org/10.1016/S1473-3099(16)30006-8. PMID: 27068488.
[105] Zammarchi L, Stella G, Mantella A, et al. Zika virus infections imported to Italy: clinical, immunological and virological findings, and public health implications. J Clin Virol 2015;63:32–5.
[106] Musso D, de Pina JJ, Nhan TX, Deparis X. Uncommon presentation of Zika fever or co-infection? Lancet 2016;387:1812–3.
[107] Sun LH. First Zika virus-related death reported in U.S. in Puerto Rico. Wash Post 2016. April 29, 2016.
[108] Sharp TM, Munoz-Jordan J, Perez-Padilla J, et al. Zika virus infection associated with severe thrombocytopenia. Clin Infect Dis 2016;63:1198–201.
[109] Ioos S, Mallet HP, Leparc Goffart I, Gauthier V, Cardoso T, Herida M. Current Zika virus epidemiology and recent epidemics. Med Mal Infect 2014;44:302–7.
[110] Adams L, Bello-Pagan M, Lozier M, et al. Update: ongoing Zika virus transmission—Puerto Rico, November 1, 2015-July 7, 2016. MMWR Morb Mortal Wkly Rep 2016;65:774–9.
[111] Chraibi S, Najioullah F, Bourdin C, et al. Two cases of thrombocytopenic purpura at onset of Zika virus infection. J Clin Virol 2016;83:61–2.
[112] Calvet GA, Santos FB, Sequeira PC. Zika virus infection: epidemiology, clinical manifestations and diagnosis. Curr Opin Infect Dis 2016;29:459–66.
[113] Boyer Chammard T, Schepers K, Breurec S, et al. Severe thrombocytopenia after Zika virus infection, Guadeloupe, 2016. Emerg Infect Dis 2017;23:696–8.
[114] Swaminathan S, Schlaberg R, Lewis J, Hanson KE, Couturier MR. Fatal Zika virus infection with secondary nonsexual transmission. N Engl J Med 2016;375(19):1907–9. PMID: 27681699. PMCID: PMC5267509.
[115] Brazilian Ministry of Health. Informe epidemiologico no 02/2015-Semana epidemiologica 47 (22 a 28/11/2015): Monitoramento dos casos de microcefalias no Brasil. (The Public Health Emergency Operations Center report on microcephaly. Epidemiological week 47 of 2015.); 2015.
[116] Azevedo RS, Araujo MT, Martins Filho AJ, et al. Zika virus epidemic in Brazil. I. Fatal disease in adults: clinical and laboratorial aspects. J Clin Virol 2016;85:56–64.
[117] Soares CN, Brasil P, Carrera RM, et al. Fatal encephalitis associated with Zika virus infection in an adult. J Clin Virol 2016;83:63–5.
[118] Roze B, Najioullah F, Signate A, et al. Zika virus detection in cerebrospinal fluid from two patients with encephalopathy, Martinique, February 2016. Euro Surveill 2016;21.
[119] Carteaux G, Maquart M, Bedet A, et al. Zika virus associated with meningoencephalitis. N Engl J Med 2016;374:1595–6.

[120] Mécharles S, Herrmann C, Poullain P, Tran TH, Deschamps N, Mathon G, Landais A, Breurec S, Lannuzel A. Acute myelitis due to Zika virus infection. Lancet 2016;387(10026):1481. https://doi.org/10.1016/S0140-6736(16)00644-9. PMID: 26946926.
[121] Cao-Lormeau VM, Blake A, Mons S, et al. Guillain-Barre syndrome outbreak associated with Zika virus infection in French Polynesia: a case-control study. Lancet 2016;387:1531–9.
[122] Fontes CA, Dos Santos AA, Marchiori E. Magnetic resonance imaging findings in Guillain-Barré syndrome caused by Zika virus infection. Neuroradiology 2016;58(8):837–8. https://doi.org/10.1007/s00234-016-1687-9. PMID: 27067205.
[123] Oehler E, Watrin L, Larre P, et al. Zika virus infection complicated by Guillain-Barre syndrome—case report, French Polynesia, December 2013. Euro Surveill 2014;19.
[124] Kassavetis P, Joseph JM, Francois R, Perloff MD, Berkowitz AL. Zika virus-associated Guillain-Barré syndrome variant in Haiti. Neurology 2016;87(3):336–7. https://doi.org/10.1212/WNL.0000000000002759. PMID: 27164708.
[125] Medina MT, England JD, Lorenzana I, et al. Zika virus associated with sensory polyneuropathy. J Neurol Sci 2016;369:271–2.
[126] Zambrano H, Waggoner JJ, Almeida C, Rivera L, Benjamin JQ, Pinsky BA. Zika virus and chikungunya virus coinfections: a series of three cases from a single center in Ecuador. Am J Trop Med Hyg 2016;95:894–6.
[127] Siu R, Bukhari W, Todd A, Gunn W, Huang QS, Timmings P. Acute Zika infection with concurrent onset of Guillain-Barre syndrome. Neurology 2016;87:1623–4.
[128] Septfons A, Leparc-Goffart I, Couturier E, et al. Travel-associated and autochthonous Zika virus infection in mainland France, 1 January to 15 July 2016. Euro Surveill 2016;21.
[129] Asadi-Pooya AA. Zika virus-associated seizures. Seizure 2016;43:13.
[130] Arias A, Torres-Tobar L, Hernandez G, et al. Guillain-Barre syndrome in patients with a recent history of Zika in Cucuta, Colombia: a descriptive case series of 19 patients from December 2015 to March 2016. J Crit Care 2016;37:19–23.
[131] Parra B, Lizarazo J, Jimenez-Arango JA, et al. Guillain-Barre syndrome associated with Zika virus infection in Colombia. N Engl J Med 2016;375:1513–23.
[132] Roze B, Najioullah F, Ferge JL, et al. Zika virus detection in urine from patients with Guillain-Barre syndrome on Martinique, January 2016. Euro Surveill 2016;21.
[133] Dos Santos T, Rodriguez A, Almiron M, et al. Zika virus and the Guillain-Barre syndrome—case series from seven countries. N Engl J Med 2016;375:1598–601.
[134] Dirlikov E, Major CG, Mayshack M, et al. Guillain-Barre syndrome during ongoing Zika virus transmission—Puerto Rico, January 1-July 31, 2016. MMWR Morb Mortal Wkly Rep 2016;65:910–4.
[135] do Rosario MS, de Jesus PA, Vasilakis N, et al. Guillain-Barre syndrome after Zika virus infection in Brazil. Am J Trop Med Hyg 2016;95:1157–60.
[136] Munoz LS, Barreras P, Pardo CA. Zika virus-associated neurological disease in the adult: Guillain-Barre syndrome, encephalitis, and myelitis. Semin Reprod Med 2016;34:273–9.
[137] Fabrizius RG, Anderson K, Hendel-Paterson B, Kaiser RM, Maalim S, Walker PF. Guillain-Barre syndrome associated with Zika virus infection in a traveler returning from Guyana. Am J Trop Med Hyg 2016;95:1161–5.
[138] Nicastri E, Castilletti C, Balestra P, Galgani S, Ippolito G. Zika virus infection in the central nervous system and female genital tract. Emerg Infect Dis 2016;22:2228–30.
[139] Zucker J, Neu N, Chiriboga CA, et al. Zika virus-associated cognitive impairment in adolescent, 2016. Emerg Infect Dis 2017;23:1047–8.
[140] Gold CA, Josephson SA. Anticipating the challenges of Zika virus and the incidence of Guillain-Barré syndrome. JAMA Neurol 2016;73(8):905–6. https://doi.org/10.1001/jamaneurol.2016.1268. PMID: 27272118.

[141] Yung CF, Thoon KC. Guillain-Barré syndrome and Zika virus: estimating attributable risk to inform intensive care capacity preparedness. Clin Infect Dis 2016;63(5):708–9. https://doi.org/10.1093/cid/ciw355. PMID: 27225243.

[142] Smith DW, Mackenzie J. Zika virus and Guillain-Barre syndrome: another viral cause to add to the list. Lancet 2016;387:1486–8.

[143] Zika virus disease in the United States, 2015–2016. Accessed June 26, 2016, at, http://www.cdc.gov/zika/geo/united-states.html; 2016.

[144] Ferreira da Silva IR, Frontera JA, Moreira do Nascimento OJ. News from the battlefront: Zika virus-associated Guillain-Barre syndrome in Brazil. Neurology 2016;87:e180–1.

[145] Paploski IA, Prates AP, Cardoso CW, et al. Time lags between exanthematous illness attributed to Zika virus, Guillain-Barre syndrome, and microcephaly, Salvador, Brazil. Emerg Infect Dis 2016;22:1438–44.

[146] Moore CA, Staples JE, Dobyns WB, Pessoa A, Ventura CV, Fonseca EB, Ribeiro EM, Ventura LO, Neto NN, Arena JF, Rasmussen SA. Characterizing the pattern of anomalies in congenital Zika syndrome for pediatric clinicians. JAMA Pediatr 2017;171(3):288–95. https://doi.org/10.1001/jamapediatrics.2016.3982. PMID: 27812690. PMCID: PMC5561417.

[147] PAHO/WHO. Zika cases and congenital syndrome associated with Zika virus reported by countries and territories in the Americas, 2015–2017: cumulative cases 2017. August 10, 2017.

[148] Microcephaly Epidemic Research Group. Microcephaly in infants, Pernambuco State, Brazil, 2015. Emerg Infect Dis 2016;22:1090–3.

[149] Meneses JDA, Ishigami AC, de Mello LM, et al. Lessons learned at the epicenter of Brazil's congenital Zika epidemic: evidence from 87 confirmed cases. Clin Infect Dis 2017;64:1302–8.

[150] Del Campo M, Feitosa IM, Ribeiro EM, et al. The phenotypic spectrum of congenital Zika syndrome. Am J Med Genet A 2017;173:841–57.

[151] van der Linden V, Filho EL, Lins OG, et al. Congenital Zika syndrome with arthrogryposis: retrospective case series study. BMJ 2016;354:i3899.

[152] Leal MC, Muniz LF, Ferreira TS, et al. Hearing loss in infants with microcephaly and evidence of congenital Zika virus infection—Brazil, November 2015-May 2016. MMWR Morb Mortal Wkly Rep 2016;65:917–9.

[153] Leal MC, Muniz LF, Caldas Neto SD, van der Linden V, Ramos RC. Sensorineural hearing loss in a case of congenital Zika virus. Braz J Otorhinolaryngol 2016;https://doi.org/10.1016/j.bjorl.2016.06.001. (e-pub ahead of print). PMID: 27444419.

[154] Valentine G, Marquez L, Pammi M. Zika virus-associated microcephaly and eye lesions in the newborn. J Pediatric Infect Dis Soc 2016;5:323–8.

[155] de Paula Freitas B, de Oliveira Dias JR, Prazeres J, Sacramento GA, Ko AI, Maia M, Belfort Jr R. Ocular findings in infants with microcephaly associated with presumed Zika virus congenital infection in Salvador, Brazil. JAMA Ophthalmol 2016;134(5):529–35. https://doi.org/10.1001/jamaophthalmol.2016.0267. PMID: 26865554; PMCID: PMC5444996.

[156] Ventura CV, Maia M, Bravo-Filho V, Gois AL, Belfort Jr R. Zika virus in Brazil and macular atrophy in a child with microcephaly. Lancet 2016;387:228.

[157] Ventura CV, Maia M, Travassos SB, et al. Risk factors associated with the ophthalmoscopic findings identified in infants with presumed Zika virus congenital infection. JAMA Ophthalmol 2016;134:912–8.

[158] Verçosa I, Carneiro P, Verçosa R, Girão R, Ribeiro EM, Pessoa A, Almeida NG, Verçosa P, Tartarella MB. The visual system in infants with microcephaly related to presumed congenital Zika syndrome. J AAPOS 2017;21(4):300–304.e1. https://doi.org/10.1016/j.jaapos.2017.05.024. PMID: 28652051.

[159] Ventura LO, Ventura CV, Lawrence L, van der Linden V, van der Linden A, Gois AL, Cavalcanti MM, Barros EA, Dias NC, Berrocal AM, Miller MT. Visual impairment in children with congenital Zika syndrome. J AAPOS 2017;21(4):295–299.e2. https://doi.org/10.1016/j.jaapos.2017.04.003. PMID: 28450178.

[160] Yepez JB, Murati FA, Pettito M, et al. Ophthalmic manifestations of congenital Zika syndrome in Colombia and Venezuela. JAMA Ophthalmol 2017;135:440–5.

[161] Miranda 2nd HA, Costa MC, Frazão MAM, Simão N, Franchischini S, Moshfeghi DM. Expanded spectrum of congenital ocular findings in microcephaly with presumed Zika infection. Ophthalmology 2016;123(8):1788–94. https://doi.org/10.1016/j.ophtha.2016.05.001. PMID: 27236271.

[162] Ventura CV, Maia M, Ventura BV, et al. Ophthalmological findings in infants with microcephaly and presumable intra-uterus Zika virus infection. Arq Bras Oftalmol 2016;79:1–3.

[163] Zin AA, Tsui I, Rossetto J, Vasconcelos Z, Adachi K, Valderramos S, Halai UA, Pone MVDS, Pone SM, Silveira Filho JCB, Aibe MS, da Costa ACC, Zin OA, Belfort Jr R, Brasil P, Nielsen-Saines K, Moreira MEL. Screening criteria for ophthalmic manifestations of congenital Zika virus infection. JAMA Pediatr 2017;171(9):847–54. https://doi.org/10.1001/jamapediatrics.2017.1474. PMID: 28715527. PMCID: PMC5710409.

[164] Cavalcanti DD, Alves LV, Furtado GJ, et al. Echocardiographic findings in infants with presumed congenital Zika syndrome: retrospective case series study. PLoS One 2017;12:e0175065.

[165] Orofino DHG, Passos SRL, de Oliveira RVC, de Farias CVB, Leite MFMP, Pone SM, Pone MVDS, Teixeira Mendes HAR, Moreira MEL, Nielsen-Saines K. Cardiac findings in infants with in utero exposure to Zika virus—a cross sectional study. PLoS Negl Trop Dis 2018;12(3):e0006362.

[166] Ramalho FS, Yamamoto AY, da Silva LL, Figueiredo LTM, Rocha LB, Neder L, Teixeira SR, Apolinário LA, Ramalho LNZ, Silva DM, Coutinho CM, Melli PP, Augusto MJ, Santoro LB, Duarte G, Mussi-Pinhata MM. Congenital Zika virus infection induces severe spinal cord injury. Clin Infect Dis 2017;65(4):687–90. https://doi.org/10.1093/cid/cix374. PMID: 28444144.

[167] van der Linden V, Pessoa A, Dobyns W, et al. Description of 13 infants born during October 2015-January 2016 with congenital Zika virus infection without microcephaly at birth—Brazil. MMWR Morb Mortal Wkly Rep 2016;65:1343–8.

[168] Kapogiannis BG, Chakhtoura N, Hazra R, Spong CY. Bridging knowledge gaps to understand how Zika virus exposure and infection affect child development. JAMA Pediatr 2017;171:478–85.

[169] Moura da Silva AA, Ganz JS, Sousa PD, et al. Early growth and neurologic outcomes of infants with probable congenital Zika virus syndrome. Emerg Infect Dis 2016;22:1953–6.

[170] Petribu NCL, Aragao MFV, van der Linden V, et al. Follow-up brain imaging of 37 children with congenital Zika syndrome: case series study. BMJ 2017;359:j4188.

[171] Alves LV, Mello MJG, Bezerra PG, Alves JGB. Congenital Zika syndrome and infantile spasms: case series study. J Child Neurol 2018;33:664–6.

[172] Pinato L, Ribeiro EM, Leite RFP, et al. Sleep findings in Brazilian children with congenital Zika syndrome. Sleep 2018;41.

[173] Raymond A, Jakus J. Cerebral infarction and refractory seizures in a neonate with suspected Zika virus infection. Pediatr Infect Dis J 2018;37:e112–4.

[174] Cunha AJ, de Magalhaes-Barbosa MC, Lima-Setta F, Medronho RA, Prata-Barbosa A. Microcephaly case fatality rate associated with Zika virus infection in Brazil: current estimates. Pediatr Infect Dis J 2017;36:528–30.

[175] Melo AS, Aguiar RS, Amorim MM, Arruda MB, Melo FO, Ribeiro ST, Batista AG, Ferreira T, Dos Santos MP, Sampaio VV, Moura SR, Rabello LP, Gonzaga CE, Malinger G, Ximenes R, de Oliveira-Szejnfeld PS, Tovar-Moll F, Chimelli L, Silveira PP, Delvechio R, Higa L, Campanati L, Nogueira RM, Filippis AM, Szejnfeld J, Voloch CM, Ferreira Jr OC, Brindeiro RM, Tanuri A. Congenital Zika virus infection: beyond

neonatal microcephaly. JAMA Neurol 2016;73(12):1407–16. https://doi.org/10.1001/jamaneurol.2016.3720. PMID: 27695855.
[176] Satterfield-Nash A, Kotzky K, Allen J, et al. Health and development at age 19-24 months of 19 children who were born with microcephaly and laboratory evidence of congenital Zika virus infection during the 2015 Zika virus outbreak—Brazil, 2017. MMWR Morb Mortal Wkly Rep 2017;66:1347–51.
[177] Wheeler AC. Development of infants with congenital Zika syndrome: what do we know and what can we expect? Pediatrics 2018;141:S154–60.
[178] Gostin LO, Hodge Jr JG. Zika virus and global health security. Lancet Infect Dis 2016;16:1099–100.
[179] ARA A, Trussell J. Women and children are political pawns in the Zika funding battle. Time 2016. July 5, 2016.
[180] Hotez PJ. What does Zika virus mean for the children of the Americas? JAMA Pediatr 2016;170(8):787–9. https://doi.org/10.1001/jamapediatrics.2016.1465. Erratum in: JAMA Pediatr 2016;170 (8):811. PMID: 27322491.
[181] Broussard CS, Shapiro-Mendoza CK, Peacock G, et al. Public health approach to addressing the needs of children affected by congenital Zika syndrome. Pediatrics 2018;141:S146–53.
[182] World Health Organization. Zika situation Report: neurological syndrome and congenital anomalies. World Health Organization; 2016.
[183] Lazear HM, Govero J, Smith AM, et al. A mouse model of Zika virus pathogenesis. Cell Host Microbe 2016;19:720–30.
[184] Makhluf H, Shresta S. Development of Zika virus vaccines. Vaccines (Basel) 2018;6.
[185] WHO Vaccine pipeline tracker. Accessed July 2018, at http://www.who.int/immunization/research/vaccine_pipeline_tracker_spreadsheet/en/, 2018.
[186] Tripp RA, Ross TM. Development of a Zika vaccine. Expert Rev Vaccines 2016;15:1083–5.
[187] Gostin LO, Hodge Jr JG. Is the United States prepared for a major Zika virus outbreak? JAMA 2016;315:2395–6.
[188] Mier YT-RL, Delorey MJ, Sejvar JJ, Johansson MA. Guillain-Barre syndrome risk among individuals infected with Zika virus: a multi-country assessment. BMC Med 2018;16:67.
[189] de Oliveira WK, de Franca GVA, Carmo EH, Duncan BB, de Souza KR, Schmidt MI. Infection-related microcephaly after the 2015 and 2016 Zika virus outbreaks in Brazil: a surveillance-based analysis. Lancet 2017;390:861–70.
[190] Halai UA, Nielsen-Saines K, Moreira ML, et al. Maternal Zika virus disease severity, virus load, prior dengue antibodies, and their relationship to birth outcomes. Clin Infect Dis 2017;65:877–83.
[191] Richner JM, Jagger BW, Shan C, et al. Vaccine mediated protection against Zika virus-induced congenital disease. Cell 2017;170:273–83. e12.
[192] Shan C, Muruato AE, Jagger BW, et al. A single-dose live-attenuated vaccine prevents Zika virus pregnancy transmission and testis damage. Nat Commun 2017;8:676.
[193] Wang R, Liao X, Fan D, et al. Maternal immunization with a DNA vaccine candidate elicits specific passive protection against post-natal Zika virus infection in immunocompetent BALB/c mice. Vaccine 2018;36:3522–32.
[194] Zhu X, Li C, Afridi SK, et al. E90 subunit vaccine protects mice from Zika virus infection and microcephaly. Acta Neuropathol Commun 2018;6:77.
[195] Wilder-Smith A, Vannice K, Durbin A, et al. Zika vaccines and therapeutics: landscape analysis and challenges ahead. BMC Med 2018;16:84.
[196] WHO. Zika virus (ZIKV) vaccine target product profile (TPP): vaccine to protect against congenital Zika virus syndrome for use during an emergency. at, http://www.who.int/immunization/research/development/WHO_UNICEF_Zikavac_TPP_Feb2017.pdf; 2016.
[197] Lyerly AD, Robin SG, Jaffe E. Rubella and Zika vaccine research—a cautionary tale about caution. JAMA Pediatr 2017;171:719–20.

[198] Leineweber B, Grote V, Schaad UB, Heininger U. Transplacentally acquired immunoglobulin G antibodies against measles, mumps, rubella and varicella-zoster virus in preterm and full term newborns. Pediatr Infect Dis J 2004;23:361–3.
[199] Joguet G, Mansuy JM, Matusali G, et al. Effect of acute Zika virus infection on sperm and virus clearance in body fluids: a prospective observational study. Lancet Infect Dis 2017;17:1200–8.
[200] Nicastri E, Castilletti C, Liuzzi G, Iannetta M, Capobianchi MR, Ippolito G. Persistent detection of Zika virus RNA in semen for six months after symptom onset in a traveller returning from Haiti to Italy, February 2016. Euro Surveill 2016;21.
[201] Singh MV, Weber EA, Singh VB, Stirpe NE, Maggirwar SB. Preventive and therapeutic challenges in combating Zika virus infection: are we getting any closer? J Neurovirol 2017;23:347–57.
[202] Omer SB, Beigi RH. Pregnancy in the time of Zika: addressing barriers for developing vaccines and other measures for pregnant women. JAMA 2016;315:1227–8.
[203] Johansson MA, Mier-y-Teran-Romero L, Reefhuis J, Gilboa SM, Hills SL. Zika and the risk of microcephaly. N Engl J Med 2016;375:1–4.
[204] Goncalves APR, Chu MC, Gubler DJ, de Silva AM, Harris E, Murtagh M, Chua A, Rodriguez W, Kelly C. Innovative and new approaches to laboratory diagnosis of Zika and dengue: a meeting report. J Infect Dis 2017;217:1060–8.
[205] Barrett ADT. Current status of Zika vaccine development: Zika vaccines advance into clinical evaluation. NPJ Vaccines 2018;3:24.
[206] Barrett ADT. Zika vaccine candidates progress through nonclinical development and enter clinical trials. NPJ Vaccines 2016;1:16023.
[207] Lin HH, Yip BS, Huang LM, Wu SC. Zika virus structural biology and progress in vaccine development. Biotechnol Adv 2018;36:47–53.
[208] Kalkeri R, Murthy KK. Zika virus reservoirs: Implications for transmission, future outbreaks, drug and vaccine development. F1000Res 2017;6:1850.
[209] Barzon L, Palu G. Current views on Zika virus vaccine development. Expert Opin Biol Ther 2017;17:1185–92.
[210] Coller BA, Clements DE, Bett AJ, Sagar SL, Ter Meulen JH. The development of recombinant subunit envelope-based vaccines to protect against dengue virus induced disease. Vaccine 2011;29:7267–75.
[211] Katzelnick LC, Gresh L, Halloran ME, et al. Antibody-dependent enhancement of severe dengue disease in humans. Science 2017;358:929–32.
[212] Dejnirattisai W, Supasa P, Wongwiwat W, et al. Dengue virus sero-cross-reactivity drives antibody-dependent enhancement of infection with zika virus. Nat Immunol 2016;17:1102–8.
[213] Barba-Spaeth G, Dejnirattisai W, Rouvinski A, et al. Structural basis of potent Zika-dengue virus antibody cross-neutralization. Nature 2016;536:48–53.
[214] Charles AS, Christofferson RC. Utility of a dengue-derived monoclonal antibody to enhance Zika infection in vitro. PLoS Curr 2016;8.
[215] Stettler K, Beltramello M, Espinosa DA, et al. Specificity, cross-reactivity, and function of antibodies elicited by Zika virus infection. Science 2016;353:823–6.
[216] Barouch DH, Thomas SJ, Michael NL. Prospects for a Zika virus vaccine. Immunity 2017;46:176–82.
[217] Barzon L, Palu G. Current views on Zika virus vaccine development. Expert Opin Biol Ther 2017;1–8.
[218] Richner JM, Himansu S, Dowd KA, et al. Modified mRNA vaccines protect against Zika virus infection. Cell 2017;168:1114–25. e10.
[219] Yang M, Dent M, Lai H, Sun H, Chen Q. Immunization of Zika virus envelope protein domain III induces specific and neutralizing immune responses against Zika virus. Vaccine 2017;35:4287–94.

[220] Magnani DM, Rogers TF, Beutler N, et al. Neutralizing human monoclonal antibodies prevent Zika virus infection in macaques. Sci Transl Med 2017;9.
[221] Gopichandran V. Controlled human infection models for vaccine development: Zika virus debate. Indian J Med Ethics 2018;3:51–5.
[222] Durbin AP, Whitehead SS. Zika vaccines: role for controlled human infection. J Infect Dis 2017;216:S971–5.
[223] Marston HD, Lurie N, Borio LL, Fauci AS. Considerations for developing a Zika virus vaccine. N Engl J Med 2016;375:1209–12.
[224] Gruber MF, Krause PR. Regulating vaccines at the FDA: development and licensure of Zika vaccines. Expert Rev Vaccines 2017;16:525–7.
[225] Modjarrad K, Lin L, George SL, et al. Preliminary aggregate safety and immunogenicity results from three trials of a purified inactivated Zika virus vaccine candidate: phase 1, randomised, double-blind, placebo-controlled clinical trials. Lancet 2018;391:563–71.
[226] Gaudinski MR, Houser KV, Morabito KM, et al. Safety, tolerability, and immunogenicity of two Zika virus DNA vaccine candidates in healthy adults: randomised, open-label, phase 1 clinical trials. Lancet 2018;391:552–62.
[227] Tebas P, Roberts CC, Muthumani K, et al. Safety and immunogenicity of an anti-Zika virus DNA vaccine—preliminary report. N Engl J Med 2017;https://doi.org/10.1056/NEJMoa1708120. (e-pub ahead of print). PMID: 28976850.
[228] Kim E, Erdos G, Huang S, Kenniston T, Falo Jr LD, Gambotto A. Preventative vaccines for Zika virus outbreak: preliminary evaluation. EBioMedicine 2016;13:315–20.
[229] Boigard H, Alimova A, Martin GR, Katz A, Gottlieb P, Galarza JM. Zika virus-like particle (VLP) based vaccine. PLoS Negl Trop Dis 2017;11:e0005608.
[230] Guy B, Guirakhoo F, Barban V, Higgs S, Monath TP, Lang J. Preclinical and clinical development of YFV 17D-based chimeric vaccines against dengue, West Nile and Japanese encephalitis viruses. Vaccine 2010;28:632–49.
[231] Durbin AP, McArthur JH, Marron JA, et al. rDEN2/4Delta30(ME), a live attenuated chimeric dengue serotype 2 vaccine is safe and highly immunogenic in healthy dengue-naive adults. Hum Vaccin 2006;2:255–60.
[232] Abbink P, Larocca RA, De La Barrera RA, et al. Protective efficacy of multiple vaccine platforms against Zika virus challenge in rhesus monkeys. Science 2016;353:1129–32.
[233] Sumathy K, Kulkarni B, Gondu RK, et al. Protective efficacy of Zika vaccine in AG129 mouse model. Sci Rep 2017;7:46375.
[234] Richner JM, Himansu S, Dowd KA, et al. Modified mRNA vaccines protect against Zika virus infection. Cell 2017;169:176.
[235] Pardi N, Weissman D. Nucleoside modified mRNA vaccines for infectious diseases. Methods Mol Biol 2017;1499:109–21.
[236] Pardi N, Hogan MJ, Pelc RS, et al. Zika virus protection by a single low-dose nucleoside-modified mRNA vaccination. Nature 2017;543:248–51.
[237] Kool M, Fierens K, Lambrecht BN. Alum adjuvant: some of the tricks of the oldest adjuvant. J Med Microbiol 2012;61:927–34.
[238] Ramsauer K, Schwameis M, Firbas C, et al. Immunogenicity, safety, and tolerability of a recombinant measles-virus-based chikungunya vaccine: a randomised, double-blind, placebo-controlled, active-comparator, first-in-man trial. Lancet Infect Dis 2015;15:519–27.
[239] The Lancet Infectious Diseases. Vaccine against Zika virus must remain a priority. Lancet Infect Dis 2017;17:1003.
[240] Brooks JT, Friedman A, Kachur RE, LaFlam M, Peters PJ, Jamieson DJ. Update: interim guidance for prevention of sexual transmission of Zika virus—United States, July 2016. MMWR Morb Mortal Wkly Rep 2016;65:745–7.
[241] Silva JVJJ, Lopes TRR, Oliveira-Filho EF, Oliveira R, Gil L. Perspectives on the Zika outbreak: herd immunity, antibody-dependent enhancement and vaccine. Rev Inst Med Trop Sao Paulo 2017;59:e21.

CHAPTER

14

Malaria

Patrick E. Duffy, Sara Healy, J. Patrick Gorres, Michal Fried

Laboratory of Malaria Immunology and Vaccinology, National Institute of Allergy and Infectious Diseases, National Institutes of Health, Bethesda, MD, United States

Introduction

Plasmodium parasites cause malaria, and no vaccine is currently licensed for use against any human parasite including *Plasmodium*. Malaria vaccines have been pursued for generations, but only in recent years have specific malaria vaccine candidates proven to be safe and at least partially efficacious in humans. Compared to viruses and bacteria, parasites including *Plasmodia* have a complex biology that hinders vaccine development. *Plasmodium* has a large genome and multiple life cycle stages in both mosquito and human hosts with distinct antigenic repertoires that are targeted by different immune mediators (Fig. 1) [1]. Human malaria is caused by five species of *Plasmodium* (*Plasmodium falciparum* (Pf), *Plasmodium vivax, Plasmodium malariae, Plasmodium ovale*, and *Plasmodium knowlesi*), but Pf commands special attention due to its high incidence and virulence, causing more than 98% of malaria mortality, and has the unique characteristic of sequestering in the placenta.

Pregnant women, and women of child-bearing age, require special consideration for malaria vaccine development. Malaria susceptibility increases during pregnancy, rendering these women subject to poor maternal and perinatal outcomes, and making pregnant women an important parasite reservoir in the community. Despite their susceptibility to malaria infection, most women in areas of stable transmission are semi-immune and consequently may have few or no acute symptoms, and hence remain unaware of their infection. Further, Pf parasite-infected erythrocytes sequester in the human placenta where they inflict maternofetal damage, but in many cases

https://doi.org/10.1016/B978-0-12-814582-1.00015-2

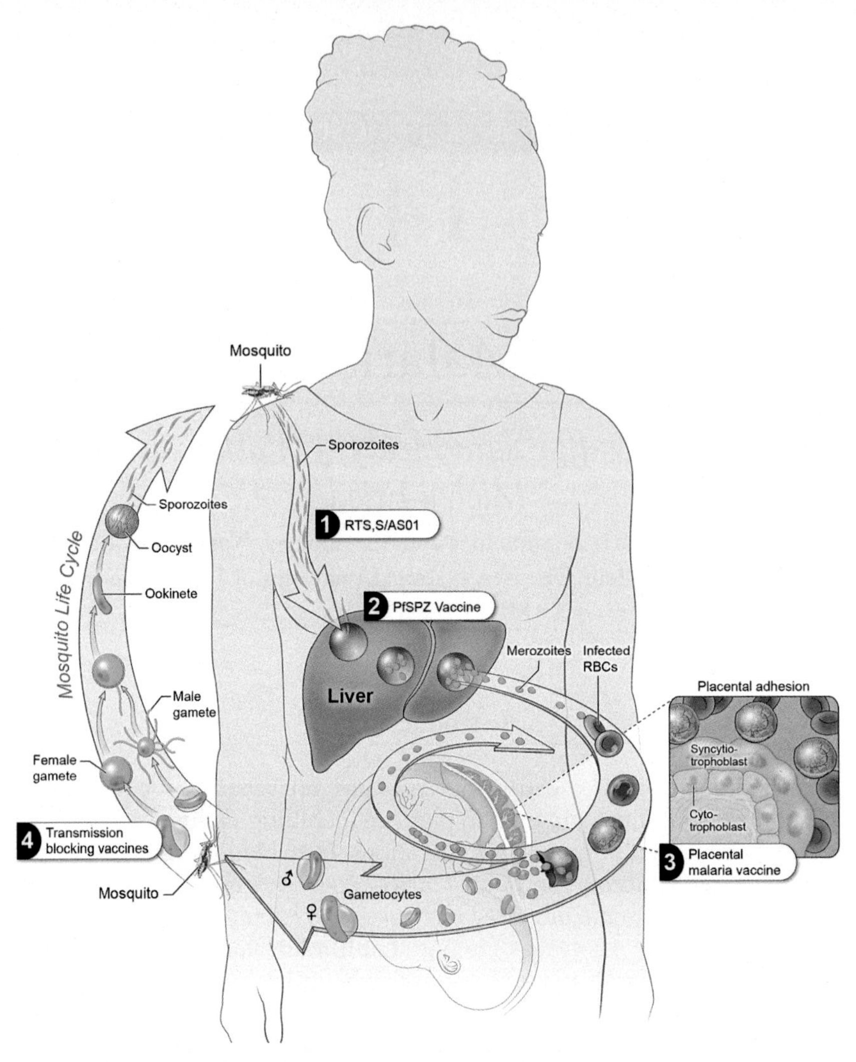

FIG. 1 Life cycle of *P. falciparum* as well as protective immune responses and vaccines that target each stage. The malaria parasite life cycle is generally divided into pre-erythrocytic, erythrocytic, and sexual-mosquito stages. The pre-erythrocytic stage commences when a female Anopheline mosquito inoculates sporozoites that rapidly pass through the bloodstream of the host to the liver and invade hepatocytes, where each sporozoite develops into 10,000–40,000 merozoites over about a week. The erythrocytic or blood stage begins with the release of merozoites into the blood stream, each of which can invade an erythrocyte, initiating the cycle of intraerythrocytic-stage development, rupture, and reinvasion that leads to a 10–20-fold increase in parasite numbers every 2 days and to all the clinical manifestations and pathology of malaria. Sexual-stage parasites develop first as intraerythrocytic gametocytes from a subset of infected erythrocytes, which then emerge from erythrocytes and form gametes when ingested by another blood-feeding mosquito to initiate a new mosquito infection; within 2 weeks a new crop of infectious sporozoites have accumulated in the mosquito salivary glands, waiting to be inoculated into the next human victim. Among the malaria vaccines currently in development, RTS,S/AS01 and PfSPZ Vaccine target the pre-erythrocytic stages by inducing antibodies and CD8 T cells, respectively, while placental malaria vaccines and transmission blocking vaccines seek to induce antibodies that target surface proteins of infected erythrocytes or of gametes, respectively. *Source: Illustration designed by Alan Hoofring, NIH Medical Arts.*

peripheral blood smears that are used for diagnosis are negative owing to very low peripheral blood parasite levels. This distinct biology and clinical presentation of *P. falciparum* in semi-immune pregnant women interferes with diagnosis, rendering targeted interventions ineffective for control. Finally, concerns for teratogenicity and embryotoxicity complicate the use of drugs, vaccines, or anti-vector measures in women of reproductive age, and complicate clinical development of new interventions.

Malaria parasitology and mechanisms of placental malaria

A histologic hallmark of *P. falciparum* malaria in pregnant women is placental sequestration of infected erythrocytes (Fig. 2). Parasites that infect pregnant mothers and sequester in the placenta have a distinct binding phenotype: *P. falciparum* infected erythrocytes (IE) from pregnant women bind to chondroitin sulfate A (CSA) found in placental intervillous spaces and on the syncytiotrophoblast, whereas parasites from non-pregnant individuals bind to other receptors such as CD36 [2]. Further, primigravidae have a higher parasite burden in the placenta than multigravidae, which can be explained by the acquisition of specific anti-adhesion antibodies over successive pregnancies [3]. The acquisition of these anti-adhesion antibodies has been associated with a ~400 g increase in birthweight among secundigravid pregnancies in Kenya [4]. Thus, women experience parity-dependent susceptibility to pregnancy malaria that, from a host perspective, is explained by the acquisition over successive pregnancies of functional antibodies associated with protection.

From a parasite perspective, parity-dependent susceptibility is rooted in the ability of the parasite to vary the display of IE surface antigen(s) that adhere to receptors on the vascular endothelium and thereby mediate IE sequestration in deep vascular beds. During pregnancy, the placenta presents a new receptor for IE adhesion, which has been confirmed to be CSA in numerous studies conducted at different sites [3,5–8]. CSA is a glycosaminoglycan composed of repeats of the disaccharide D-glucuronic acid (GlcUA) and *N*-acetyl-D-galactosamine (GalNAc).

Adhesion and sequestration of IE in the placenta selects a CSA-binding parasite subpopulation. Women are naïve to the CSA-binding parasites before their first pregnancy, making first-time mothers most susceptible to infection and poor outcomes. With successive pregnancies, women develop specific antibodies to CSA-binding placental IE that are associated with improved control of placental infection [3] and improved pregnancy outcomes [4,9].

IE sequestration in the placenta is followed by the infiltration of macrophages and B cells into the intervillous spaces. The intensity of the inflammatory immune infiltrate varies between women, and is inversely

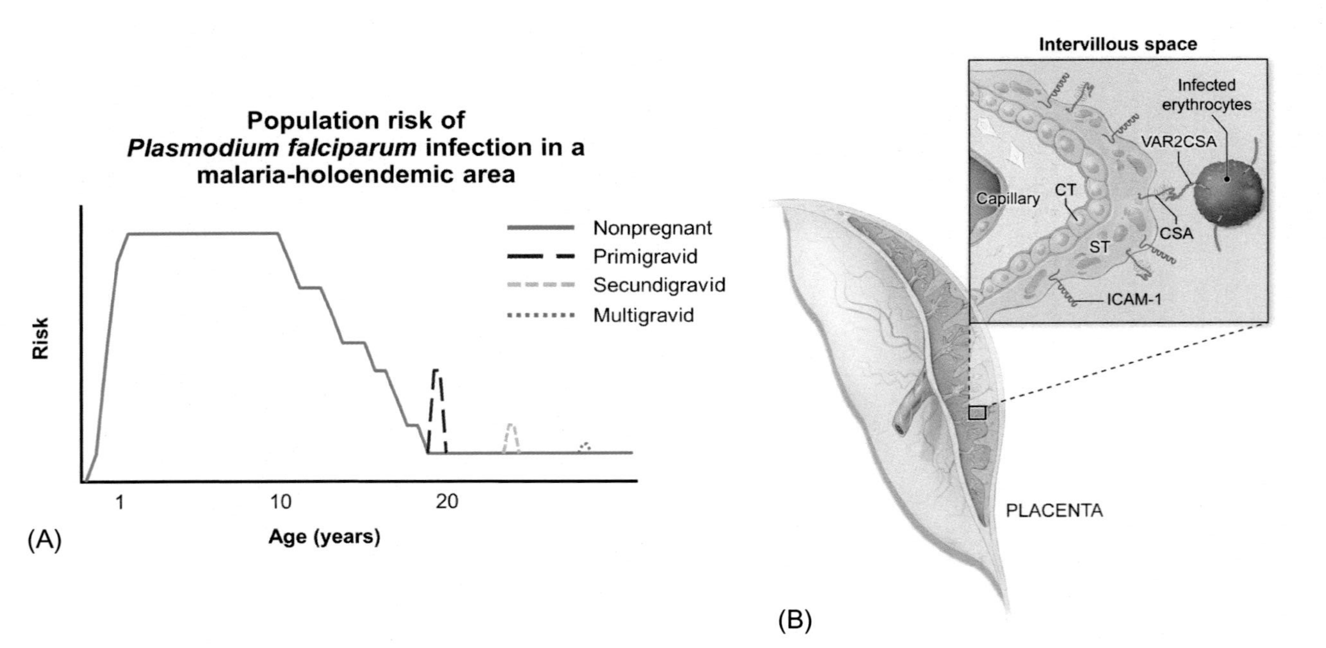

FIG. 2 Epidemiology and pathogenesis of pregnancy malaria in areas of stable transmission. Malaria is more common in pregnant women than other adults, and among pregnant women in areas of stable transmission, primigravidae are at highest risk (panel A). The pathological hallmark of *P. falciparum* pregnancy malaria is placental sequestration of infected erythrocytes, which is mediated by adhesion of the parasite surface antigen VAR2CSA to chondroitin sulfate A (CSA) on syncytiotrophoblast and in the intervillous spaces (panel B). Women become clinically resistant to placental malaria over successive pregnancies as they acquire antibodies against CSA-binding parasites including antibodies that block adhesion. In areas of low or unstable transmission, clinical resistance is acquired more slowly over multiple pregnancies. *Source: Illustration by Alan Hoofring, NIH Medical Arts.*

related to acquired immunity: macrophages are more commonly observed in placentas from primigravidae who lack specific immunity to placental IE than in those from multigravidae [10,11]. In histologic studies, the inflammatory infiltrate has a stronger relationship to poor outcomes like severe maternal anemia than does parasitemia itself [12].

The cytokine milieu in a healthy uninfected placenta displays a bias toward type 2 cytokines [13]. Placental malaria leads to increased levels of several inflammatory mediators, particularly pronounced in the placenta but also reflected in peripheral blood, including tumor necrosis factor alpha (TNF-α), interferon gamma (IFN-γ), interleukin-10 (IL-10), monocyte chemoattractant protein 1, macrophage inflammatory protein 1 (MIP-1a and MIP-1b), CXC ligand 8, CXC ligand 9 and CXC ligand 13 [13–20]. Increased placental levels of TNF-α and IFN-γ, and the chemokine CXCL9 that is upregulated by IFN-γ, have been associated with low birthweight deliveries especially among primigravidae [14,17,19,20]. Similarly, CXCL13, CXCL9 and CCL18 transcript levels negatively correlate with birthweight, while upregulation of IL-8 and TNF-α transcription in the placenta has been associated with intrauterine growth retardation [11,13]. These findings suggest that inflammatory mediators contribute to placental malaria sequelae, but animal models that reproduce placental sequestration and inflammation are needed to test these mechanistic hypotheses.

Clinical burden of disease and epidemiology

Despite progress made possible by new investments in disease control, malaria remains an enormous public health problem, with an estimated 219 million cases and 435,000 malaria-related deaths in 2017 [21], mostly attributable to *P. falciparum* and *P. vivax*. After a decade of decline however, the decrease of malaria mortality may have stalled or even reversed in countries that carry a disproportionate burden of disease [21]. In areas of stable malaria transmission, the incidence of malaria peaks in early childhood and declines thereafter, at a rate that varies with malaria transmission. In general, adults enjoy semi-immunity that limits parasitemia and controls symptoms.

However, *P. falciparum* malaria susceptibility increases again during first pregnancy, and then declines with each subsequent pregnancy unless the woman is immunosuppressed as in human immunodeficiency virus (HIV) infection [22,23]. Pregnancy malaria due to *P. falciparum* may lead to poor pregnancy outcomes such as miscarriage, stillbirth, and neonatal death; and the offspring is at increased risk of fetal growth restriction, preterm birth and low birthweight. Meta-analyses of trials have suggested that malaria chemoprevention reduces perinatal mortality by more than 25% [Risk Ratio (RR) 0.73, 95% confidence intervals [CI] 0.53 to 0.99] among first and second time mothers in areas of stable transmission [24]. Further, low birthweight and preterm delivery increase infant mortality risks, resulting in an estimated 100,000 infant deaths in Africa each year [25]. Infected pregnant women may suffer severe anemia, preeclampsia, and death. Malaria was the cause of 7.4% of 605 maternal deaths across Ghana [26] and 7% of 57 maternal deaths in Maputo, Mozambique [27]. Malaria commonly leads to anemia or severe anemia, potentially contributing to the 24.5% of maternal deaths (or an estimated 321,000 maternal deaths between 2003 and 2009) in sub-Saharan Africa attributed to hemorrhage [28].

Since pregnant women and children bear the greatest burden of malaria morbidity and mortality, they are the greatest beneficiaries of improved malaria control. One hundred twenty five million women in malaria-endemic areas become pregnant each year [29], and require protection from infection to avoid disease and death for themselves and their offspring. Antimalarial drugs and anti-vector agents are currently the primary tools for malaria control and confer benefits but are cumbersome to implement and losing effectiveness as resistant parasites and mosquitoes spread. Malaria control has been substantially strengthened in many but not all areas in recent years, as a consequence of increased resources and scale-up of established tools, including insecticide-treated bed nets and intermittent preventative treatments during pregnancy (IPTp). However, malaria resurges when control efforts are scaled back, and therefore new tools for malaria elimination have received growing attention. Even when the tools are properly implemented and retain their efficacy, many or most women are infected at some point during their pregnancy, with rates that vary between sites depending on transmission intensity [30]. On average, about 1 in 4 African women in areas of stable transmission have evidence of infection at delivery [25].

Pregnancy malaria epidemiology and presentations look very different in areas of low unstable transmission versus high stable transmission, though overall disease burden in different transmission zones may be similarly heavy in the absence of preventive measures. Where malaria transmission is low and unstable, residents are infected infrequently and therefore have low immunity and often rapidly progress to severe disease syndromes once infected. Pregnant women have higher risk of severe malaria and death than their non-pregnant counterparts during *P. falciparum* infection [31], and are more likely to develop syndromes like respiratory distress and cerebral malaria [32]. In these areas, women of all parities have increased susceptibility to malaria [32], and should be routinely screened and promptly treated for infection to prevent the risk of severe disease and death.

In areas of stable *P. falciparum* malaria transmission, where approximately 50 million pregnancies occur each year, women are semi-immune and often carry their infections with few or no symptoms, even during pregnancy. Disease for mother and offspring often develops as an insidious process, and this can make it difficult to relate outcomes such as severe maternal anemia, perinatal mortality, or low birthweight back to the infection that caused these sequelae. In these areas, primigravid women are at greatest risk, and over successive pregnancies women naturally acquire resistance to *P. falciparum* that reduces parasite density and prevents disease. Resistance is associated with the acquisition of antibodies against the CSA-binding IE that sequester in the placenta [3]. In communities where malaria control has improved (incidence of malaria decreased), the incidence of *P. falciparum* pregnancy malaria also decreased, but malaria-specific antibodies waned over time and the parasite burden and sequelae during any individual pregnancy malaria infection increased [33].

Even in the presence of preventive measures such as IPTp, which entails periodic full antimalarial treatment after the first trimester, pregnancy malaria still occurs at a high rate. Perhaps surprisingly to many, women generally do not receive any preventive measures during their first trimester, when most antimalarial drugs are contraindicated. Consequently, many women develop malaria before their first IPTp dose, which is only administered when women present for their first antenatal visit, typically around 20 weeks gestation in many communities. Less than 20% of African mothers receive the recommended 3 doses of IPTp during pregnancy [21]. Other recommended interventions such as insecticide-treated bed nets and case management of intercurrent malaria infections are variably implemented, but even in areas of relatively good uptake the rate of pregnancy malaria remains high. Further, drug-resistant parasites and insecticide-resistant mosquitoes are inexorably spreading, which will further erode the benefits of existing interventions. Thus, a vaccine to specifically reduce pregnancy malaria or a universal vaccine to reduce all malaria, is urgently needed to improve pregnancy outcomes. To have maximum benefit an effective vaccine should ideally be administered before first pregnancies.

Challenges of developing a vaccine for malaria during pregnancy

All malaria parasites have a complex life cycle and significant antigenic diversity. Hence, there are many different approaches to the development of malaria vaccines [1]. The multi-stage life cycle of the malaria parasite displays a continuously changing antigenic repertoire that requires distinct immune mediators to attack the parasite at different developmental stages. For example, protective antibodies against sporozoites injected by mosquitoes do not recognize the asexual erythrocytic stage parasites that cause human disease. Thus, a single sporozoite that evades vaccine-induced protective antibodies could ultimately lead to an uncontrolled blood stage infection. This complexity makes it more difficult to develop a vaccine for parasites than for viruses and bacteria [34]. In fact, there is no vaccine on the market for prevention of any human parasitic infection.

Various approaches to malaria vaccine development entail induction of antigen specific protective antibodies, CD4+ T cells, and/or CD8+ T cells. Some vaccines are directed against the sporozoite and/or liver stages of the life cycle (pre-erythrocytic stages), some against the asexual erythrocytic stages, and some against the sexual erythrocytic and mosquito stages. Pre-erythrocytic vaccine candidates aim to prevent infection and thereby prevent all disease, while asexual erythrocytic stage vaccines often seek to limit parasite growth in the bloodstream by blocking red blood cell invasion by the parasite. A special example of an erythrocytic stage vaccine is one that targets the IE surface antigens of placenta-sequestering parasites, such as the variant surface antigen VAR2CSA that binds the placental CSA receptor, to specifically control placental parasitemia and prevent poor pregnancy outcomes. Sexual stage vaccines aim to induce antibodies that prevent parasite transmission to mosquitoes, hence referred to as transmission-blocking vaccines. Both transmission-blocking and anti-infection vaccines are expected to be components of vaccines that target multiple life cycle stages of the parasite to interrupt malaria transmission [35], which could be useful for malaria elimination and eradication programs. Any of these vaccine types might be useful for pregnant women. A vaccine targeting placenta-sequestering parasites would aim to protect pregnant women exclusively.

While two vaccines (influenza and pertussis) have been recommended by many regulatory bodies to be administered during pregnancy, no vaccine has a specific indication approved by the United States (US) Food and Drug Administration to be used in pregnant women. Furthermore, there has never been a trial of a malaria vaccine in pregnant women. Trials in pregnant women have traditionally faced many obstacles. While many of these issues also pertain in malaria endemic areas, malaria vaccine trials in pregnant women will face special considerations, including the likely need to enroll poorer rural populations where malaria incidence is high, poor pregnancy outcomes such as miscarriage and stillbirth are relatively common, and specialized obstetric facilities and care are often absent. Additional considerations are the social and cultural mores that surround pregnancy and can hinder participation in research studies, and these can vary between communities even in the same country. Recent progress in the US is lowering hurdles for vaccine trials in pregnant women, and this may also favorably impact the prospects for trials of malaria vaccines in endemic areas.

Malaria vaccines in development

A number of candidate vaccines currently in clinical development could be considered for testing in pregnant women (see Table 1). For example, vaccines that prevent pregnancy malaria will be tested in pregnant women to demonstrate their safety and efficacy, as should vaccines that will be incorporated into mass vaccination programs for malaria elimination.

Currently, the leading candidate antigen for a vaccine that will specifically protect against pregnancy malaria is VAR2CSA, a member of the *var* gene or *Plasmodium falciparum* erythrocyte membrane protein 1(PfEMP1) protein family that is upregulated in CSA-binding and placental parasites [36,37]. The PfEMP1 family is the immunodominant variant surface antigen on *P. falciparum* infected red blood cells, encoded by approximately 60 alleles in each parasite haploid genome, and mediates parasite adhesion and deep vascular sequestration associated with severe malarial syndromes [38]. VAR2CSA is a unique PfEMP1 family member because it is identifiable in all *P. falciparum* genomes, albeit VAR2CSA alleles display substantial sequence variation between isolates. VAR2CSA is comprised of 6 extracellular Duffy binding-like (DBL) domains, interspersed with several inter-domain (ID) regions, forming a ~350 kD protein too large to be manufactured as an intact molecule with current recombinant expression systems. Therefore, the immunogens currently being considered for product development incorporate one or a few domains, with or without adjacent interdomain regions. The first vaccine products to move into clinical trials have focused on the N-terminal region around DBL-2 that has been implicated as the minimal binding domain for adhesion to the placental receptor CSA (Fig. 3) [40].

TABLE 1 Malaria vaccines in clinical development that could be tested in pregnant women.

Vaccine (type)	Status	Intended activity	Indication	Developer
RTS,S/ Mosquirix (subunit)	Phase III complete	Prevent disease	Reduce childhood disease	GSK
PfSPZ Vaccine (whole organism)	Phase II	Prevent infection	Prevent infection	Sanaria
VAR2CSA (subunit)	Phase I	Control infection	Prevent placental disease	Various (EVI)
Pfs25/Pfs230 (subunit)	Phase I	Block transmission	Elimination	NIAID

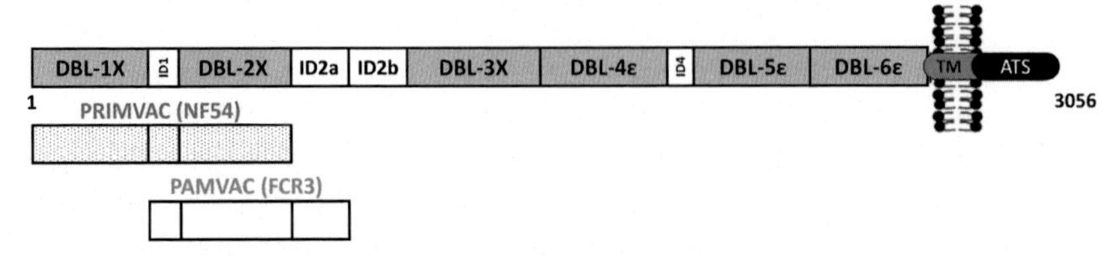

FIG. 3 VAR2CSA structure and current Placental Malaria Vaccine candidates. VAR2CSA structure (3D7/NF54 allele) displaying 6 DBL domains interspersed with several ID regions. Current vaccine candidates contain the DBL1-ID1-DBL2 fragment of 3D7 allele (PRIMVAC) [39] or the ID1-DBL2-ID2 fragment of the FCR3 allele (PAMVAC) [40].

To predict the VAR2CSA fragments that might be effective vaccine components, several studies compared susceptible primigravid to resistant multigravid women for their seroreactivity to different VAR2CSA domains (Fig. 2), but evidence for parity-dependent acquisition of VAR2CSA antibodies has varied [9,41,42]. This could reflect differences in the recombinant proteins being used for seroreactivity surveys, such as expression platform, allelic variant or domain boundaries, or differences in study populations such as transmission intensity, as well as gestational age or prevalence of infection at time of serum sampling [43]. Regarding the time of sampling, antibody boosting during malaria infection can confound attempts to distinguish between protective antibodies and markers of exposure [42].

Similarly, studies that examine relationships between VAR2CSA antibodies and pregnancy outcomes have also yielded differing results. Among infected Kenyan women, higher DBL5 antibody levels correlated with reduced risk of low birthweight deliveries [44]. Among infected Mozambican women, higher antibody levels to DBL3, DBL6 and the unrelated merozoite antigen AMA1, were associated with increased birthweight and gestational age [45]. In Benin, high antibody levels targeting DBL3 and full-length VAR2CSA were associated with a reduced number of infections, and high DBL1-ID1-DBL2 antibody levels measured during the first two trimesters reduced the risk of low birthweight [9]. In Mali, however, antibody levels early in pregnancy to DBL1, DBL2, DBL3, DBL4, and DBL5 were not associated with reduced risks of pregnancy loss, preterm delivery, or low birthweight [42]. Additional studies that define relationships between specific antibody and protection will advance the development of a placental malaria vaccine.

Malaria vaccines that protect the general population can also be expected to benefit pregnant women and improve pregnancy outcomes for their offspring. The PfSPZ Vaccine product being developed by Sanaria, Inc. is an advanced malaria candidate that is being strongly considered for trials in pregnant women, owing in part to its highly favorable safety profile [34]. The vaccine is comprised of aseptic, metabolically active, non-replicating, purified, cryopreserved *P. falciparum* sporozoites that meet regulatory requirements and are suitable for parenteral inoculation. In multiple double-blind, placebo-controlled trials, there have been no differences in adverse events between vaccines versus controls who received normal saline [34,46,47]. PfSPZ Vaccine induces immune responses to the sporozoite and liver stages of parasite development in the human host and prevents progression to blood stage parasitemia. Licensure is being pursued with prevention of infection as an initial indication. Preventing blood stage infection will also avert disease sequelae, and therefore there is a compelling rationale to test PfSPZ Vaccine for its benefits in pregnant women. Finally, by preventing blood stage parasitemia, PfSPZ Vaccine is envisioned for mass vaccination as part of malaria elimination programs, a scenario in which pregnant women are likely to receive PfSPZ Vaccine whether intended or not.

Proof-of-concept clinical trials involving volunteers from the US, Mali, Tanzania, and Equatorial Guinea demonstrated that PfSPZ Vaccine is extremely well-tolerated. As of a 2015 report [48], 2155 doses of PfSPZ had been administered to 824 adults by different routes, with an excellent safety and tolerability profile, and numerous trials have been completed or started subsequently with similar results. Importantly, when efficacy was measured by controlled human malaria infection (CHMI), the vaccine achieved short-term homologous protection (87–100%) and short-term heterologous protection (80%), with lower levels of long-term protection against CHMI and against natural exposure in Mali for up to 14 and 6 months respectively. In Malian adults, PfSPZ Vaccine reduced the risk of *P. falciparum* infection across the transmission season by 52% in time-to-event analysis [47].

Preclinical studies have not suggested any safety concerns for PfSPZ Vaccine. Toxicology studies performed in rabbits and biodistribution studies in mice have been unremarkable [49]. In humans, inoculation of PfSPZ Vaccine has not been associated with breakthrough infection [48]. ReproTox studies should include inoculation at different stages of pregnancy and should investigate whether PfSPZ can cross the placenta. After an infectious mosquito bite, sporozoites circulate in blood no longer than 30–60 min; any interaction with the placenta would need to occur during this period. Mosquitoes inoculate malaria sporozoites into millions of individuals every day, including pregnant women, and there is no evidence from these natural inoculations that sporozoites cross the placenta. The possibility of liver damage may increase with pregnancy related conditions (e.g., preeclampsia) and therefore should be monitored during trials involving pregnant women.

Other vaccines that are in clinical development include the RTS,S/AS01 vaccine developed at GlaxoSmithKline, and transmission blocking vaccines like Pfs25 or Pfs230 that are envisioned for use in malaria elimination programs. RTS,S/AS01 is the most advanced malaria vaccine product having completed phase III trials and is currently being evaluated in implementation trials in 3 African countries. However, while RTS,S/AS01 was shown in phase III trials to reduce clinical malaria in children, no studies have yet shown that this vaccine product confers significant efficacy in African adults against naturally occurring infection. Trials of this product in pregnant women may not be justified before demonstrating benefits in nonpregnant adults. Malaria transmission blocking vaccines will be deployed as part of mass vaccination programs, meaning that pregnant women are likely to be exposed to the product. Hence, clinical trials in pregnant women would be of great value for ensuring safety and activity. However, transmission blocking vaccines do not provide a direct benefit and therefore evidence that vaccination of communities reduces their incidence of malaria infection are warranted before including pregnant women in trials of these products. Pfs25 has completed safety and immunogenicity testing in US and African adults, as a protein-protein conjugate vaccine formulated in Alhydrogel, and the results suggested improvements would be needed before advancing this product further [50,51]. Pfs230 protein-protein conjugate vaccines are currently undergoing trials in the US and Mali (clinicaltrials.gov: NCT02334462 and NCT02942277).

Efficacy endpoints for clinical trials of malaria vaccines

When the goal is to prevent infection, the primary endpoints for a vaccine trial in pregnant women will be placental and peripheral parasitemia. The secondary endpoints will be those known to be sequelae of malaria, including low birthweight, preterm delivery, miscarriage, stillbirth, neonatal death, and severe maternal anemia.

Parasitemia can be detected through a number of diagnostic modalities, such as microscopy, rapid diagnostic tests (RDT), molecular tools [i.e., polymerase chain reaction (PCR)], and Loop Mediated Isothermal Amplification (LAMP). These tools can be applied to peripheral blood samples collected antenatally or at delivery, or to placental blood sampled by tissue impression, mechanical extraction, or incision at delivery. Peripheral blood samples are less sensitive for diagnosing malaria in pregnant women, and placental blood samples are more likely to yield a positive diagnosis. This is largely explained by placental sequestration of parasites, although the physiological hemodilution that occurs in pregnant women may also contribute. Initial evaluation of a newer ultrasensitive RDT reported a slight improvement in the detection of parasites in peripheral blood from pregnant women [52,53]. PCR diagnosis requires infrastructure, reagents, trained staff, and careful controls, but yields more positive diagnoses than peripheral blood microscopy, as does LAMP which requires less infrastructure. Notably, the performance of microscopy, RDT, and PCR can vary between sites depending on transmission intensity.

Placental histology is particularly valuable for defining clinical trial endpoints. Placental histology can detect parasites and is considered the gold standard for its sensitivity and specificity. Placental histopathology also detects malaria pigment, which indicates past infection in the absence of parasites, thus adding additional data to assess vaccine efficacy. One concern for histology is false positives, particularly due to formalin pigment (also known as acid hematin) which can be mistaken as malaria pigment. These errors can confound study interpretation but can be avoided by using buffered formalin for fixation and by formal external quality assurance.

Available evidence of effectiveness/efficacy in pregnant women

No efficacy data are currently available to understand and compare the benefits of malaria vaccination during pregnancy. Clinical trials of PfSPZ Vaccine in pregnant women are being planned, and a pregnancy registry is ongoing in Mali to gain background data on maternal/fetal outcomes in the target population in anticipation of testing PfSPZ Vaccine. These outcomes include miscarriages, stillbirths, pregnancy complications, fever, and acute illnesses. Rates of these outcomes will inform halting rules when the proposed clinical trials are in progress. The clinical development plan for PfSPZ Vaccine in pregnant women envisions a stepwise approach to studies including safety and efficacy studies in non-pregnant women of child-bearing potential, studies of the safety and efficacy of a primary immunization series in all trimesters, and finally studies to evaluate the efficacy of boosting during pregnancy.

Phase I safety and immunogenicity trials of VAR2CSA-based vaccines to prevent placental malaria have also been initiated in non-pregnant populations (PAMVAC: NCT02647489; PRIMVAC: NCT02658253). Results of the first PAMVAC trial have recently been published, concluding that the vaccine was safe, well-tolerated and induced functionally active antibodies [40].

Ideal timing of vaccination (infant/adolescent/pre-conception/early pregnancy)

The ideal timing for a vaccine that provides direct protection against malaria is preconception as well as early in pregnancy. Current antenatal practices fail to protect the pregnant mother and her unborn child from malaria. The primary intervention for malaria prevention, as recommended by the

World Health Organization, is the IPTp, consisting of monthly dose of the antimalarial drug combination sulfadoxine-pyrimethamine (SP) starting after first trimester. The first and most harmful gap in this strategy is that most women receive no protective interventions in the first half of pregnancy, a time when maternal malaria infections may peak and the fetus is developmentally at its most vulnerable. IPTp with SP is contraindicated during first trimester, and African women typically do not seek antenatal care until mid-pregnancy in many areas. The second gap occurs in the second half of pregnancy: by one estimate, only 19% of pregnant African women receive all three recommended IPTp treatments during the remainder of their term [2]. For these reasons, a malaria vaccination for pregnancy program is a better solution than an antimalarial drug program to close these gaps, if the vaccine is determined to be safe and efficacious.

During mass vaccination programs, women are likely to receive vaccine at any time during pregnancy. Therefore, clinical trials are needed to assess both safety and immunogenicity of these products in all three trimesters of pregnancy.

Conclusion

In the future, pregnant women will receive a malaria vaccine either because they will receive direct benefits or because they are included in mass vaccination programs aimed at malaria elimination. In the case of a malaria vaccine providing a direct benefit, three approaches to vaccination can be considered: (1) immunize women of child-bearing age before conception to protect through pregnancy, (2) immunize women of child-bearing age before pregnancy and boost as early in pregnancy as possible, or (3) administer the primary vaccine series to pregnant women starting at 1st antenatal visit. For each of these approaches, interventional trials can be designed to collect safety and efficacy data at antenatal visits.

Mass administration of malaria vaccines for elimination programs will likely only be feasible if pregnant women can also be vaccinated. Toward this end, clinical and regulatory pathways to demonstrate safety in women of child-bearing age and pregnant women are needed, and success will hinge on more education and dialogue among researchers, regulators, health ministers, and populations.

References

[1] Hoffman SL, Miller LH. Perspectives on malaria vaccine development. In: Hoffman SL, editor. Malaria vaccine development: a multi-immune response approach. Washington, DC: ASM Press; 1996. p. 1–13.

[2] Fried M, Duffy PE. Adherence of Plasmodium falciparum to chondroitin sulfate A in the human placenta. Science 1996;272(5267):1502–4.

[3] Fried M, Nosten F, Brockman A, Brabin BJ, Duffy PE. Maternal antibodies block malaria. Nature 1998;395(6705):851–2.
[4] Duffy PE, Fried M. Antibodies that inhibit Plasmodium falciparum adhesion to chondroitin sulfate A are associated with increased birth weight and the gestational age of newborns. Infect Immun 2003;71(11):6620–3.
[5] Beeson JG, Brown GV, Molyneux ME, Mhango C, Dzinjalamala F, Rogerson SJ. Plasmodium falciparum isolates from infected pregnant women and children are associated with distinct adhesive and antigenic properties. J Infect Dis 1999;180(2):464–72.
[6] Maubert B, Fievet N, Tami G, Boudin C, Deloron P. Cytoadherence of Plasmodium falciparum-infected erythrocytes in the human placenta. Parasite Immunol 2000;22(4):191–9.
[7] Fried M, Domingo GJ, Gowda CD, Mutabingwa TK, Duffy PE. Plasmodium falciparum: chondroitin sulfate A is the major receptor for adhesion of parasitized erythrocytes in the placenta. Exp Parasitol 2006;113(1):36–42.
[8] Muthusamy A, Achur RN, Valiyaveettil M, Botti JJ, Taylor DW, Leke RF, et al. Chondroitin sulfate proteoglycan but not hyaluronic acid is the receptor for the adherence of Plasmodium falciparum-infected erythrocytes in human placenta, and infected red blood cell adherence up-regulates the receptor expression. Am J Pathol 2007;170(6):1989–2000.
[9] Ndam NT, Denoeud-Ndam L, Doritchamou J, Viwami F, Salanti A, Nielsen MA, et al. Protective antibodies against placental malaria and poor outcomes during pregnancy, Benin. Emerg Infect Dis 2015;21(5):813–23.
[10] Garnham PCC. The placenta in malaria with special reference to reticulo-endothelial immunity. Trans R Soc Trop Med Hyg 1938;32(1):13–34.
[11] Muehlenbachs A, Fried M, Lachowitzer J, Mutabingwa TK, Duffy PE. Genome-wide expression analysis of placental malaria reveals features of lymphoid neogenesis during chronic infection. J Immunol 2007;179(1):557–65.
[12] Duffy PE, Fried M. Immunity to malaria during pregnancy: different host, different parasite. In: Fried M, editor. Malaria in pregnancy: deadly parasite, susceptible host. London and New York: Taylor & Francis Publishers; 2001. [chapter 4].
[13] Moormann AM, Sullivan AD, Rochford RA, Chensue SW, Bock PJ, Nyirenda T, et al. Malaria and pregnancy: placental cytokine expression and its relationship to intrauterine growth retardation. J Infect Dis 1999;180(6):1987–93.
[14] Fried M, Muga RO, Misore AO, Duffy PE. Malaria elicits type 1 cytokines in the human placenta: IFN-gamma and TNF-alpha associated with pregnancy outcomes. J Immunol 1998;160(5):2523–30.
[15] Abrams ET, Brown H, Chensue SW, Turner GD, Tadesse E, Lema VM, et al. Host response to malaria during pregnancy: placental monocyte recruitment is associated with elevated beta chemokine expression. J Immunol 2003;170(5):2759–64.
[16] Chaisavaneeyakorn S, Moore JM, Mirel L, Othoro C, Otieno J, Chaiyaroj SC, et al. Levels of macrophage inflammatory protein 1 alpha (MIP-1 alpha) and MIP-1 beta in intervillous blood plasma samples from women with placental malaria and human immunodeficiency virus infection. Clin Diagn Lab Immunol 2003;10(4):631–6.
[17] Rogerson SJ, Brown HC, Pollina E, Abrams ET, Tadesse E, Lema VM, et al. Placental tumor necrosis factor alpha but not gamma interferon is associated with placental malaria and low birth weight in Malawian women. Infect Immun 2003;71(1):267–70.
[18] Suguitan Jr AL, Cadigan TJ, Nguyen TA, Zhou A, Leke RJ, Metenou S, et al. Malaria-associated cytokine changes in the placenta of women with pre-term deliveries in Yaounde, Cameroon. Am J Trop Med Hyg 2003;69(6):574–81.
[19] Kabyemela ER, Fried M, Kurtis JD, Mutabingwa TK, Duffy PE. Fetal responses during placental malaria modify the risk of low birth weight. Infect Immun 2008;76(4):1527–34.

[20] Dong S, Kurtis JD, Pond-Tor S, Kabyemela E, Duffy PE, Fried M. CXC ligand 9 response to malaria during pregnancy is associated with low-birth-weight deliveries. Infect Immun 2012;80(9):3034–8.
[21] WHO. World malaria report 2018. Geneva: World Health Organizaiton; 2018.
[22] McGregor I. Malaria—recollections and observations. Trans R Soc Trop Med Hyg 1984;78(1):1–8.
[23] van Eijk AM, Ayisi JG, ter Kuile FO, Misore AO, Otieno JA, Rosen DH, et al. HIV increases the risk of malaria in women of all gravidities in Kisumu, Kenya. AIDS 2003;17(4):595–603.
[24] Garner P, Gulmezoglu AM. Drugs for preventing malaria in pregnant women. Cochrane Database Syst Rev 2006;4:CD000169.
[25] Desai M, ter Kuile FO, Nosten F, McGready R, Asamoa K, Brabin B, et al. Epidemiology and burden of malaria in pregnancy. Lancet Infect Dis 2007;7(2):93–104.
[26] Asamoah BO, Moussa KM, Stafstrom M, Musinguzi G. Distribution of causes of maternal mortality among different socio-demographic groups in Ghana; a descriptive study. BMC Public Health 2011;11:159.
[27] Castillo P, Hurtado JC, Martinez MJ, Jordao D, Lovane L, Ismail MR, et al. Validity of a minimally invasive autopsy for cause of death determination in maternal deaths in Mozambique: an observational study. PLoS Med 2017;14(11):e1002431.
[28] Say L, Chou D, Gemmill A, Tuncalp O, Moller AB, Daniels J, et al. Global causes of maternal death: a WHO systematic analysis. Lancet Glob Health 2014;2(6):e323–33.
[29] Dellicour S, Tatem AJ, Guerra CA, Snow RW, ter Kuile FO. Quantifying the number of pregnancies at risk of malaria in 2007: a demographic study. PLoS Med 2010;7(1):e1000221.
[30] Berry I, Walker P, Tagbor H, Bojang K, Coulibaly SO, Kayentao K, et al. Seasonal dynamics of malaria in pregnancy in West Africa: evidence for carriage of infections acquired before pregnancy until first contact with antenatal care. Am J Trop Med Hyg 2018;98(2):534–42.
[31] Duffy PE, Desowitz RS. Pregnancy malaria throughout history: dangerous labors. In: Duffy PE, Fried M, editors. Malaria in pregnancy: deadly parasite, susceptible host. New York: Taylor and Francis; 2001. p. 1–25.
[32] Nosten F, ter Kuile F, Maelankirri L, Decludt B, White NJ. Malaria during pregnancy in an area of unstable endemicity. Trans R Soc Trop Med Hyg 1991;85(4):424–9.
[33] Mayor A, Bardaji A, Macete E, Nhampossa T, Fonseca AM, Gonzalez R, et al. Changing trends in P. falciparum burden, immunity, and disease in pregnancy. N Engl J Med 2015;373(17):1607–17.
[34] Hoffman SL, Vekemans J, Richie TL, Duffy PE. The march toward malaria vaccines. Vaccine 2015;33(Suppl 4):D13–23.
[35] malERA Consultative Group on Vaccines. A research agenda for malaria eradication: vaccines. PLoS Med 2011;8(1):e1000398.
[36] Tuikue Ndam NG, Salanti A, Bertin G, Dahlback M, Fievet N, Turner L, et al. High level of var2csa transcription by Plasmodium falciparum isolated from the placenta. J Infect Dis 2005;192(2):331–5.
[37] Salanti A, Staalsoe T, Lavstsen T, Jensen AT, Sowa MP, Arnot DE, et al. Selective upregulation of a single distinctly structured var gene in chondroitin sulphate A-adhering Plasmodium falciparum involved in pregnancy-associated malaria. Mol Microbiol 2003;49(1):179–91.
[38] Hviid L, Jensen AT. PfEMP1—a parasite protein family of key importance in Plasmodium falciparum malaria immunity and pathogenesis. Adv Parasitol 2015;88:51–84.
[39] Chene A, Gangnard S, Dechavanne C, Dechavanne S, Srivastava A, Tetard M, et al. Down-selection of the VAR2CSA DBL1-2 expressed in E. coli as a lead antigen for placental malaria vaccine development. NPJ Vaccines 2018;3:28.

[40] Mordmuller B, Sulyok M, Egger-Adam D, Resende M, de Jongh WA, Jensen MH, et al. First-in-human, randomized, double-blind clinical trial of differentially adjuvanted PAMVAC, a vaccine candidate to prevent pregnancy-associated malaria. Clin Infect Dis 2019; https://doi.org/10.1093/cid/ciy1140. (e-pub ahead of print).
[41] Tutterrow YL, Avril M, Singh K, Long CA, Leke RJ, Sama G, et al. High levels of antibodies to multiple domains and strains of VAR2CSA correlate with the absence of placental malaria in Cameroonian women living in an area of high Plasmodium falciparum transmission. Infect Immun 2012;80(4):1479–90.
[42] Fried M, Kurtis JD, Swihart B, Morrison R, Pond-Tor S, Barry A, et al. Antibody levels to recombinant VAR2CSA domains vary with Plasmodium falciparum parasitaemia, gestational age, and gravidity, but do not predict pregnancy outcomes. Malar J 2018;17(1):106.
[43] Fried M, Duffy PE. Malaria during pregnancy. Cold Spring Harb Perspect Med 2017;7(6):1–24. https://doi.org/10.1101/cshperspect.a025551.
[44] Salanti A, Dahlback M, Turner L, Nielsen MA, Barfod L, Magistrado P, et al. Evidence for the involvement of VAR2CSA in pregnancy-associated malaria. J Exp Med 2004;200(9):1197–203.
[45] Mayor A, Kumar U, Bardaji A, Gupta P, Jimenez A, Hamad A, et al. Improved pregnancy outcomes in women exposed to malaria with high antibody levels against Plasmodium falciparum. J Infect Dis 2013;207(11):1664–74.
[46] Seder RA, Chang LJ, Enama ME, Zephir KL, Sarwar UN, Gordon IJ, et al. Protection against malaria by intravenous immunization with a nonreplicating sporozoite vaccine. Science 2013;341(6152):1359–65.
[47] Sissoko MS, Healy SA, Katile A, Omaswa F, Zaidi I, Gabriel EE, et al. Safety and efficacy of PfSPZ vaccine against Plasmodium falciparum via direct venous inoculation in healthy malaria-exposed adults in Mali: a randomised, double-blind phase 1 trial. Lancet Infect Dis 2017;17(5):498–509.
[48] Richie TL, Billingsley PF, Sim BK, James ER, Chakravarty S, Epstein JE, et al. Progress with Plasmodium falciparum sporozoite (PfSPZ)-based malaria vaccines. Vaccine 2015;33(52):7452–61.
[49] Hoffman SL, Billingsley PF, James E, Richman A, Loyevsky M, Li T, et al. Development of a metabolically active, non-replicating sporozoite vaccine to prevent Plasmodium falciparum malaria. Hum Vaccin 2010;6(1):97–106.
[50] Sagara I, Healy SA, Assadou MH, Gabriel EE, Kone M, Sissoko K, et al. Safety and immunogenicity of Pfs25H-EPA/Alhydrogel, a transmission-blocking vaccine against Plasmodium falciparum: a randomised, double-blind, comparator-controlled, dose-escalation study in healthy Malian adults. Lancet Infect Dis 2018;18(9):P969–82.
[51] Talaat KR, Ellis RD, Hurd J, Hentrich A, Gabriel E, Hynes NA, et al. Safety and immunogenicity of Pfs25-EPA/Alhydrogel(R), a transmission blocking vaccine against Plasmodium falciparum: an open label study in malaria naive adults. PLoS One 2016;11(10):e0163144.
[52] Das S, Jang IK, Barney B, Peck R, Rek JC, Arinaitwe E, et al. Performance of a high-sensitivity rapid diagnostic test for Plasmodium falciparum malaria in asymptomatic individuals from Uganda and Myanmar and naive human challenge infections. Am J Trop Med Hyg 2017;97(5):1540–50.
[53] Vasquez AM, Medina AC, Tobon-Castano A, Posada M, Velez GJ, Campillo A, et al. Performance of a highly sensitive rapid diagnostic test (HS-RDT) for detecting malaria in peripheral and placental blood samples from pregnant women in Colombia. PLoS One 2018;13(8):e0201769.

PART IV

Conclusion

CHAPTER

15

Conclusion

Elke E. Leuridan[a], *Marta C. Nunes*[b,c], *Christine E. Jones*[d]

[a]Centre for the Evaluation of Vaccination, Vaccine & Infectious Diseases Institute, Faculty of Medicine and Health Sciences, University of Antwerp, Antwerp, Belgium, [b]Medical Research Council: Respiratory and Meningeal Pathogens Research Unit, Faculty of Health Sciences, University of the Witwatersrand, Johannesburg, South Africa, [c]Department of Science and Technology/National Research Foundation: Vaccine Preventable Diseases Unit, University of the Witwatersrand, Johannesburg, South Africa, [d]Faculty of Medicine and Institute for Life Sciences, University of Southampton and University Hospital Southampton NHS Foundation Trust, Southampton, United Kingdom

Maternal immunization: Past, present and future

This textbook aims to provide an overview of the current knowledge on maternal immunization. The book reviews in depth this quickly evolving topic, aims to create awareness of the current developments in this area and of the potential of maternal vaccination to improve the health of mothers and infants worldwide.

Maternal vaccination is a proven strategy to prevent infection in both the mother and the young infant. There is an increasing focus from different stakeholders to harness this approach to better protect these vulnerable groups.

Despite the substantial progress over the past two decades in reducing global under-five deaths, progress in newborn mortality has been comparatively slower. A priority of the United Nations' Sustainable Development Goals is to end preventable neonatal deaths by 2030 [1]. A significant contributor to neonatal mortality is infection, and maternal immunization has emerged as a key intervention which can begin to address this problem.

https://doi.org/10.1016/B978-0-12-814582-1.00016-4

Pregnancy and the neonatal period continue to come into focus as areas of medicine that need more research and investment. The history of women's involvement in research is evolving from the need to "protect" pregnant women from research, to recognizing the complex nature of pregnancy and emphasizing the need for research as a way to make pregnancy and the post-partum period safer (Chapter 1: The history of maternal immunization). Vaccination of pregnant women is not a new concept, but has gained renewed interest in the last decade. It is equally important to study and highlight the maternal benefit from vaccination in pregnancy as well as fetal and infant benefits.

Vaccines against tetanus toxoid, influenza and pertussis are currently recommended in pregnancy on a global or regional level, and the history of maternal immunization has led the way for research on new vaccines currently in development, such as those against respiratory syncytial virus (RSV) and Group B *Streptococcus* (GBS).

Tetanus vaccination has been recommended since the 1960s, with high coverage rates, and has been a cornerstone of elimination of maternal and neonatal tetanus in all but 14 countries worldwide (Chapter 6: Tetanus) [2]. On a global level, lessons have been learnt from the implementation of this strategy for other vaccines that can be administered in pregnancy.

Maternal immunization is a complex field that must consider multiple biological factors such as maternal immunogenicity, transplacental antibody transfer, the neonatal immune system and subsequent neonatal immunogenicity, as well as the potential of adverse effects in the mother and infant and sociological factors like acceptability, planning and implementation of this strategy.

The first section of the book (Chapters 1–4) reviews general concepts of the strategy. The first chapter offers a review of the history of maternal vaccination, highlighting the lessons learnt from the already recommended vaccines as well as the need for further investigation. The following chapters review the current knowledge on immunobiology of maternal vaccination from both the maternal and infant perspectives, as well as a general overview of the implementation and acceptance of maternal immunization in a global context.

The second and third parts of the book review in depth different aspects of disease specific vaccines that are recommended for—or are being designed for—administration during pregnancy (Chapters 6–14).

Immunobiological aspects of vaccines in pregnancy

Maternal perspective

From early in pregnancy, the immune system is adapted in order to allow immunological tolerance of the developing fetus. This modulation includes a reduction in B cell and dendritic cell numbers and alterations in the glycosylation of immunoglobulin G (IgG) (which changes the three-dimensional structure of the IgG Fc fragment and therefore the interaction with IgG-binding Fcγ receptors) [3–8]. How these immunological changes might impact the responses to vaccination during pregnancy, and whether gestational timing of vaccination has a role to play—is incompletely understood; some studies suggest similar antibody responses to influenza and pertussis vaccination [9–14] whereas other suggest that immunological responses might be attenuated in pregnancy [15–17]. While pregnancy cannot be considered as conferring a state of immunosuppression, there are infections which cause more severe disease in pregnant women compared to non-pregnant women, including influenza, varicella zoster, hepatitis E, listeria, and malaria. Immune adaptions, increased levels of estradiol and progesterone, decreased pulmonary reserve and increased cardiac output may be responsible for more severe clinical disease phenotype in pregnancy.

Transplacental transfer of IgG from mother to infant provides protection to the infant in early life against infections to which the mother has been previously exposed either by natural infection or vaccination. This process increases exponentially from the second to third trimester of pregnancy such that infant levels of maternally-derived specific IgG may be greater than the level found in the mother. Chronic maternal infections, including malaria [18–22] and Human Immunodeficiency Virus (HIV), can reduce the transplacental transfer of IgG from mother to infant [23–25]. This reduction of antibody transfer has been shown to be associated with an increased risk of hospitalization among infants born to mothers living with HIV who themselves are uninfected [25]. Maternal immunization may provide a solution to the lower concentrations of antibody at birth in these infants and reduce the immunological disadvantage of HIV-exposed, uninfected infants, however the effectiveness of this strategy may be limited by a reduction in vaccine immunogenicity and impaired transplacental transfer of antibody.

Infant perspective

High concentrations of maternally-derived vaccine-specific IgG transmitted to the infant through the placenta provide protection against vaccine-preventable infections early in life. However, IgG induced by previous maternal infection and vaccine-induced IgG are associated with inhibitory effects on the infant's own response to vaccination. The blunting effect of high concentrations of maternal measles-specific IgG on the infants response to measles vaccination has long been demonstrated and informed the timing of the infant measles vaccination schedule [26,27]. This blunting effect has also been described for tetanus, pneumococcus, influenza, *Haemophilus influenzae* type B (Hib), mumps, hepatitis A & B, rotavirus and poliovirus [28]. More recently, this has also been demonstrated in the context of maternal vaccination [13,29–33]. Maternal pertussis vaccination is associated with lower concentrations of specific antibody to pertussis antigens following primary infant immunization in most studies and also following booster vaccination in some but not all studies [30]. Moreover, since the pertussis-containing vaccines recommended for use in pregnancy in many countries are combination vaccines, which also contain tetanus, diphtheria and in some cases polio, an impact on infant responses to these antigens is also observed. Interestingly, while the response to diphtheria is reduced, the response to infant tetanus vaccination is greater in infants born to women vaccinated in pregnancy compared to infants born to women not vaccinated in pregnancy. Infant responses to vaccines conjugated with tetanus, (such as Hib) are normally also greater, while those conjugated to CRM197 (a naturally occurring diphtheria toxin variant), e.g., pneumococcal vaccines, are reduced [31,34].

The mechanisms for the blunting effect are not fully understood in humans. In the mouse model, it has been demonstrated that maternally-derived IgG-vaccine antigen complexes, cross-link the B cell receptor to the inhibitory FcγRII, thus resulting in a signal which inhibits the proliferation of B cells and the secretion of antibody [35].

The clinical effects of blunting are important to evaluate, however, the burden of severe disease and death from infections, such as pertussis, in early infancy is high and maternal immunization has been shown to be very effective at protecting young infants [36–44]. Therefore, relatively small reductions in IgG later in infancy might be acceptable in order to realize the benefits of high concentrations of maternally-derived antibody early in life. That said, an increase in pertussis in later infancy has not be observed to date [37].

Maternal antibody, in the form of secretory IgA and to a smaller extent IgG, is provided to the infant through breastfeeding. The role of breastmilk antibodies in protecting infants against infection is not well understood, however vaccination in pregnancy does induce pathogen-specific IgA and IgG in breastmilk and could potentially contribute to the protective benefits of maternal immunization [45–47]. Several studies have demonstrated significantly higher vaccine-specific secretory IgA in the breastmilk of women vaccinated against influenza, pertussis, meningococcus, and pneumococcus compared to unvaccinated women [48–51]. Further studies are required to fully understand the cellular and immunological mechanisms of breastmilk-mediated protection after vaccination in pregnancy and establish its clinical benefits.

Conducting research in maternal immunization

Including pregnant women in clinical trials of vaccines is essential to ensure that mothers and their newborns can benefit from vaccines, and the success of maternal immunization implementation as a public health strategy to reduce maternal and infant mortality. While most vaccines may directly benefit the pregnant woman, all have the potential to result in passive protection for the infant. Maternal immunization is therefore critical when no other disease prevention or treatment alternatives are available. Research in maternal immunization is essential to support the recommendation and licensure of vaccines for use during pregnancy. Substantial progress has been made since the 2009 influenza pandemic in the field of maternal immunization research.

Vaccines licensed for non-pregnant populations can be administered to pregnant women when women or their fetuses are at risk of infection, morbidity or mortality from a vaccine preventable disease for which a safe (non-live) vaccine is available. Historically, while studies of vaccines in pregnant women have been conducted since the advent of vaccines, these vaccines have not been systematically evaluated in pregnancy, with the exception of tetanus vaccine used in the global maternal-neonatal tetanus elimination strategy [52]. Vaccines are now being developed for the specific indication of administration in pregnancy, advancing maternal immunization research.

The ethical bases of maternal immunization research include the following [53–59]:

- Pregnancy should not be a deterrent for women and their infants to receive the benefits of safe and effective vaccines. This is particularly relevant in situations of outbreaks or epidemics, and when pregnant women and their infants are disproportionately affected.
- Pregnant women must be included in research of vaccines that have the potential to protect them and their infant.
- The risks of not including pregnant women in vaccine research are substantial when data on safety and effectiveness of vaccines is not generated for these susceptible populations.
- Vaccine studies in pregnancy are designed to ensure and optimize the safety of the pregnant mother and her fetus and newborn.
- Pregnant women can provide consent for participation in research for themselves and for their fetus and newborn.
- Pregnant women are no longer considered a vulnerable population for participation in research, given that their capacity to provide consent is not in question.

In general, vaccine candidates studied in pregnant women should meet the following requirements before a maternal immunization trial starts:

- Pre-clinical studies have been performed.
- Reproductive toxicology testing demonstrated no fetal toxicity.
- Phase I–II clinical trials in healthy non-pregnant adults provide guidance on dosage, safety, and immunogenicity.

- The disease to prevent poses a special risk to the mother and/or fetus.
- The vaccine is unlikely to cause harm to the mother or fetus.

In addition, the utilization of vaccines in pregnancy, including existing vaccines or new vaccines as they become available, needs to be supported by ensuring the following:

- Strong health information and surveillance systems to ensure that maternal and newborn health, as well as obstetric outcomes and vaccine safety are captured, monitored, and reported.
- Continued collection and reporting of safety and effectiveness data at every available opportunity, including for routinely recommended vaccines, vaccines inadvertently administered to pregnant women, and vaccines administered in special situations determined by risk-benefit considerations, such as during outbreak responses. Establishing systems to actively or passively, prospectively or retrospectively, collect and report safety and effectiveness data supports the successful implementation of maternal immunization.
- Activities focusing on vaccine education, awareness, and effective communication to promote confidence in maternal immunization among health care providers and the public.

Global implementation and acceptance of maternal immunization

For maternal immunization to be effective, successful implementation is needed and acceptance by both the target population as well as the health care workers responsible for delivering the vaccines to pregnant women. The setting and facilities for vaccinating pregnant women differ widely, depending on existing infrastructure and background data [60,61].

First of all, careful planning with national and local leadership involvement are crucial. Maternal tetanus vaccination programs were the first ever implemented to consistently deliver vaccines to pregnant women and have provided an opportunity to understand the operational challenges that may hamper achieving high vaccine coverage [2]. Ongoing evaluation of programs and global implementation research, help policy makers to better understand the opportunities and challenges detected during the implementation phases in various settings, to improve maternal tetanus vaccination efforts as a public health strategy [62].

Secondly, burden of disease data are necessary in informing the introduction of different maternal immunizations. Operational challenges, along with significant gaps in local burden of disease data, have become evident with maternal influenza vaccine programs that are not yet widely implemented in low and middle-income countries (LMICs) [63]. This requires support focused on informing countries' decisions to introduce the vaccine, based on the local potential benefit and reassuring safety profile of the vaccine.

Identifying the optimal service delivery is another key factor for successful implementation [64]. The optimal service delivery needs to account for the optimal timing of vaccination, format of the service, and careful assessment of already existing delivery capacity services. With an increasing number of maternal vaccines becoming available, there is urgent need to ensure their equitable operationalization across all countries. Among these, resource constrained contexts require consideration of local capacity and delivery options to ensure the maximization of the potential benefits of maternal immunization and to avoid overburdening of existing health systems.

With the emergence of new vaccines for use during pregnancy, such as GBS and RSV, key issues common to all maternal vaccines need to be considered during implementation. Such issues include education, training and communication of key messages targeted at pregnant women and health workers [65]. Finally, ensuring safety vigilance, and evaluation of the proportional benefit to the mother and/or newborn is essential to guarantee safe and sustainable use of vaccines in pregnant women [66].

Furthermore, it is crucial to identify and address questions and concerns among pregnant women and health care workers. Clear barriers have been identified in the pregnant population and in the health care providers: information and its sources, safety concerns, the role that the health care provider plays, ignorance of the disease and unawareness of the recommendation and lack of access [67–69].

Chapter 5 provides an analysis of these potential barriers, but also describes strategies to increase confidence and uptake of vaccines in pregnancy. Arguably the most relevant group to engage are the pregnant women themselves [70]. Taking a customer-oriented approach requires being inclusive and understanding the importance, benefit and most appropriate platform to access a range of vaccines and other important health services in pregnancy. Addressing pregnant women early will help to generate demand, avoid potential misperceptions of vaccines' safety, and ultimately prevent low uptake due to suboptimal implementation. Healthcare workers as key influencers of program implementation and source of trust for pregnant women, also need to be seen as customers and need to be empowered to drive communication and recommendation efforts.

Vaccination of women in the pre-conception and post-partum periods

Some vaccines are preferably given before or after pregnancy.

During a pre-conception medical assessment, it is important to review a woman's vaccination history and where indicated perform serology to assess immunity to hepatitis B, measles, varicella and rubella [71]. Identification of non-immune women will enable appropriate "catch up" vaccines, especially live vaccines such as rubella, to be administered before pregnancy if possible, or after delivery. Varicella has large variations in epidemiology worldwide, hence routine testing is not implemented everywhere. Universal screening of hepatitis B will identify women who are chronically infected and may benefit from anti-viral treatment prior to conception [72]. Influenza vaccine, without screening, is recommended for women planning to become pregnant prior to the influenza season or if they will be pregnant during the influenza season [73].

Live attenuated vaccines are to be avoided during pregnancy because of the theoretical risk of viraemia and the vaccine strain crossing the placenta. If a non-immune woman receives a live attenuated vaccine she should avoid pregnancy for at least 1 month [74]. However, data on accidental rubella immunization in pregnancy from the Pan American Health Organisation (PAHO), indicate that there was no increased risk of congenital rubella syndrome [75].

The post-partum period offers an opportunity to catch up with vaccination, especially with live attenuated vaccines that cannot be offered during pregnancy. Rubella immunization should be performed post-partum for all non-immune women. If pertussis vaccination was not offered during gestation, post-partum vaccination, the so-called cocoon vaccination, is still recommended in many countries [76]. Following yellow fever vaccination during lactation, a few case reports describe yellow fever disease in lactating neonates, caused by the vaccine strain [77,78]. This has never been described for other live attenuated vaccines offered during lactation.

Currently recommended vaccines for use in pregnancy

In the second section of the book, disease-specific vaccines recommended for use in pregnancy are described. Chapters' structure for each vaccine include, the burden of disease, rationale for implementing maternal immunization as a way for protecting the infant, fetus and/or mother, the effectiveness of the strategy, and safety data available. Based on experiences from the tetanus, influenza and pertussis vaccination, new vaccines are in the pipeline and specific chapters tackle the progress on vaccines protecting against RSV, GBS, cytomegalovirus (CMV), Zika virus and malaria.

Tetanus

Maternal and neonatal tetanus (MNT) remains a preventable cause of maternal and neonatal deaths in 14 countries, or regions within those countries [2]. Maternal tetanus is associated with abortion, miscarriages and unhygienic delivery conditions, whereas neonatal tetanus occurs secondary to poor post-partum cord care practices [79]. When MNT occurs, it is often fatal and is characterized by muscular rigidity and spasm. The Maternal and Neonatal Tetanus Elimination program launched by the World Health Assembly aims to eliminate MNT through promotion of clean birth and neonatal care practices, as well as maternal immunization with tetanus toxoid including vaccines. The present aim for elimination is set for 2020. The effectiveness of the tetanus vaccine in preventing maternal and neonatal tetanus is documented by elimination of the disease in pregnant women and neonates well before the year 2000 in all industrialized countries. For elimination, enhanced neonatal tetanus surveillance and monitoring of the impact of the existing immunization programs are essential for identification of problems and implementation of targeted interventions [80]. Home births contribute to the underreporting of cases, and impact the accurate validation of elimination interventions. Improving local reporting systems with education of local community workers and traditional birth attendants should therefore be promoted.

Immunization of pregnant women or women of childbearing age with tetanus toxoid is an inexpensive, effective and safe strategy to prevent maternal and neonatal tetanus [81]. Active transplacental transport has been documented and maternal antibodies remain above the protective level until the first infant vaccination is planned [82]. Once elimination of maternal and neonatal tetanus has been achieved, maintaining access to clean delivery practices and sustaining high immunization coverage with tetanus toxoid in women, in addition to routine childhood vaccines are essential. The evidence accumulated on the effectiveness and safety of maternal tetanus vaccination has also contributed to the increased acceptance and uptake of other vaccines given during pregnancy.

Influenza

Influenza is an important cause of morbidity and mortality among pregnant women and young infants, and influenza infection during pregnancy may also cause adverse obstetric and birth outcomes. Pregnant women are therefore a priority group for seasonal influenza vaccination. The recommendations by the Advisory Committee on Immunization Practices in the United States for influenza vaccination of pregnant women dates to the 1960s [83]. It was, however, only after the 2009 H1N1 influenza pandemic observation that pregnant women were the most severely affected group, that prioritizing pregnant women for influenza vaccination was emphasized. Correspondingly, it is now recommended by the World Health Organization (WHO) for those countries that include influenza vaccines in their public immunization programs [73]. While the initial recommendation for maternal influenza vaccination was based on limited safety and immunogenicity data, several studies have since corroborated the safety of maternal influenza vaccination. Epidemiological studies and clinical trials have also substantiated the effectiveness of inactivated influenza vaccines (IIV) in pregnant women in protecting the women and their infants from influenza illness [84–87]. Furthermore, albeit conflicting evidence, influenza vaccination during pregnancy has been reported to positively influence fetal outcomes including reducing the risk of low-birth weight, premature and stillbirth, an effect size that might be, however, more evident during pandemic influenza virus circulation than normal seasonal epidemics [88].

Across four randomized controlled trials (RCTs) influenza vaccination of pregnant women was efficacious against laboratory-confirmed influenza illness in their young infants, with an overall vaccine efficacy of 36% (95% CI: 22–48%) being reported [89]. Accordingly, two of the RCTs, in South Africa and Mali, also reported significant efficacy against influenza-confirmed illness in the women (50% and 70%, respectively) [84,86]. Notably, however, the duration of protection against influenza illness among the infants might be more concentrated to the first 2–3 months of life, rather than extended protection up to 6 months of age [86,90].

From a public health perspective especially for LMICs, the RCTs were designed to evaluate efficacy against any influenza-confirmed illness and not specifically against influenza hospitalization. Nevertheless, the most compelling reason for maternal influenza vaccination in protection of their infants against severe disease, is evident from the RCTs being used as a probe to delineate the impact of vaccination against other biologically plausible endpoints. A pooled analysis across the RCTs reported a 20% lower rate of all-cause pneumonia in infants born to women who received IIV [91].

Concern about the safety of vaccines during pregnancy, including vaccination against influenza, is consistently among the most common reasons cited for non-vaccination. More than five decades of observations, however, have established that influenza vaccination during pregnancy is safe. Different systematic reviews on safety studies found no elevated risk of adverse fetal or neonatal outcomes associated with both seasonal and pandemic maternal influenza immunization [92]. Regarding maternal safety, retrospective studies found no association between receipt of antenatal influenza vaccine and medically attended adverse obstetric events or maternal adverse events, including among first-trimester vaccinees [93]. Nonetheless evidence base on longer term health outcomes in children following influenza vaccination during pregnancy are still sparse and future studies in different settings, with different influenza vaccine formulations should address these outcomes.

Pertussis

Neonates are more susceptible to severe pertussis disease and death than any other age group [94,95]. Despite the high vaccine coverage of pertussis-containing vaccines in all age groups globally, the incidence of whooping cough has increased during the last decades, particularly in high-income countries [96]. The reasons for this resurgence of pertussis in these countries are multi-factorial, including the switch from the use of whole cell pertussis (wP) vaccines to acellular pertussis (aP) containing vaccines [97].

Immunization of pregnant women with aP vaccines for adult use has been implemented in several high-income countries, as it is the only means to protect young infants with the currently available vaccines [98,99]. The recommendations are based on epidemiological needs to fight the increasing incidence of pertussis cases with the highest number of hospitalizations and mortality among the very young infants, too young to have received the primary immunization doses.

Data on the immunogenicity and safety of maternal immunization obtained in those countries are reassuring. The first data were gathered in the United Kingdom, after implementation of the strategy; with no obstetrical or other adverse events reported. In large studies from the United States, interrogation of the Vaccines Safety Data databases confirmed safety of aP vaccination during pregnancy [100,101].

Pregnant women respond in a similar way compared to non-pregnant women with regards to humoral immune responses. The antibody transplacental transport is efficient for several *B. pertussis*-specific antigens and maternal antibodies remain at detectable levels in the neonates at least until the priming vaccination scheme is started, usually at the age of 8 weeks [102]. Therefore, the maternal antibodies close the susceptibility gap in the first weeks of life. In order to elicit high titers of maternal antibodies to be transported to the fetus, vaccination is recommended for every pregnancy. Indeed, antibodies in the women wane very rapidly, leaving less maternal antibodies to be transported with extending the interval between vaccination and pregnancy.

Effectiveness data from the United Kingdom, United States and other countries, indicate that maternal immunization with an aP containing vaccine is highly effective in preventing severe pertussis disease and mortality among young infants. However, suppression of infant responses to primary and booster vaccination after maternal immunization (the so-called blunting effect) has been demonstrated. Yet, current surveillance data, mainly from the United Kingdom, show no increased risk of pertussis disease in the first years of life, what could be attributable to the blunting effect of the infant vaccine responses [37]. Although the use of tetanus-diphtheria-aP (Tdap) vaccine in pregnancy has been shown to also blunt responses to infant pneumococcal vaccines, there was no impact on the actual percentage of children with putative protective levels of the pneumococcal antibodies [31,34].

Future studies using different vaccines or vaccines with more restricted pertussis antigens for use during pregnancy, and different immunization schedules in infants are needed to evaluate if the impact on infant immune responses are also reduced.

Vaccines recommended in specific circumstances

In addition to those vaccines for which there are national recommendations for use in pregnancy, there are other vaccines that might be considered for use in pregnant women in particular circumstances. The protective benefits of a vaccine must be carefully weighed against the potential risks to the mother, fetus or infant. This risk assessment can be challenging, as for many of these vaccines, there is paucity of clinical trial data and these vaccines have not been widely administered during pregnancy.

Vaccination against *Streptococcus pneumoniae* during pregnancy might be a useful strategy to reduce the burden of pneumococcal infection in young infants, prior to receipt of infant primary vaccination [103]. There is evidence from multiple clinical trials that polysaccharide vaccination in pregnancy is safe for the mother and fetus and can increase the concentration of specific antibody in the infant, however this was not associated with a reduction in disease in early infancy [104,105]. Pneumococcal conjugate vaccination in pregnancy has been less studied to date, however there is evidence that this is more likely to impact on disease reduction in the infant [106]. *Haemophilus influenzae* type B vaccines are also immunogenic in pregnant women and specific antibody is transferred across the placenta, resulting in higher concentrations of Hib-specific antibody in infants of vaccinated mothers, however, as yet, there is no evidence that this is effective in reducing disease in the infant [107].

In situations where the risk of infection is significantly increased, for example in regions where meningococcal infection is endemic, vaccination should be considered for the protection of the mother and the infant [108].

Vaccination may also be strongly considered when a pregnant woman has been exposed to infection, such as rabies where the mortality associated with the infection is high [58]. The risk assessment is more nuanced for pre-exposure vaccination against rabies, however since no evidence of harm has been shown for post-exposure vaccination, this approach might be considered in some circumstances [109]. Vaccination against typhoid, Japanese encephalitis, yellow fever, hepatitis A, cholera, tick borne encephalitis may be considered for women traveling to high-risk regions, however there is little evidence upon which to base recommendations of these vaccines in pregnancy. Where possible, live vaccines should be avoided in pregnancy and inactivated vaccines used in preference.

Future vaccines which may be recommended for use pre-conception or during pregnancy

Robust disease burden and cost-effectiveness estimates for RSV, GBS and CMV are being collected and will serve to inform expectations of vaccine program impact. These vaccines are also being developed with WHO programmatic suitability criteria in mind.

Respiratory syncytial virus

Respiratory syncytial virus is the most common cause of viral lower respiratory tract infection in infants globally and is associated with high rates of hospitalization in young infants [110]. The majority of deaths associated with RSV occur in LMICs and therefore the prevention of severe RSV disease is a high priority in these settings [111,112]. The burden of infection in pregnant women is less well characterized, however it is emerging that it may well be a significant pathogen in this population with one study showing an attack rate of laboratory-confirmed RSV of 10% among ambulatory pregnant women [113], suggesting that vaccination in pregnancy may provide protection to the woman herself as well as the infant. Other studies show a lower burden of illness but are limited by low rates of testing for respiratory viruses in this population.

Maternal IgG transferred to the infant across the placenta is protective against RSV disease, with infants with low concentrations being at the highest risk of severe disease [114–116] and hospitalization. The aim of vaccinating the mothers is to provide high concentrations of RSV-specific neutralizing antibody to the infants during the period of highest risk of severe disease, in particular IgG targeting the RSV fusion (F) protein. The F-protein is highly conserved and is expressed on the virion membrane and aids attachment to host cells and therefore is critical to viral entry. Passive immunization of infants with humanized monoclonal antibody specific for the F-protein of RSV (Palivizumab) demonstrates that high titres of IgG targeting the F-protein are effective at preventing severe RSV disease, however this approach is limited by the high cost and need for monthly intramuscular injection and therefore it is restricted to use in the most vulnerable preterm infants.

There are multiple vaccine candidates progressing through the development pipeline, the most advanced of those being developed for a maternal indication, targeting the F-protein (GlaxoSmithKline [GSK]

Investigational Vaccines and Novavax, Inc.) In 600 non-pregnant women, the GSK vaccine has been shown to be immunogenic and safe and is entering phase I/II clinical trials in pregnant women in 2020. The Novavax manufactured vaccine has been evaluated in phase II and phase III clinical trials in pregnancy and has been demonstrated to be well-tolerated and safe, inducing high concentrations of RSV-specific antibodies which peaked in mothers at 14 days post vaccination and was efficiently transferred to the infant [117,118]. The phase III trial, which included 4636 women, failed to demonstrate efficacy against the primary endpoint of medically significant RSV lower respiratory tract infections (LRTI) in infants, however efficacy against RSV-associated LRTI hospitalization in the first 90 days and up to 180 days after birth was demonstrated [119]. When the data collected as part of routine clinical care was included, vaccine efficacy against medically significant RSV LRTI was 41% (95% CI: 16–58%) and 60% (95% CI: 32–76%) against RSV LRTI with severe hypoxemia. While these results have not allowed progression to licensure, this pivotal study has allowed a significant leap forward for the field of maternal vaccination and shows that vaccination in pregnancy to prevent severe RSV infection early in life is within our grasp.

Group B Streptococcus

Streptococcus agalactiae (Group B *Streptococcus*) remains a leading cause of neonatal sepsis and meningitis in many countries. GBS infection has also been demonstrated as a cause of stillbirths and preterm births [120]. Although in low-middle income settings, the disease burden remains uncertain, estimates from 2015 revealed that there were at least 33,000 pregnant or post-partum women with GBS sepsis and 57,000 stillbirths due to GBS disease [121]. Intrapartum antibiotic prophylaxis (IAP) strategies have reduced the incidence of early-onset neonatal GBS disease (EOD) in several countries but have had no impact on late onset GBS disease (LOD). As disease may be rapidly fulminating, cases might be missed before appropriate samples are obtained and this may lead to underestimation of the true burden. Given the rapid onset of disease and progression within hours of birth as well as the deficiencies in IAP strategies and absence of a solution for preventing LOD, it is reasonable to speculate that administration of a suitable vaccine in pregnancy would provide a better solution in all settings; it should also be cost effective.

Vaccine manufacturers are increasingly committed to invest in GBS vaccines and several vaccine candidates are in active development. The current leading vaccine candidates are capsular polysaccharide protein conjugate vaccines but protein-based vaccines are also being pursued [122]. Advantages of maternal vaccination over IAP include the possible leverage of existing antenatal care platforms, the reduction of adverse outcomes also in pregnant and postnatal women, unborn babies, and infants, and prevention of the less clearly measured burden from non-invasive disease, including, preterm birth [123].

Maternally derived GBS antibodies are associated with protection from EOD, however, since a correlate of efficacy has not been defined, phase III efficacy trials may be required for licensure. But because of the large number of participants needed for those type of trials, alternate pathways to licensure could be explored, e.g., identification of serological correlates of protection with subsequent phase IV studies establishing vaccine-effectiveness against invasive GBS disease [124].

Cytomegalovirus

Congenital cytomegalovirus (cCMV) infection is asymptomatic in the majority of infants at birth, however for those who have signs and symptoms at birth—such as growth restriction, microcephaly, petechiae, blueberry muffin rash, hepatosplenomegaly, jaundice, seizures—the risk of neurodevelopmental sequelae such as sensorineural hearing loss (SNHL), cerebral palsy, seizure disorders and learning difficulties are significantly increased. However, even those who are apparently unaffected at birth, up to a quarter of these infants will develop SNHL in childhood [125–128]. The risks of adverse neurodevelopmental sequelae are greatest in those infants who are infected in the first trimester of pregnancy and the risk is substantially less as pregnancy progresses [129–132]. Therefore, a vaccine to prevent CCMV infection would need to be given prior to pregnancy to prevent the most severe complications of cCMV.

There are multiple challenges in the development of an effective vaccine to prevent cCMV. Immunological correlates of protection are as yet unknown and a vaccine to prevent CMV needs to be better than natural immunity, which does not fully protect against re-infection or re-activation. The optimal timing of vaccination remains a matter of debate. Potential options are women of childbearing age, teenagers and toddlers—to prevent transmission to pregnant family members from young children who commonly excrete the virus in urine and saliva for prolonged periods and are a common source of CMV infection in pregnancy [133,134].

There are several vaccines that have been tested in clinical trials including live, attenuated or disabled, single cycle (DISC) vaccines (AD169 vaccine, Towne and Toledo' strains and V160) and subunit (recombinant/vectored) vaccines (the majority of which are based on recombinant forms of glycoprotein B protein, either alone or in combination with other proteins such as 65-kilodalton phosphoprotein or the major immediate early protein 1—or employ vectors or plasmids expressing these proteins). At the current time, only the efficacy of the MF59-adjuvanted purified recombinant glycoprotein B protein vaccine has been demonstrated in clinical trials, with a 50% efficacy against acquisition of primary CMV infection in seronegative women [131]. This is a seminal clinical trial, as was the first to demonstrate efficacy. There have been no concerning safety signals from any of the studies of CMV vaccines performed to date. To continue the progress in the field, not only do the immunological and logistical issues need to be addressed, but it is important to raise awareness about CMV for all stakeholders to ensure that a vaccine, when available, is acceptable and achieves high uptake. These efforts will not only benefit the field of CMV vaccines, but improved awareness will also benefit primary prevention in pregnancy by hygiene based interventions and improved timely diagnosis of CMV in pregnancy and in the neonatal period to allow infants the opportunity to benefit from treatment and monitoring.

Zika virus

Zika virus is transmitted by bite of the *Aedes aegypti* mosquito and less commonly through sexual contact and blood transfusion. Zika virus infection generally causes mild disease and can even be asymptomatic in pregnant women and adults. It can, however, also manifest as dengue-like symptoms, including rash, myalgias, fever, and conjunctivitis [135]. In 2015, reports of association of Zika infection with increased incidence of Guillain-Barre syndrome and other neurologic conditions led to worldwide alarm. Further evidence of the Zika virus 2015–16 pandemic suggested that infants born to Zika-infected women can develop microcephaly, limb contractures, ocular and hearing problems and other central nervous system abnormalities [136]. Zika infection during pregnancy can result in placental infection and damage, which allows the virus to cross the placenta and infect the developing fetus. The risk of congenital abnormalities is particularly increased with infection acquired in the first trimester [137].

There is currently no vaccine for Zika, so prevention against infection includes control of mosquito breeding sites and protection against bites. Treatment is supportive based on symptoms. Similarly, there is no specific treatment for infants with congenital Zika syndrome other than supportive care. Although Zika cases have declined since 2016, the threat of new outbreaks emphasizes the urgent need for a vaccine to prevent the symptoms and complications of Zika virus infection, mainly to prevent congenital Zika syndrome. The goal of a Zika virus vaccine will be to elicit protective antibodies against the virus to prevent infection and severe disease. The challenges in developing a safe and effective vaccine, as for other vaccines, include the lack of a defined correlate of protection. Zika infection is mostly asymptomatic, therefore clinical endpoints for vaccine trials are impractical. If immunological endpoints are to be used for vaccine authorization, vaccine effectiveness will still need to be demonstrated in follow-up studies. Another difficulty in the development of Zika vaccines, is that since dengue virus is closely related to Zika virus, the vaccine needs to minimize the possibility of antibody-dependent enhancement of dengue virus infection [138]. There are presently multiple Zika virus vaccine candidates in different stages of development; including nucleic acid vaccines, inactivated whole virus vaccines, a live attenuated measles vector vaccine and a vaccine that targets mosquito salivary antigens (a universal mosquito-borne disease vaccine) [139–141]. Common targets for these vaccines include the viral pre-membrane and envelope proteins, which represent multiple surface epitopes easily accessible to antibodies. During outbreaks, it is possible that live Zika vaccines would be allowed at 6 months of age. Alternatively, nucleic acid or inactivated vaccines, could be given as early as 6 weeks of age as these are normally safe. Nonetheless, most important is that individuals are vaccinated before they become sexually active. In cases where vaccination is needed during pregnancy, it would be important to vaccinate as early as possible.

Malaria

Plasmodium falciparum malaria severely affects pregnant women and children. Despite immunity through lifelong exposure to malaria, pregnant women become susceptible to infections that may lead to poor pregnancy outcomes such as miscarriage, stillbirth, and neonatal death; and the offspring is at increased risk of fetal growth restriction, preterm birth and low-birth weight. Infected pregnant women may suffer severe anemia, preeclampsia, and death [142].

Pregnant women experience massive accumulation of infected erythrocytes (IE) in the placenta. Adhesion of IE to host endothelial receptors is mediated by members of a large diverse protein family called *P. falciparum* erythrocyte membrane protein 1 (PfEMP1) [143]. Pregnancy malaria is generally associated with the emergence of a distinct subset of parasites expressing a unique PfEMP1 that binds to the host-receptor chondroitin sulfate A (CSA) found in placental intervillous spaces and on the syncytiotrophoblast [144]. Resistance to pregnancy malaria is associated with the acquisition of antibodies that block IE binding to placental CSA. The absence (or rare occurrence) of CSA-binding parasites in malaria patients (children, men and non-pregnant women) suggests that these parasites become virulent only during pregnancy. Primigravidae women are at higher risk of placental malaria than multigravidae, which can be explained by the acquisition of specific anti-adhesion antibodies over successive pregnancies [145].

Currently, antimalarial drugs and anti-mosquitoes agents are the primary tools for malaria control; however, drug-resistant parasites and insecticide-resistant mosquitoes are emerging. Since pregnant women and children have the highest burden of malaria morbidity and mortality, they will be the greatest beneficiaries of a malaria vaccine. Although clinical trials on malaria vaccines have been conducted, none included pregnant women. At the moment the leading vaccine candidate to specifically protect against pregnancy malaria is based on the PfEMP1 that binds to CSA, but protein sequence variation needs to be carefully studied. This vaccine should ideally be administered before first pregnancies. Malaria vaccines that protect the general population are, however, also expected to benefit pregnant women [146].

The primary endpoints for a vaccine trial in pregnant women should be placental and peripheral parasitemia; and the secondary endpoints those known to be sequelae of malaria, including poor birth outcomes, and severe maternal anemia.

Conclusion

Maternal immunization is a safe and effective strategy to protect women and their infants, during and after pregnancy. Examples are available of successful implementation of maternal immunization and lessons can be learnt for future vaccines to be specifically developed for administration during pregnancy. The research on this topic is quickly evolving and offers exciting new insights in both basic research on the mechanism of protection as well as on applied research for acceptance and implantation of the strategy on a global level.

Acknowledgments

Editors would like to express their gratitude to Dr Flor M. Munoz for her help on writing the paragraph on **Conducting research in maternal immunization**.

References

[1] United Nations. Sustainable development goal 3; 2019.
[2] World Health Organisation. Maternal and neonatal tetanus elimination (MNTE); 2019.
[3] Mahmoud F, Abul H, Omu A, Al-Rayes S, Haines D, Whaley K. Pregnancy-associated changes in peripheral blood lymphocyte subpopulations in normal Kuwaiti women. Gynecol Obstet Investig 2001;52(4):232–6.
[4] Watanabe M, Iwatani Y, Kaneda T, Hidaka Y, Mitsuda N, Morimoto Y, et al. Changes in T, B, and NK lymphocyte subsets during and after normal pregnancy. Am J Reprod Immunol 1997;37(5):368–77.
[5] Zimmer JP, Garza C, Butte NF, Goldman AS. Maternal blood B-cell (CD19+) percentages and serum immunoglobulin concentrations correlate with breast-feeding behavior and serum prolactin concentration. Am J Reprod Immunol 1998;40(1):57–62.

[6] Matthiesen L, Berg G, Ernerudh J, Hakansson L. Lymphocyte subsets and mitogen stimulation of blood lymphocytes in normal pregnancy. Am J Reprod Immunol 1996;35(2):70–9.

[7] Valdimarsson H, Mulholland C, Fridriksdottir V, Coleman DV. A longitudinal study of leucocyte blood counts and lymphocyte responses in pregnancy: a marked early increase of monocyte-lymphocyte ratio. Clin Exp Immunol 1983;53(2):437–43.

[8] Moore MP, Carter NP, Redman CW. Lymphocyte subsets defined by monoclonal antibodies in human pregnancy. Am J Reprod Immunol 1983;3(4):161–4.

[9] Hulka JF. Effectiveness of polyvalent influenza vaccine in pregnancy. Report of a controlled study during an outbreak of Asian influenza. Obstet Gynecol 1964;23:830–7.

[10] Murray DL, Imagawa DT, Okada DM, St Geme Jr JW. Antibody response to monovalent A/New Jersey/8/76 influenza vaccine in pregnant women. J Clin Microbiol 1979;10(2):184–7.

[11] Christian LM, Porter K, Karlsson E, Schultz-Cherry S, Iams JD. Serum proinflammatory cytokine responses to influenza virus vaccine among women during pregnancy versus non-pregnancy. Am J Reprod Immunol 2013;70(1):45–53.

[12] Kay AW, Bayless NL, Fukuyama J, Aziz N, Dekker CL, Mackey S, et al. Pregnancy does not attenuate the antibody or plasmablast response to inactivated influenza vaccine. J Infect Dis 2015;212(6):861–70.

[13] Munoz FM, Bond NH, Maccato M, Pinell P, Hammill HA, Swamy GK, et al. Safety and immunogenicity of tetanus diphtheria and acellular pertussis (Tdap) immunization during pregnancy in mothers and infants: a randomized clinical trial. JAMA 2014;311(17):1760–9.

[14] Huygen K, Cabore RN, Maertens K, Van Damme P, Leuridan E. Humoral and cell mediated immune responses to a pertussis containing vaccine in pregnant and nonpregnant women. Vaccine 2015;33(33):4117–23.

[15] Schlaudecker EP, McNeal MM, Dodd CN, Ranz JB, Steinhoff MC. Pregnancy modifies the antibody response to trivalent influenza immunization. J Infect Dis 2012;206(11):1670–3.

[16] Schlaudecker EP, Ambroggio L, McNeal MM, Finkelman FD, Way SS. Declining responsiveness to influenza vaccination with progression of human pregnancy. Vaccine 2018;36(31):4734–41.

[17] Bischoff AL, Folsgaard NV, Carson CG, Stokholm J, Pedersen L, Holmberg M, et al. Altered response to A(H1N1)pnd09 vaccination in pregnant women: a single blinded randomized controlled trial. PLoS ONE 2013;8(4):e56700.

[18] de Moraes-Pinto MI, Verhoeff F, Chimsuku L, Milligan PJ, Wesumperuma L, Broadhead RL, et al. Placental antibody transfer: influence of maternal HIV infection and placental malaria. Arch Dis Child Fetal Neonatal Ed 1998;79(3):F202–5.

[19] Okoko BJ, Wesuperuma LH, Ota MO, Banya WA, Pinder M, Gomez FS, et al. Influence of placental malaria infection and maternal hypergammaglobulinaemia on materno-foetal transfer of measles and tetanus antibodies in a rural west African population. J Health Popul Nutr 2001;19(2):59–65.

[20] Brair ME, Brabin BJ, Milligan P, Maxwell S, Hart CA. Reduced transfer of tetanus antibodies with placental malaria. Lancet 1994;343(8891):208–9.

[21] Ogolla S, Daud II, Asito AS, Sumba OP, Ouma C, Vulule J, et al. Reduced transplacental transfer of a subset of Epstein-Barr virus-specific antibodies to neonates of mothers infected with Plasmodium falciparum malaria during pregnancy. Clin Vaccine Immunol 2015;22(11):1197–205.

[22] Cumberland P, Shulman CE, Maple PA, Bulmer JN, Dorman EK, Kawuondo K, et al. Maternal HIV infection and placental malaria reduce transplacental antibody transfer and tetanus antibody levels in newborns in Kenya. J Infect Dis 2007;196(4):550–7.

[23] Abu-Raya B, Smolen KK, Willems F, Kollmann TR, Marchant A. Transfer of maternal antimicrobial immunity to HIV-exposed uninfected newborns. Front Immunol 2016;7:338.
[24] Jones CE, Naidoo S, De Beer C, Esser M, Kampmann B, Hesseling AC. Maternal HIV infection and antibody responses against vaccine-preventable diseases in uninfected infants. JAMA 2011;305(6):576–84.
[25] Goetghebuer T, Smolen KK, Adler C, Das J, McBride T, Smits G, et al. Initiation of antiretroviral therapy before pregnancy reduces the risk of infection-related hospitalization in human immunodeficiency virus-exposed uninfected infants born in a high-income country. Clin Infect Dis 2019;68(7):1193–203.
[26] Albrecht P, Ennis FA, Saltzman EJ, Krugman S. Persistence of maternal antibody in infants beyond 12 months: mechanism of measles vaccine failure. J Pediatr 1977;91(5):715–8.
[27] Leuridan E, Hens N, Hutse V, Ieven M, Aerts M, Van Damme P. Early waning of maternal measles antibodies in era of measles elimination: longitudinal study. BMJ 2010;340:c1626.
[28] Siegrist CA. Mechanisms by which maternal antibodies influence infant vaccine responses: review of hypotheses and definition of main determinants. Vaccine 2003;21(24):3406–12.
[29] Maertens K, Cabore RN, Huygen K, Vermeiren S, Hens N, Van Damme P, et al. Pertussis vaccination during pregnancy in Belgium: follow-up of infants until 1 month after the fourth infant pertussis vaccination at 15 months of age. Vaccine 2016;34(31):3613–9.
[30] Maertens K, Hoang TT, Nguyen TD, Cabore RN, Duong TH, Huygen K, et al. The effect of maternal pertussis immunization on infant vaccine responses to a booster pertussis-containing vaccine in Vietnam. Clin Infect Dis 2016;63(Suppl. 4): S197–204.
[31] Ladhani SN, Andrews NJ, Southern J, Jones CE, Amirthalingam G, Waight PA, et al. Antibody responses after primary immunization in infants born to women receiving a pertussis-containing vaccine during pregnancy: single arm observational study with a historical comparator. Clin Infect Dis 2015;61(11):1637–44.
[32] Halperin SA, Langley JM, Ye L, MacKinnon-Cameron D, Elsherif M, Allen VM, et al. A randomized controlled trial of the safety and immunogenicity of tetanus, diphtheria, and acellular pertussis vaccine immunization during pregnancy and subsequent infant immune response. Clin Infect Dis 2018;67(7):1063–71.
[33] Hardy-Fairbanks AJ, Pan SJ, Decker MD, Johnson DR, Greenberg DP, Kirkland KB, et al. Immune responses in infants whose mothers received Tdap vaccine during pregnancy. Pediatr Infect Dis J 2013;32(11):1257–60.
[34] Maertens K, Burbidge P, Van Damme P, Goldblatt D, Leuridan E. Pneumococcal immune response in infants whose mothers received tetanus, diphtheria and acellular pertussis vaccination during pregnancy. Pediatr Infect Dis J 2017;36(12):1186–92.
[35] Kim D, Huey D, Oglesbee M, Niewiesk S. Insights into the regulatory mechanism controlling the inhibition of vaccine-induced seroconversion by maternal antibodies. Blood 2011;117(23):6143–51.
[36] Dabrera G, Amirthalingam G, Andrews N, Campbell H, Ribeiro S, Kara E, et al. A case-control study to estimate the effectiveness of maternal pertussis vaccination in protecting newborn infants in England and Wales, 2012-2013. Clin Infect Dis 2015;60(3):333–7.
[37] Amirthalingam G, Campbell H, Ribeiro S, Fry NK, Ramsay M, Miller E, et al. Sustained effectiveness of the maternal pertussis immunization program in England 3 years following introduction. Clin Infect Dis 2016;63(Suppl. 4):S236–43.
[38] Amirthalingam G, Letley L, Campbell H, Green D, Yarwood J, Ramsay M. Lessons learnt from the implementation of maternal immunization programs in England. Hum Vaccin Immunother 2016;12(11):2934–9.

[39] Byrne L, Campbell H, Andrews N, Ribeiro S, Amirthalingam G. Hospitalisation of preterm infants with pertussis in the context of a maternal vaccination programme in England. Arch Dis Child 2018;103(3):224–9.
[40] Bellido-Blasco J, Guiral-Rodrigo S, Miguez-Santiyan A, Salazar-Cifre A, Gonzalez-Moran F. A case-control study to assess the effectiveness of pertussis vaccination during pregnancy on newborns, Valencian community, Spain, 1 March 2015 to 29 February 2016. Euro Surveill 2017;22(22).
[41] Baxter R, Bartlett J, Fireman B, Lewis E, Klein NP. Effectiveness of vaccination during pregnancy to prevent infant pertussis. Pediatrics 2017;139(5).
[42] Skoff TH, Blain AE, Watt J, Scherzinger K, McMahon M, Zansky SM, et al. Impact of the US maternal tetanus, diphtheria, and acellular pertussis vaccination program on preventing pertussis in infants <2 months of age: a case-control evaluation. Clin Infect Dis 2017;65(12):1977–83.
[43] Winter K, Cherry JD, Harriman K. Effectiveness of prenatal tetanus, diphtheria, and acellular pertussis vaccination on pertussis severity in infants. Clin Infect Dis 2017;64(1):9–14.
[44] Winter K, Nickell S, Powell M, Harriman K. Effectiveness of prenatal versus postpartum tetanus, diphtheria, and acellular pertussis vaccination in preventing infant pertussis. Clin Infect Dis 2017;64(1):3–8.
[45] Marchant A, Sadarangani M, Garand M, Dauby N, Verhasselt V, Pereira L, et al. Maternal immunisation: collaborating with mother nature. Lancet Infect Dis 2017;17(7):e197–208.
[46] Caballero-Flores G, Sakamoto K, Zeng MY, Wang Y, Hakim J, Matus-Acuna V, et al. Maternal immunization confers protection to the offspring against an attaching and effacing pathogen through delivery of IgG in breast milk. Cell Host Microbe 2019;25(2):313–323.e4.
[47] Shao HY, Chen YC, Chung NH, Lu YJ, Chang CK, Yu SL, et al. Maternal immunization with a recombinant adenovirus-expressing fusion protein protects neonatal cotton rats from respiratory syncytia virus infection by transferring antibodies via breast milk and placenta. Virology 2018;521:181–9.
[48] Schlaudecker EP, Steinhoff MC, Omer SB, McNeal MM, Roy E, Arifeen SE, et al. IgA and neutralizing antibodies to influenza a virus in human milk: a randomized trial of antenatal influenza immunization. PLoS ONE 2013;8(8):e70867.
[49] De Schutter S, Maertens K, Baerts L, De Meester I, Van Damme P, Leuridan E. Quantification of vaccine-induced antipertussis toxin secretory IgA antibodies in breast milk: comparison of different vaccination strategies in women. Pediatr Infect Dis J 2015;34(6):e149–52.
[50] Shahid NS, Steinhoff MC, Hoque SS, Begum T, Thompson C, Siber GR. Serum, breast milk, and infant antibody after maternal immunisation with pneumococcal vaccine. Lancet 1995;346(8985):1252–7.
[51] Shahid NS, Steinhoff MC, Roy E, Begum T, Thompson CM, Siber GR. Placental and breast transfer of antibodies after maternal immunization with polysaccharide meningococcal vaccine: a randomized, controlled evaluation. Vaccine 2002;20(17–18):2404–9.
[52] World Health Organization. Maternal and neonatal tetanus elimination (MNTE) - the initiative and challenges, http://www.who.int/immunization/diseases/MNTE_initiative/en.
[53] Munoz FM, Sheffield JS, Beigi RH, Read JS, Swamy GK, Jevaji I, et al. Research on vaccines during pregnancy: protocol design and assessment of safety. Vaccine 2013;31(40):4274–9.
[54] Beigi RH, Omer SB, Thompson KM. Key steps forward for maternal immunization: policy making in action. Vaccine 2018;36(12):1521–3.
[55] Overcoming barriers and identifying opportunities for developing maternal immunizations: recommendations from the National Vaccine Advisory Committee. Public Health Rep 2017;132(3):271–84.

[56] American College of Obstetricians and Gynecologists. Ethical considerations for including women as research participants. ACOG Committee Opinion No 646; 2015.
[57] Chamberlain AT, Lavery JV, White A, Omer SB. Ethics of maternal vaccination. Science 2017;358(6362):452–3.
[58] Centers for Disease Control and Prevention. Guidelines for vaccinating pregnant women. Available from: https://www.cdc.gov/vaccines/pregnancy/hcp/guidelines.html; 2019.
[59] Krubiner CB, Faden RR, Karron RA, Little MO, Lyerly AD, Abramson JS, et al. Pregnant women & vaccines against emerging epidemic threats: ethics guidance for preparedness, research, and response. Vaccine 2019;. (in press) https://doi.org/10.1016/j.vaccine.2019.01.011.
[60] World Health Organisation. Maternal immunization research and implementation portfolio. 2019.
[61] World Health Organisation. Maternal Immunization and Antenatal Care Situation Analysis (MIACSA); 2019.
[62] Zuber PLF, Moran AC, Chou D, Renaud F, Halleux C, Pena-Rosas JP, et al. Mapping the landscape of global programmes to evaluate health interventions in pregnancy: the need for harmonised approaches, standards and tools. BMJ Glob Health 2018;3(5):e001053.
[63] World Health Organisation. Manual for estimating disease burden associated with seasonal influenza. World Health Organization; 2015.
[64] Krishnaswamy S, Wallace EM, Buttery J, Giles ML. Strategies to implement maternal vaccination: a comparison between standing orders for midwife delivery, a hospital based maternal immunisation service and primary care. Vaccine 2018;36(13):1796–800.
[65] Wilson RJ, Paterson P, Jarrett C, Larson HJ. Understanding factors influencing vaccination acceptance during pregnancy globally: a literature review. Vaccine 2015;33(47):6420–9.
[66] Bonhoeffer J, Kochhar S, Hirschfeld S, Heath PT, Jones CE, Bauwens J, et al. Global alignment of immunization safety assessment in pregnancy - the GAIA project. Vaccine 2016;34(49):5993–7.
[67] Healy CM, Rench MA, Montesinos DP, Ng N, Swaim LS. Knowledge and attitudes of pregnant women and their providers towards recommendations for immunization during pregnancy. Vaccine 2015;33(41):5445–51.
[68] Wilson RJCTLS, Paterson P, Larson H. The patient – healthcare worker relationship: how does it affect patient views towards vaccination during pregnancy? In: Health and health care concerns among women and racial and ethnic minorities. Research in the sociology of health care, vol. 35. Emerald Publishing Limited; 2017.
[69] SAGE. Report of the SAGE working group on vaccine hesitancy. Available from: http://www.who.int/immunization/sage/meetings/2014/october/1_Report_WORKING_GROUP_vaccine_hesitancy_final.pdf; 2014.
[70] Yudin MH, Salripour M, Sgro MD. Impact of patient education on knowledge of influenza and vaccine recommendations among pregnant women. J Obstet Gynaecol Can 2010;32(3):232–7.
[71] Coonrod DV, Jack BW, Boggess KA, Long R, Conry JA, Cox SN, et al. The clinical content of preconception care: immunizations as part of preconception care. Am J Obstet Gynecol 2008;199(6 Suppl. 2):S290–5.
[72] Henderson JT, Webber EM, Bean SI. Screening for hepatitis B infection in pregnant women: updated evidence report and systematic review for the US Preventive Services Task Force. JAMA 2019;322(4):360–2.
[73] Vaccines against influenza WHO position paper - November 2012. Wkly Epidemiol Rec 2012;87(47):461–76.
[74] Centers for Disease Control and Prevention. Revised ACIP recommendation for avoiding pregnancy after receiving a rubella-containing vaccine. MMWR Morb Mortal Wkly Rep 2001;50(49):1117.

[75] Castillo-Solorzano C, Reef SE, Morice A, Vascones N, Chevez AE, Castalia-Soares R, et al. Rubella vaccination of unknowingly pregnant women during mass campaigns for rubella and congenital rubella syndrome elimination, the Americas 2001-2008. J Infect Dis 2011;204(Suppl. 2):S713–7.
[76] Healy CM, Rench MA, Wootton SH, Castagnini LA. Evaluation of the impact of a pertussis cocooning program on infant pertussis infection. Pediatr Infect Dis J 2015;34(1):22–6.
[77] Traiber C, Coelho-Amaral P, Ritter VR, Winge A. Infant meningoencephalitis caused by yellow fever vaccine virus transmitted via breastmilk. J Pediatr 2011;87(3):269–72.
[78] Kuhn S, Twele-Montecinos L, MacDonald J, Webster P, Law B. Case report: probable transmission of vaccine strain of yellow fever virus to an infant via breast milk. CMAJ 2011;183(4):E243–5.
[79] Hatheway CL, Johnson EA. Topley & Wilson's Microbiology and Microbial Infections. Wiley; 1998731–82.
[80] World Health Organisation. WHO-recommended surveillance standard of neonatal tetanus; n.d. Available from: http://www.who.int/immunization/monitoring_surveillance/burden/vpd/surveillance_type/active/NT_Standards/en. [Accessed 5 October 2019].
[81] Zheteyeva YA, Moro PL, Tepper NK, Rasmussen SA, Barash FE, Revzina NV, et al. Adverse event reports after tetanus toxoid, reduced diphtheria toxoid, and acellular pertussis vaccines in pregnant women. Am J Obstet Gynecol 2012;207(1). 59.e1–7.
[82] Gendrel D, Richard-Lenoble D, Massamba MB, Picaud A, Francoual C, Blot P. Placental transfer of tetanus antibodies and protection of the newborn. J Trop Pediatr 1990;36(6):279–82.
[83] Burney LE. Influenza immunization: Statement. Public Health Rep 1960;75(10):944.
[84] Madhi SA, Cutland CL, Kuwanda L, Weinberg A, Hugo A, Jones S, et al. Influenza vaccination of pregnant women and protection of their infants. N Engl J Med 2014;371(10):918–31.
[85] Steinhoff MC, Katz J, Englund JA, Khatry SK, Shrestha L, Kuypers J, et al. Year-round influenza immunisation during pregnancy in Nepal: a phase 4, randomised, placebo-controlled trial. Lancet Infect Dis 2017;17(9):981–9.
[86] Tapia MD, Sow SO, Tamboura B, Teguete I, Pasetti MF, Kodio M, et al. Maternal immunisation with trivalent inactivated influenza vaccine for prevention of influenza in infants in Mali: a prospective, active-controlled, observer-blind, randomised phase 4 trial. Lancet Infect Dis 2016;16(9):1026–35.
[87] Zaman K, Roy E, Arifeen SE, Rahman M, Raqib R, Wilson E, et al. Effectiveness of maternal influenza immunization in mothers and infants. N Engl J Med 2008;359(15):1555–64.
[88] Nunes MC, Aqil AR, Omer SB, Madhi SA. The effects of influenza vaccination during pregnancy on birth outcomes: a systematic review and meta-analysis. Am J Perinatol 2016;33(11):1104–14.
[89] Nunes MC, Madhi SA. Influenza vaccination during pregnancy for prevention of influenza confirmed illness in the infants: a systematic review and meta-analysis. Hum Vaccin Immunother 2017;14(3):758–66.
[90] Nunes MC, Cutland CL, Jones S, Hugo A, Madimabe R, Simoes EA, et al. Duration of infant protection against influenza illness conferred by maternal immunization: secondary analysis of a randomized clinical trial. JAMA Pediatr 2016;170(9):840–7.
[91] Omer SB, Clark DR, Aqil AR, Tapia MD, Nunes MC, Kozuki N, et al. Maternal influenza immunization and prevention of severe clinical pneumonia in young infants: analysis of randomized controlled trials conducted in Nepal, Mali, and South Africa. Pediatr Infect Dis J 2018;37(5):436–40.
[92] Munoz FM. Safety of influenza vaccines in pregnant women. Am J Obstet Gynecol 2012;207(3 Suppl):S33–7.

[93] McMillan M, Porritt K, Kralik D, Costi L, Marshall H. Influenza vaccination during pregnancy: a systematic review of fetal death, spontaneous abortion, and congenital malformation safety outcomes. Vaccine 2015;33(18):2108–17.
[94] Paddock CD, Sanden GN, Cherry JD, Gal AA, Langston C, Tatti KM, et al. Pathology and pathogenesis of fatal Bordetella pertussis infection in infants. Clin Infect Dis 2008;47(3):328–38.
[95] Centers for Disease Control and Prevention. Pertussis (whooping cough): surveillance and reporting. 2018.
[96] Yeung KHT, Duclos P, Nelson EAS, Hutubessy RCW. An update of the global burden of pertussis in children younger than 5 years: a modelling study. Lancet Infect Dis 2017;17(9):974–80.
[97] Edwards KM. Unraveling the challenges of pertussis. Proc Natl Acad Sci U S A 2014;111(2):575–6.
[98] ACIP. Updated recommendations for use of tetanus toxoid, reduced diphtheria toxoid, and acellular pertussis vaccine (Tdap) in pregnant women. MMWR Morb Mortal Wkly Rep 2013;62:131–5.
[99] Public Health England. Vaccination against pertussis (whooping cough) for pregnant women-2016: Information for healthcare professionals. Available from: https://www.gov.uk/government/publications/vaccination-against-pertussis-whooping-cough-for-pregnant-women; 2016.
[100] Donegan K, King B, Bryan P. Safety of pertussis vaccination in pregnant women in UK: observational study. BMJ 2014;349:g4219.
[101] Kharbanda EO, Vazquez-Benitez G, Lipkind HS, Klein NP, Cheetham TC, Naleway A, et al. Evaluation of the association of maternal pertussis vaccination with obstetric events and birth outcomes. JAMA 2014;312(18):1897–904.
[102] Maertens K, Cabore RN, Huygen K, Hens N, Van Damme P, Leuridan E. Pertussis vaccination during pregnancy in Belgium: results of a prospective controlled cohort study. Vaccine 2016;34(1):142–50.
[103] Clarke E, Kampmann B, Goldblatt D. Maternal and neonatal pneumococcal vaccination - where are we now? Expert Rev Vaccines 2016;15(10):1305–17.
[104] Quiambao BP, Nohynek H, Kayhty H, Ollgren J, Gozum L, Gepanayao CP, et al. Maternal immunization with pneumococcal polysaccharide vaccine in the Philippines. Vaccine 2003;21(24):3451–4.
[105] Lopes CR, Berezin EN, Ching TH, de Souza Canuto J, Costa VO, Klering EM. Ineffectiveness for infants of immunization of mothers with pneumococcal capsular polysaccharide vaccine during pregnancy. Braz J Infect Dis 2009;13(2):104–6.
[106] Daly KA, Scott Giebink G, Lindgren BR, Knox J, Haggerty BJ, Nordin J, et al. Maternal immunization with pneumococcal 9-valent conjugate vaccine and early infant otitis media. Vaccine 2014;32(51):6948–55.
[107] Salam RA, Das JK, Dojo Soeandy C, Lassi ZS, Bhutta ZA. Impact of Haemophilus influenzae type B (Hib) and viral influenza vaccinations in pregnancy for improving maternal, neonatal and infant health outcomes. Cochrane Database Syst Rev 2015;6:CD009982.
[108] Centers for Disease Control and Prevention. Meningococcal disease; n.d. Available from: https://www.cdc.gov/meningococcal/index.html [Accessed 5 October 2019].
[109] Sudarshan MK, Madhusudana SN, Mahendra BJ. Post-exposure prophylaxis with purified vero cell rabies vaccine during pregnancy--safety and immunogenicity. J Commun Dis 1999;31(4):229–36.
[110] Hall CB, Weinberg GA, Iwane MK, Blumkin AK, Edwards KM, Staat MA, et al. The burden of respiratory syncytial virus infection in young children. N Engl J Med 2009;360(6):588–98.
[111] Shi T, McAllister DA, O'Brien KL, Simoes EAF, Madhi SA, Gessner BD, et al. Global, regional, and national disease burden estimates of acute lower respiratory infections

due to respiratory syncytial virus in young children in 2015: a systematic review and modelling study. Lancet 2017;390(10098):946–58.
[112] Scheltema NM, Gentile A, Lucion F, Nokes DJ, Munywoki PK, Madhi SA, et al. Global respiratory syncytial virus-associated mortality in young children (RSV GOLD): a retrospective case series. Lancet Glob Health 2017;5(10):e984–91.
[113] Hause AM, Avadhanula V, Maccato ML, Pinell PM, Bond N, Santarcangelo P, et al. Clinical characteristics and outcomes of respiratory syncytial virus infection in pregnant women. Vaccine 2019;37(26):3464–71.
[114] Eick A, Karron R, Shaw J, Thumar B, Reid R, Santosham M, et al. The role of neutralizing antibodies in protection of American Indian infants against respiratory syncytial virus disease. Pediatr Infect Dis J 2008;27(3):207–12.
[115] Glezen WP, Paredes A, Allison JE, Taber LH, Frank AL. Risk of respiratory syncytial virus infection for infants from low-income families in relationship to age, sex, ethnic group, and maternal antibody level. J Pediatr 1981;98(5):708–15.
[116] Chu HY, Steinhoff MC, Magaret A, Zaman K, Roy E, Langdon G, et al. Respiratory syncytial virus transplacental antibody transfer and kinetics in mother-infant pairs in Bangladesh. J Infect Dis 2014;210(10):1582–9.
[117] August A, Glenn GM, Kpamegan E, Hickman SP, Jani D, Lu H, et al. A phase 2 randomized, observer-blind, placebo-controlled, dose-ranging trial of aluminum-adjuvanted respiratory syncytial virus F particle vaccine formulations in healthy women of childbearing age. Vaccine 2017;35(30):3749–59.
[118] Munoz FM, Swamy GK, Hickman SP, Agrawal S, Piedra PA, Glenn GM, et al. Safety and immunogenicity of a respiratory syncytial virus fusion (F) protein nanoparticle vaccine in healthy third-trimester pregnant women and their infants. J Infect Dis 2019;. https://doi.org/10.1093/infdis/jiz390. [Epub ahead of print].
[119] Novavax. 2019. Available from: http://ir.novavax.com/news-releases/news-release-details/novavax-announces-topline-results-phase-3-preparetm-trial#.
[120] Lawn JE, Bianchi-Jassir F, Russell NJ, Kohli-Lynch M, Tann CJ, Hall J, et al. Group B Streptococcal disease worldwide for pregnant women, stillbirths, and children: why, what, and how to undertake estimates? Clin Infect Dis 2017;65(Suppl. 2):S89–99.
[121] Seale AC, Bianchi-Jassir F, Russell NJ, Kohli-Lynch M, Tann CJ, Hall J, et al. Estimates of the burden of group B streptococcal disease worldwide for pregnant women, stillbirths, and children. Clin Infect Dis 2017;65(Suppl. 2):S200–19.
[122] Dzanibe S, Madhi SA. Systematic review of the clinical development of group B streptococcus serotype-specific capsular polysaccharide-based vaccines. Expert Rev Vaccines 2018;17(7):635–51.
[123] Heath PT. Status of vaccine research and development of vaccines for GBS. Vaccine 2016;34(26):2876–9.
[124] Madhi SA, Dangor Z, Heath PT, Schrag S, Izu A, Sobanjo-Ter Meulen A, et al. Considerations for a phase-III trial to evaluate a group B Streptococcus polysaccharide-protein conjugate vaccine in pregnant women for the prevention of early- and late-onset invasive disease in young-infants. Vaccine 2013;31(Suppl. 4):D52–7.
[125] Fowler KB, Boppana SB. Congenital cytomegalovirus (CMV) infection and hearing deficit. J Clin Virol 2006;35(2):226–31.
[126] Goderis J, De Leenheer E, Smets K, Van Hoecke H, Keymeulen A, Dhooge I. Hearing loss and congenital CMV infection: a systematic review. Pediatrics 2014;134(5):972–82.
[127] Goderis J, Keymeulen A, Smets K, Van Hoecke H, De Leenheer E, Boudewyns A, et al. Hearing in children with congenital cytomegalovirus infection: results of a longitudinal study. J Pediatr 2016;172:110–5. [e2].
[128] Dahle AJ, Fowler KB, Wright JD, Boppana SB, Britt WJ, Pass RF. Longitudinal investigation of hearing disorders in children with congenital cytomegalovirus. J Am Acad Audiol 2000;11(5):283–90.

[129] Faure-Bardon V, Magny JF, Parodi M, Couderc S, Garcia P, Maillotte AM, et al. Sequelae of congenital cytomegalovirus (cCMV) following maternal primary infection are limited to those acquired in the first trimester of pregnancy. Clin Infect Dis 2018;69(9):1526–32.
[130] Lipitz S, Yinon Y, Malinger G, Yagel S, Levit L, Hoffman C, et al. Risk of cytomegalovirus-associated sequelae in relation to time of infection and findings on prenatal imaging. Ultrasound Obstet Gynecol 2013;41(5):508–14.
[131] Pass RF, Duliege AM, Boppana S, Sekulovich R, Percell S, Britt W, et al. A subunit cytomegalovirus vaccine based on recombinant envelope glycoprotein B and a new adjuvant. J Infect Dis 1999;180(4):970–5.
[132] Pass RF, Fowler KB, Boppana SB, Britt WJ, Stagno S. Congenital cytomegalovirus infection following first trimester maternal infection: symptoms at birth and outcome. J Clin Virol 2006;35(2):216–20.
[133] Amin MM, Stowell JD, Hendley W, Garcia P, Schmid DS, Cannon MJ, et al. CMV on surfaces in homes with young children: results of PCR and viral culture testing. BMC Infect Dis 2018;18(1):391.
[134] Pass RF, Hutto C, Ricks R, Cloud GA. Increased rate of cytomegalovirus infection among parents of children attending day-care centers. N Engl J Med 1986;314(22):1414–8.
[135] Brasil P, Pereira Jr JP, Moreira ME, Ribeiro Nogueira RM, Damasceno L, Wakimoto M, et al. Zika virus infection in pregnant women in Rio de Janeiro. N Engl J Med 2016;375(24):2321–34.
[136] Brasil P, Nielsen-Saines K. More pieces to the microcephaly-Zika virus puzzle in Brazil. Lancet Infect Dis 2016;16(12):1307–9.
[137] Honein MA, Dawson AL, Petersen EE, Jones AM, Lee EH, Yazdy MM, et al. Birth defects among fetuses and infants of US women with evidence of possible Zika virus infection during pregnancy. JAMA 2017;317(1):59–68.
[138] Dejnirattisai W, Supasa P, Wongwiwat W, Rouvinski A, Barba-Spaeth G, Duangchinda T, et al. Dengue virus sero-cross-reactivity drives antibody-dependent enhancement of infection with zika virus. Nat Immunol 2016;17(9):1102–8.
[139] Boigard H, Alimova A, Martin GR, Katz A, Gottlieb P, Galarza JM. Zika virus-like particle (VLP) based vaccine. PLoS Negl Trop Dis 2017;11(5):e0005608.
[140] Durbin AP, McArthur JH, Marron JA, Blaney JE, Thumar B, Wanionek K, et al. rDEN2/4Delta30(ME), a live attenuated chimeric dengue serotype 2 vaccine is safe and highly immunogenic in healthy dengue-naive adults. Hum Vaccin 2006;2(6):255–60.
[141] Abbink P, Larocca RA, De La Barrera RA, Bricault CA, Moseley ET, Boyd M, et al. Protective efficacy of multiple vaccine platforms against Zika virus challenge in rhesus monkeys. Science 2016;353(6304):1129–32.
[142] Desai M, ter Kuile FO, Nosten F, McGready R, Asamoa K, Brabin B, et al. Epidemiology and burden of malaria in pregnancy. Lancet Infect Dis 2007;7(2):93–104.
[143] Fried M, Domingo GJ, Gowda CD, Mutabingwa TK, Duffy PE. Plasmodium falciparum: chondroitin sulfate A is the major receptor for adhesion of parasitized erythrocytes in the placenta. Exp Parasitol 2006;113(1):36–42.
[144] Nunes MC, Scherf A. Plasmodium falciparum during pregnancy: a puzzling parasite tissue adhesion tropism. Parasitology 2007;134(Pt 13):1863–9.
[145] Muehlenbachs A, Fried M, Lachowitzer J, Mutabingwa TK, Duffy PE. Genome-wide expression analysis of placental malaria reveals features of lymphoid neogenesis during chronic infection. J Immunol 2007;179(1):557–65.
[146] Hoffman SL, Vekemans J, Richie TL, Duffy PE. The march toward malaria vaccines. Vaccine 2015;33(Suppl. 4):D13–23.

Index

Note: Page numbers followed by *f* indicate figures and *t* indicate tables.

D

E

F

G

H

I

Q

R

S

V

W

Y

Z

Printed in the United States
By Bookmasters